BY THE LAW
OF NATURE

BY THE LAW OF NATURE

Form and Value in Nineteenth-Century America

HOWARD HORWITZ

New York   Oxford
OXFORD UNIVERSITY PRESS
1991

Oxford University Press

Oxford New York Toronto
Delhi Bombay Calcutta Madras Karachi
Petaling Jaya Singapore Hong Kong Tokyo
Nairobi Dar es Salaam Cape Town
Melbourne Auckland

and associated companies in
Berlin Ibadan

Library of Congress Cataloging-in-Publication Data
Horwitz, Howard.
By the law of nature : form and value in nineteenth-century
America / Howard Horwitz.
p. cm.
Includes bibliographical references.
ISBN 0-19-506227-2
1. American literature—19th century—History and criticism.
2. Economics in literature. 3. Nature in literature. 4. Value
in literature. 5. Literary form. I. Title.
PS217.E35H67 1990
810'.9003—dc20   89-26654 CIP

Portions of this book have appeared in previous publications and are
reprinted here with the permission of the respective publishers. Parts
of Chapter 3 were adapted in "'Ours by the Law of Nature': Romance
and Independents on Mark Twain's River," *boundary 2: An International Journal of Literature and Culture* 17 (Spring 1990), copyright © 1990 by Duke University Press. An earlier version of Chapter
5 appeared as "'To Find the Value of *X*': *The Pit* as a Renunciation
of Romance," in *American Realism: New Essays,* ed. Eric J. Sundquist (Baltimore: The Johns Hopkins University Press, 1982). An earlier version of Chapter 8 was published as "*O Pioneers!* and the Paradox of Property: Cather's Aesthetics of Divestment," in *Prospects:
An Annual of American Cultural Studies,* vol. 13, ed. Jack Salzman
(New York: Cambridge University Press, 1988). Portions of Chapters
2 and 6 first appeared in "The Standard Oil Trust as Emersonian
Hero," *Raritan: A Quarterly Review* 6 (Spring 1987), copyright ©
1987 by *Raritan.*

2 4 6 8 9 7 5 3 1

Printed in the United States of America
on acid-free paper

For my mother, Rosalind

Preface

This book consists of three parts, and may be said to investigate three intersecting problems. First, it explores in detail the formal problems raised by a number of different American texts and paintings in the nineteenth and early twentieth centuries. Second, it considers the relations (both affinities and disjunctions) between these formal problems and developments in both forms of economic practice and conceptions of economic value. Third, it examines what these intersections illustrate about the liberal tradition and ethos in which the various documents and practices treated in this book were produced, sometimes in order to contest. One goal, then, of studying aesthetic artifacts in conjunction with economic and political documents and developments is to elucidate conceptions of value and identity in liberal culture.

The aesthetic performances and public debates and events I study here invariably invoked nature as the ground of value (both economic and moral), and this rhetorical strategy was characteristic of the liberal ethos. Yet this pervasive justificatory strategy performed multifarious work because the character, availability, and authority of nature were themselves disputed, and consequently nature was invoked in support of diverse and competing projects. Thus, liberalism and its constitutive rhetorical strategies were not monolithic or static but dynamic, constantly being defined and forged in various arenas under pressure of conflicts arising within the liberal regime itself. But liberalism is dynamic in another way, I argue, since both self and value were understood as appearing and emerging only in a network of relations and, indeed, transactions. More specifically, I conclude that the evolving notions of value and self patterning American liberalism were generally paradoxical, although often this paradoxical, indeed quite contradictory, dynamic was not crippling to, but rather constitutive of, the liberal ethos.

In juxtaposing a variety of documents, artifacts, and events (for example, fiction, paintings, pamphlets, legislative debate, textbooks, the corporate trust), I deliberately do not treat the literary or aesthetic merely as a representation or refraction of culture. The aesthetic, rather, is part of what makes up culture, as both a material and structuring element. Much of the literary work I study here was expressly social practice, not just *about* the social but often intended to intervene in the social in some way. Moreover, aesthetic categories and rhetorical strategies, I will show, deeply inform the semantic and conceptual field in which historical debates and developments occur. Therefore, this book explores not just economies of representation in literary forms but also the rhetorical economies informing other historical practices.

Part One, "Nature and the Problem of Value," introduces the problems I address throughout. In the Introduction and Chapter 1, I delineate the ubiquitous but multifarious invocation of nature as the ground of value, and explore the relation between notions of representation (specifically pictorial representation) and understandings of the character, status, and social force of nature (as either a prescriptive agent or a resource to be appropriated, often both). Chapter 2 examines the understanding of value in the mid-nineteenth century, mainly through the work of Marx, American economists, and Emerson. Value was conceived and experienced as a problem of representation, and I explore the ramifications of this association in some detail, especially the fact that value was considered a sublime relation, which representation (through money, market prices, and so on) scarcely expresses.

In Part Two, "Realism and Romances of Freedom," I examine specific literary works that thematize and investigate the problem of value. Each work regards value as a problem of representation, and thus, in addressing certain developments in the economic fabric, considers the economy of its own artistic undertaking. The three works on which I focus—by Twain, Howells, and Norris—can be readily classified as examples of realistic representation, yet each author, even Howells, ultimately embraces romance as the genre most appropriate to his efforts. The attraction to romance is typical of what Part One begins to characterize as protectionist economies of self and representation, and this impulse both registers contradictions in the liberal ethos and seeks, with varying degrees of success, either to resolve or elide or, again in the case of Howells, confront those difficulties.

By century's end, the enduring triumph and sanction of corporations accompanied and was accompanied by a related transformation in notions (or even the psychology) of self and value that had been under way for some time. The antinomies of liberalism—between, for example, reason and desire, self and other, individual and collective—were not quite resolved but in real ways traversed and surmounted by a logic and structure of feeling that I term "transcendent agency," the title of Part Three. The most spectacular and controversial development displaying and deploying this logic of value, self, and agency was the corporate trust, specifically the Standard Oil Company. I link the Standard's conceptual and rhetorical basis to a conspicuous and influential place where this logic—part of a Christian and Enlightenment heritage—was earlier being articulated and forged, the work of Emerson. In the logic of transcendent agency, I argue, we see at once the supersession and perfection of liberalism and the self. The next chapter tests the appropriateness of this notion to Dreiser's portrayal of the financier and to the period's most prominent voice for collectivism, Eugene Debs. In this framework, the final chapter explores Willa Cather's *O Pioneers!* as a meditation on the nature of property and on the relation between property and self, and also as a measure of the limits of understandings of property at the turn of the century.

A central feature of transcendent agency, as I derive the idea from the work of Emerson and the careers of Debs and the Standard, is that even the most flagrantly individual act is conditioned, authorized, and empowered by forces greater than itself. All of us who do intellectual and academic work happily attest to this fact—that really our individual work is already collective—and I now have my opportunity to acknowledge those without whom I could not have worked alone.

First, I must express my debt and gratitude to institutions that have supported this project. A fellowship from the National Endowment for the Humanities enabled me to complete this book, and I am especially grateful to Kathleen Keyes and Karen Fuglie of the NEH, who arranged for me to defer the tenure of my grant. George Mason University and Columbia University generously contributed summer funding, Columbia through its University Council for Research and Junior Faculty Development. The research staffs at the Bancroft Library, the Huntington Library, Columbia University's Rare Book and Special Collections Room, and the Library of Congress were very helpful. Substantial portions of my research were conducted at the Library of Congress, especially at night, after classes. During much of the 1980s, funding reductions restricted the Library's hours, and thus the access of working scholars to its resources. The Library's hours have recently been restored, and I hope that those who allocate public funds will continue to remember that this great national resource is valuable only if widely available to public use.

The efforts of the editors and staff at Oxford University Press helped this book reach its final form. I wish to thank in particular William Sisler, Paul Schlotthauer, and Betty Seaver.

This book began as a dissertation at the University of California at Berkeley, and Part Two is adapted from some of its chapters. The dissertation was directed by Walter Benn Michaels, a model director, whose acute attention to my work not only illuminated its strengths and limitations but inspired unforeseen inquiries.

I have discussed the ideas in this book with many students, colleagues, and friends, some of whom have commented on portions of it. Discussions of various materials and issues with students at George Mason University and Columbia University have been quite formative. The departments and my colleagues at both institutions provided excellent atmospheres in which to pursue my research and ideas. This book and the experience of writing it have benefitted immeasurably from the contributions of Quentin Anderson, Jonathan Arac, Hans Bergmann, Mitch Breitwieser, Richard Bridgman, Richard Burt, Eric Cheyfitz, Terry Comito, Andrew Delbanco, Stanley Fish, Jeff Gordon, Edward Hutchinson, Richard Hutson, Amy Kaplan, Deborah Kaplan, William Kelly, Carol Kay, Karl Kroeber, Jessica Lane, Mary Ann O'Farrell, Barbara Packer, Donald Pease, Richard Poirier, Ross Possnock, Jack Salzman, Eric Sundquist, Brook Thomas, Jane Tompkins, Priscilla Wald, and Lynn Wardley.

I could not separate the evolution of this book from the companionship and support of Stuart Culver and Richard Grusin. Deep affection, too, to Sienna and Thurmonette. My most enduring supporter has always been my mother,

Rosalind, to whom this book is dedicated. Gillian Brown's care, criticism, and love have been indispensable to what is best about this book and what was best about writing it.

Berkeley, California   H. H.
April 1990

Contents

PART ONE

NATURE AND THE PROBLEM OF VALUE

—

Mixed Instrumentality: Nature as Protean Ground

Let's Not Hear Any More about Nature

In the House of Representatives on 9 July 1846, W. W. Payne of Alabama opened the day's discussion of a bill for distributing public lands this way:

> The time has passed when an argument addressed to this House or to the country, should be made founded upon the natural right of man to his equitable part of the soil upon which Providence has cast his destiny.[1]

Rather than assessing the merits of the bill under discussion (the Graduation Bill, not passed until 1854), Payne begins by attempting to foreclose one available, obviously popular, form of justification: "Let's stop appealing to natural right and natural law (and Providence) to back up our partisan pursuits!" Payne's embellishment registers both the prevalence of appeals to nature in political argument and also the manifold rhetorical service that the law of nature performed. Appeals to nature—often elaborate descriptions of natural processes or of the evolution of land cultivators into nations—were regularly used to ground positions for slavery, against slavery, for and against railroad interests or other concentrations of capital, and, as on the occasion of Payne's speech, for and against the cheap (or free) distribution of public land to cultivators. So protean was nature's service as the ground of values and argument that were nature (or the law of nature or natural right) an intentional agent, its ubiquitous support of competing public positions would open it to charges of capriciousness or venality.

Payne's speech is one instance of what we might call, adapting Emerson, the "mixed instrumentalities" of nature and Providence.[2] Fifteen years later, a debate about the Homestead Act of 1860 (vetoed by President Buchanan) presents another. The context is different, with amplified political tensions, and so the rhetorical gesture has different valences, but it is nevertheless typical. Here, elaborate professions to study and obey natural law support diametrically opposed policies. Senator Doolittle of Wisconsin begins a speech by concurring with Senator Chestnut from South Carolina, who has just advised legislators to "become truer and better students of nature." "We should learn the laws stamped by the Creator upon the human race, and upon the physical world," Chestnut has urged, and

3

"we should endeavor to shape our course, as legislators and statesmen, in accordance with those laws." Like Emerson, who is only typical when he adverts to "The Method of Nature" or to "the Beautiful Necessity" of the law of nature (McQuade 696), Doolittle recommends: "Ay, sir, study the laws of nature—those higher laws, which God, the Almighty, has stamped upon this earth, and stamped upon us. . . . The laws of nature are stronger than the will of man and he must yield to their necessity" (*Globe,* 36th Cong., 1st sess., 10 April 1860, 1629). The method of nature is typically threefold here: natural law stamps physical phenomena, it stamps human nature, and its stamp is God's. There is, then, something superfluous or redundant about Chestnut's and Doolittle's call to study nature. If human nature were imprinted by divine, natural law, inquiry into natural law would be unnecessary; indeed such imprinting should preclude disagreement and conflict, since then all belief and action are simply the yielding to necessity. Yet here the claim to obey natural necessity initiates and signals a combat of wills and a sectional conflict. Predictably, Wisconsin's Doolittle agrees with Chestnut's methodological counsel only in order to deride the South Carolinian's conclusion "that one of the races of mankind has superior natural and political rights," and that therefore slavery should be permitted on homesteaded land and blacks be ineligible to homestead. Nor is it surprising that Doolittle's supporting examples, expounding natural processes and the meaning of the Declaration of Independence, succeed mainly in precipitating debate over natural law and Jefferson's intent.

Payne's admonition against invoking natural law seems sensible, then, because such invocations do not resolve debate but rather relocate the interpretive contest. The dispute shifts from practical questions to questions of moral justification. Payne's own caution against appealing to nature in fact launches just such an appeal. Payne endorses the bill under debate, the Graduation Bill, because it is more consonant with natural right. Other proposals, like homesteading, whereby the government grants public land to cultivators, rob the public of its right to land and curtail each citizen's right "to supply his necessities." Like many, especially many Southerners, the Alabaman Payne denied that the federal government owned the public domain or even held it in trust for citizens. The public domain belongs to citizens to accumulate according to need. In this instance, Payne is willing to compromise the argument from right, since it is a legal fact that the government holds the land. He will therefore concede some form of public-land policy, but insists that natural value must be obliged. The Graduation Bill would auction public land at a graduated scale based on the land's real (market) value rather than grant it outright or at the famous fixed government rate of $1.25/acre.

Payne's turn to nature suggests, first, that his preamble rhetorically deploys a (possibly cynical) commonplace about the ubiquity of appeals to nature. But it indicates as well one reason nature is a protean rhetorical instrument available to competing positions: its value and moral import are not transparent. A just policy for distributing land, Payne says, would respect "the *intrinsic* value of land," but intrinsic value "is *varied*" by two factors, land's yield and its proximity to markets (my emphasis). If nature's intrinsic value is subject to contingencies, however, the law of nature may be obscure, and "there can be no uniform intrinsic value in land." In Payne's view, the Graduation Bill's sliding scale would apportion land in proportions more consonant with its so-called intrinsic value. Proponents of more liberal land policies like homesteading criticized the Graduation Bill precisely because it valued land in market terms rather than according to the exigencies of cultivation. In other words, in their view the Graduation Bill valued land according to properties insufficiently intrinsic to nature. The policy dispute is finally a dispute about the intrinsic value of nature and about how to ascertain and articulate nature's intrinsic quality.

The public-land policy debate to which Payne contributes, then, negotiated two inseparable, though not identical, problems: the law of nature needs to be both discerned and articulated, interpreted and represented. As Hobbes wrote, "All Laws, written, and unwritten, have need of Interpretation." Written laws are "easily mis-interpreted," because words have "divers significations," varying over time and place, and because of "self love." Consequently, governance is afflicted with continual "contradictions." Knowledge of the "unwritten Law of Nature" is therefore necessary, since, created by God, it is the repository of "finall causes." But if we cannot interpret written signs, and since self-love is not something Hobbes thinks we will shed anytime soon, the unwritten law of nature is perhaps "the most obscure," and has "the greatest need of able Interpreters" (322–23). Some judge is needed to interpret nature's laws, Hobbes distinctively concludes, and a "Soveraigne" to enforce them. The various documents and debates I will be examining, spanning several periods of the nineteenth and early twentieth centuries, inherit this Enlightenment sense that, as J.G.A. Pocock has argued (1985, 37), society and law must be organized around laws of nature. Many of these writers, legislators, and advocates recognized, with Hobbes, that these laws are unwritten, remote, or even illegible; more specifically, they are written (or known) only through acts of human legislation and production. This insight does not always provoke Hobbes's dismay. But it often precipitates debate about the authority of aesthetic, economic, or legislative practice.

The specific aspect of this problem of judgment and governance that Payne delineates—What is the intrinsic value of nature?—is synecdochic for a larger problem, whose fortunes through several decades it is this book's ambition to sketch: to articulate nature, or to produce articulations consistent with natural law, we must know the relation between form and value. In other words, in invoking nature, advocates must not only identify manifest phenomena but also persuade others how these point to a law or governing principle of action or conduct. Indeed, nature's physical features arouse interest precisely because of the law or value they portend. The issue is at once formal and practical, and is also at once ontological and epistemological. It is ontological because the determinate form we experience differs from the intrinsic value it purportedly embodies or represents (or conceals). Payne, for example, is probing whether intrinsic value precipitates its phenomenal form or form precipitates intrinsic value. That is, what is the generative principle of value, intrinsic substance or extrinsic and contingent factors? The issue is epistemological because often what people dispute is the basis and means for determining the magnitude, shape, and significance of the form (money, market price, law, literature, and so on) in which value enters the cognitive and cultural arenas. The ontological and epistemological questions overlap, because the answer to one may affect the answer to the other.

The problem Payne's speech illustrates—and which this book will explore throughout—has some relation to the problem of the relation between form and content long familiar to literary criticism. (Poststructuralist critiques of the classical conception of representation have problematized but not deposed this relation as a topic of inquiry.) I revise the latter term of the formulation to "value," however, for historical reasons. When various actors (authors, philosophers, legislators, policy advocates, and others) considered *form,* they did so always for practical reasons, in order to ascertain the value of that form and thereby also define or justify or reform social conduct. Moreover, if in modern literary debate content tends to be *inside* form, in the nineteenth century, value was not generally inside form, but usually the *management* of form (or sometimes superadded to it). Always, finally, the question of value (both economic and, more broadly, moral), whatever its particular relation to form, was posed in terms of nature and natural law; conversely, to ask a question about nature was to ask a question about value.

This fact—that inquiries into value and nature were inextricable—implies this study's still broader historical and intellectual concern: the liberal tradition, under whose auspices debate about natural law took place and the United States was constituted. Locke's remark that "in the beginning all the world was America" was well known,[3] and had rather mythological import: since the continent had no prior (European) civil institutions, the nation could be imagined (eliding or eliminating the prior presence of non-Europeans) as the aboriginal liberal experiment in contractarian politics and in creating property from wilderness. The rare voice, like that of Payne, discouraging the invocation of nature, went largely ignored, even by iconoclastic critics of liberalism like Thorstein Veblen, and Payne's caution was disingenuous in his own speech. All actors whom this book will look at saw their specific work (literary, economic, legislative, polemical, political) as continuing, reforming, or overturning liberalism, one of whose founding premises was that the human apprehension of nature is transactive, an activity of discovering, disclosing, and producing value. It is because questions of nature, form, and value were so densely affiliated that this book examines developments not just in literary form but in economic discourse and organization—and in the literary, economic, and legal understandings of these developments—in order to apprehend the aesthetics of value and natural law, which we might call American liberal aesthetics.

Self-Production

This book will examine some of the conceptual and practical relations (both similarities and differences) between the historically differentiated arenas of literary activity and social or political, and especially economic, activity. More specifically, it will trace the trajectory of what I take to be the transcendentalist (which is related but not equivalent to the idealist) component of liberalism, a development that culminates in the literary movement called naturalism; in the scandalous consummation of corporate development, the trust; and in the (in some sectors) equally scandalous development among American laborers, industrial unionism, led by Eugene Debs. Two interrelated questions central to liberal justifications of action and social form will continually arise, both of which are finally moral, and raise questions pertinent to a Christian culture: in generating value, what is the proper way to apprehend, appropriate, and represent nature? And, What is the nature of the self that engages and is engaged in this activity?

In the liberal vision, nature is preeminently a site of human productivity. This premise is in strong ways materialist, and was central to liberalism's innovation to the Aristotelian and scholastic traditions. For Aristotle, *homo economicus* does not produce value but rather manages resources ready to hand. "Nature intends and provides" value (*Politics* 18–19); man is value's steward. The Aristotelian notion that value inheres in nature informed economic thought through Aquinas and the physiocrats, but began to give way during the early stages of capitalism, when human and especially individual agency in the production of wealth could hardly be denied and came to be held a liberating virtue. Locke, a mercantilist (meaning that he focused on the accumulation of wealth), redefined value as the product of human labor modifying nature. Natural resources, either unusable or widely available and ready for consumption, have no value.[4] Value arises only in the transformation of nature. In converting alien matter into consumable form, labor "put[s] the difference of value on everything" (141). Value, then, is a narrative principle of differentiation distinguishing human artifact from alien nature (mere matter).

Value is not the only artifact labor fabricates. In investing nature with value, labor converts it into property. This effect and affect of labor—appropriation—follows from the essentialist corollary to Locke's materialist premise, one that led C. B. Macpherson to christen liberalism the tradition of "possessive individualism." Labor engenders property (appropriates nature) because the transaction between labor and nature realizes personhood, for, in Locke's view, labor is simultaneously the vessel and constitutive element of self. When one "hath mixed his labour with" nature, he has "joined to it something that is his own, and thereby makes it his property." Here labor is one's "own." In scholastic and feudal thought, labor does not belong to the self. It does for Locke, though, because it expresses the reason granted humankind by God. This definitive property infusing the body and all action is, because divine in origin, a constitutive principle of self-possession: "every man has a property in his own person; this nobody has any right to but himself." It is because each man owns himself that "the labour of his body and the work of his hands . . . are properly his"; and labor appropriates nature because it infuses nature with the self: I own my possessions because my labor "hath fixed my property in them" (134–35). Thus human productivity, liberalism's occasion and invention, is a doubly nuanced concept, since by their labor persons not only produce value but are produced as value-producers. Innate self-proprietorship converts nature to property, but nature is also the arena where personhood emerges, insofar as it achieves determinate and perceptible form only through appropriating nature.

The liberal self is "possessive," then, because it is based in a property of self that is "unquestionable" and "inalienable." This self not only creates property but is realized and known only through property.[5] Locke's justification of property is, then, to use Pocock's intriguing term, hylozoistic,[6] because property is property only if animate, already invested with self. Locke's parable of property rehearses a psychological formation that survives today, but as a justification (of property or self) it is perilous because it is axiomatic, even tautological. Self owns its property because it owns itself; at same time, the self owns itself only because realized by and vested in property. Self animates its property because property vests the self: property is labor is self is self-possession is property, and so forth. The relation is at all points representational and artifactual, since any term is identifiable and palpable only when represented by another. Specifically, inalienable qualities are known and realized through the alienable attributes that represent them. But it is not clear whether the relation is metaphorical, metonymic, or synecdochic, which element governs the relation, nor what the verifying mechanism is.

This problem will be probed further in chapters 2 and 3, and is the principal question treated in chapter 8, but for now it is enough to observe that Locke's ontology of property entails a metaphysics and an aesthetics of self-possession, both of which potentially assume or elide the linkage between the terms *form* and *value, materiality* and *self, self* and *self-possession.* The liberal self is dynamic, with inalienable properties experienced and identified—even acquiring substance—only through their representation by alienable attributes. Ideally, inalienable properties are independent of contingency and convention and therefore generate rather than derive from public discourse and forms of representation. The representational mode here would be synecdoche, with public forms epiphenomena of the self. But if the inalienable is known and realized only in its representation, then the relation is metonymic, a matter of contiguity, with the generative relation indefinite, and that which should be essential and abiding may well be a function (rather than the genesis) of convention.

The opening of the *Leviathan* will help elucidate this point: "life is but a motion of Limbs, the beginning whereof is in some principall part within" (81). For Hobbes, life is motion, both physical and intellectual; just as we try to interpret the unwritten law of nature, we try to ascertain the invisible principles motivating sensations and "the characters of mans [*sic*] heart" (83).[7] Therefore, every thought is "a *Representation*" of some object or desire (85). In other words, thought or consciousness is not a static condition, certainly not "the repose of a mind satisfied" (160); instead, it is the interminable motion and activity of representation. For Hobbes, thought is spatial movement—outside to inside and back again, form to principle to action—and, we might say, social movement—object to mind, physical motion to person. Hobbes calls mankind the "Artificial Animal" because the fundamental human activity is artifice—specifically, the "art" of imitating and representing nature in order to discern the laws motivating thoughts, hearts, and desires (81). Representation, finally, is the continual striving to represent us to ourselves. Hobbes is less concerned with inalienable properties than Locke, but Locke's account of property involves a similar, constitutive self-representation: property is a self-investment, the attempt to make the inalienable manifest in the forms that represent it; and this self-investment in property is also a self-production, for self is unavailable and unformed except in its representation through property.

If inalienable property (should we call this the essence of self?) is the goal of consciousness and action, the movement toward it is asymptotic, always mediated through alienable attributes. The liberal aesthetics of value and self must traverse the disparity between the alienable and inalienable, which are the materialist and idealist elements, respectively, of the liberal paradigm. In some formulations, we will see, this traversal is achieved through elision. Terms that depend on each other for definition may appear interchangeable or even identical. Each component is validated through the other: material properties and attributes are justified by appeal to an ideal self, an innate, divine property preceding any transaction with others or the world; yet this self is experienced only through cognitive and physical transactions and through material attributes, and can clearly be altered by material and contingent factors. Hence, each component is at once contingent, dependent on the other for justification, and ideal, apparently complete in itself. This mixing of materialist and idealist premises is liberalism's distinctive, though often strained, representational strategy for justifying actions and values. The stresses of change and dispute often reveal both the fissures in the logic of liberalism and also the formal, psychological, and cultural work it accomplishes. By studying a series of significant instances of this logic, this book will examine the formal, aesthetic, logical, ethical, and also political difficulties that artifacts of American liberalism address, try to resolve, and, often, elide or accept.

Nature's Nation, Of Course

If one aimed to be a firstcomer, in Emerson's and Nietzsche's term, one would eschew examining the instrumentality of nature in American literary and economic discourse. The United States is, of course, as Jefferson put it, nature's nation, conceiving national and often individual identity in terms of nature. But what this identification means has never been settled. I have already cited a few examples of nineteenth-century debate over nature's mixed instrumentality, and the question of nature's function as symbol and myth, to adapt the subtitle of Henry Nash Smith's seminal *Virgin Land,* continues to provoke academic analysis.

Scholarship has expounded, broadly, three theses about the relation between nature and human action in American literature and culture. These theses do describe patterns of thought and action in the nineteenth century, yet none alone is exhaustive. These theses revolve around two intersecting axes: one concerns whether nature is knowable or available to ground moral or political authority; the other, whether nature, however available, confirms or questions human values and institutions.

Much criticism of the 1950s and 1960s, especially that concerned with American romanticism, held nature to be immanent, a positive and determinate value or possibility, available as a model of virtue. Generally in the tradition of Frederick Jackson Turner, whom Smith characterized as a belated transmitter of the agrarian myth of free land, this view explores "the capacity of nature . . . to supply value and meaning," as Leo Marx puts it in his discussion of *Walden* (1964, 249). Such work studies pastoral withdrawal from artifice and convention and its critique of the increasing contest between wilderness life and industrialization or urbanization. Some of this work, like that of Roderick Nash, heralds the triumph of nature as a normative, moral principle influencing intellectuals and wilderness policymakers. The figure of Huck Finn lighting out for the territory to escape the constraints imposed by the Protestant culture of the Widow Douglas may serve as a trope for nature's remedial function.

More often, especially since the 1960s, the immanence of nature as a ground of reform, cleansing, or rebirth is viewed as a historically receding possibility, albeit analytically and morally recoverable through scholarship. This view of the rhetorical force of nature, retaining much of the specific content of the Turner thesis, might view Huck's removal to the territory more ironically, as Laurence Holland recently has: in heading for the territory Huck cannot leave behind middle-class conventionality, for by 1883 the territory (a federal construct rather than a natural fact) was already populated with Widow Douglases. Criticism expounding this vision (by, for example, Cecilia Tichi, Lee Clark Mitchell, and at points Nash) traces a tradition less Adamic (see R.W.B. Lewis) than oppositional, examining the efforts of some (authors and activists) to resist utilitarian developments in American culture and in Americans' treatment of the natural environment. Though less sanguine about the prospect that nature may inspire, inform, and reform conduct, this ironic and oppositional thesis is equally as progressive as the more optative view, the Emersonian mood that Matthiessen emphasized.

Critics who discern the immanent presence or possibility of nature almost uniformly adopt a "dialectical" approach to the study of nature as America's myth of identity. This approach usefully identifies confusions and contradictions in the mythology of nature and studies developments in the reconfiguration of its antinomies. The goal of this dialectical approach is progress. If historical developments have frustrated progress, criticism may achieve progress in defining virtue and value. Other critics, however, have been less optimistic about both the immanence and positive value of nature. For example, Perry Miller speaks of the "errand into the wilderness" as "a reformation which never materializes" (1956, 9). For Miller, Anglo-Americans' understanding of nature is best viewed in terms of scriptural Jeremiad, which criticizes culture as disaffected and alienated from virtue. In this view, virtue (usually designated natural, always divine) is never available to artifice, but rather is remote from human perception and convention. Indeed, its remoteness and unrealizability are the very marks of virtue's value, albeit primarily negative. This account of nature as inaccessible informs some recent work employing psychoanalytic or deconstructive premises, like Eric J. Sundquist's *Home as Found*.[8]

Let me illustrate the difference between the first two accounts of the status of nature in American discourse by contrasting Miller's traditional notion of Jeremiad with Sacvan Bercovitch's more recent analysis of the American Jeremiad.

Where Miller sees the Jeremiad as exemplifying nature's inevitable absence as a model of conduct, for Bercovitch the Jeremiad consecrates American culture by demonstrating to Americans the immanence of virtue. Bercovitch argues that America in fact revises the familiar, European Jeremiad tradition, converting it into a rhetoric of progress. The American Jeremiad transmutes the rhetoric of the wilderness's peril, with its concomitant sense of the "discrepancy between appearance and promise," into a millennial vision of the fulfillment of sacred destiny (1978, 17).

If the first of the three theses holds that nature was conceived as immanent, and the second, as remote, a third main thesis holds that Americans understood the status of nature, and consequently national and individual identity, as achievements, constantly being forged and revised. Here, the value of nature is primarily historical. Such a vision of American culture and character as forged is concisely expressed in the title to Jacob Riis's autobiography, *The Making of an American,* in which Riis, a Danish immigrant, becomes an American less by a ritual of naturalization than by the series of jobs and careers he undertakes on his way to reformism. Miller propounds this view in "The Shaping of the American Character," where he argues (using Whitman, Cooper, and Bancroft as examples) against "endeavors to fix the personality of America" "in a unitary conception." American personality is antithetic to "essence," it is "something to be achieved"; and one "abandons personality" when one seeks it "in the conditions of America's existence" (1967, 13). Although Miller does not characterize his thesis as such, it manifestly rehearses the logic of pragmatism, the American contribution to philosophy coined by Charles Sanders Peirce and popularized by William James, who argued that pragmatism "unstiffens ... theories" (1975, 32) and that truth is known in terms of, and is largely a matter of, effects. This conception of American nature and of the fortunes of the myth of nature in America influences scholars as otherwise different as Walter Benn Michaels, Richard Poirier (especially in *The Performing Self* and *The Renewal of Literature*), and Jane Tompkins.

In characterizing these three theses—that nature is (1) immanent, (2) unavailable and relentlessly critical, and (3) a value historically forged—I do not mean to suggest that Americanist scholarship on the problem of nature could or should be rigidly and exhaustively catalogued within these parameters. Most criticism incorporates components of more than one thesis. *Virgin Land,* for example, is in general historicist/pragmatist: Smith argues that the myth of nature evolves in the context of changing material and conceptual exigencies. But when Smith criticizes agrarian discourse for its contradictions, its static quality, its implicit debt to European tradition, and therefore its incapacity to enact the democratic ideals that inspired it, he is assuming the immanence of nature as a failed ideal (because compromised by conflicting premises). Similarly, if Miller sometimes sounds like a pragmatist, in "The Romantic Dilemma in American Nationalism and the Concept of Nature"[9] he propounds the immanence of nature, though in this essay immanence has sometimes positive and other times critical force. Miller traces nationalist attempts in the early romantic period to define what is characteristic about (the) American nature—its wildness—but he discovers ambivalence and uncertainty about the specific character of wildness, about how wildness should influence conduct, and about the effects and morality of what Americans called the "improvement" of nature.

I delineate these three theses, then, as helpful distinctions rather than as a formula for cataloguing. Nor are they meant to exhaust or prescribe the shape and content of all nineteenth-century American literary and economic discourse. One aim of students of culture is to discern patterns and common features in arrays of artifacts, thought, and action. But as Franco Moretti reminds us, we should curb the temptation to declare that any century or period obeyed without fail one paradigm.[10] It is not accidental that modern scholarly debate has identified more than one historical understanding of nature; in fact, Americans' conceptions of nature and their invocations of nature as the ground of value varied. If many people regarded nature as Bercovitch suggests, for example, as an immanent condition consecrating national destiny, others viewed it in the more familiar Jeremiad tradition Miller expounds. Many, as well, could shift from one rhetorical gesture to the next. I believe it is crucial to recognize that the invocation of nature and the discourse of value were heterogeneous and flexible, even inconsistent.

This book will indeed trace a line of descent, a sublime tradition of what I will call "transcendent agency," which at once supersedes liberalism and perfects it. Moreover, even in its most idealist instances, this paradigm of thought and action most resembles the third thesis I identified above. It is Emersonian, I argue, and understands the law of nature as forged and reforged through conflict. As I have suggested, this position is related to the American pragmatist tradition, and indeed many of the artifacts and documents constituting the lineage I will trace are explicitly or genetically pragmatist. But it would be a mistake to subsume all activity in a culture under one rubric, and I hope I do not succumb to this temptation too often or egregiously. In the illustrative narratives and debates I will recount—about the American School of landscape painting, protectionism, improving the Mississippi River, the status of character in an economy, speculation, the trusts, unions, homesteading, and the basis of property—we will see all three theses deployed as modes of argument by diverse competing groups. Often, too, the same argument invoked more than one account of nature, at times contradictorily.

As a result, no one argument acquired a uniform political identity or force; rather, the applications and impact of each were varied. The debates about the Graduation Bill I sketched earlier are one example of nature's mixed instrumentality. This study will introduce people with all manner of political and moral agendas claiming that they act or own "by the law of nature," as settlers along the Mississippi characterized their right of access to the river when the Spanish closed it to trade in 1802 (de Barbé-Marbois 215). The politics of the settlers' declaration are mixed. Their claim that "the Mississippi is ours by the law of nature" may seem to critics of an era called postmodern unduly to naturalize proprietorship. But since the phrase "the law of nature" in fact refers to the labor the settlers have invested in the river and its environs, the remark also champions the rights of labor against Spanish monarchial authority. If this fact makes the remark classically liberal, it would nevertheless be hard to predict other political values these settlers might hold—on slavery, for example.[11] Finally, as I will argue in chapter 3, the settlers' declaration crudely simplified Locke's logic, indicating the way liberalism splices labor-based and sovereign senses of value and authority (a point elaborated in chapter 8). In short, the settlers' declaration is multivalent, with aspects that appear to the modern eye both conservative and adversarial, some naturalizing property, but others defending a person's right to the fruits of labor. The settlers' commitment to the liberal paradigm does not determine all nuances of their beliefs and practices.

It is instructive to contrast the Mississippi settlers' declaration that they own by the law of nature with Emerson's use of the same phrase in "Self-Reliance." However multiple its resonances, the settlers' remark surely imagines that nature is immanent, available to human use, and that it sanctions (or even legislates) resistance to Spanish sovereign authority. Forty years later, in "Self-Reliance," Emerson is even more profoundly challenging the authority of convention:

> We fancy it rhetoric when we speak of eminent virtue. We do not yet see that virtue is Height, and that a man or a company of men, plastic and permeable to principles, by the law of nature must overpower and ride all cities, nations, kings, rich men, poets, who are not. (143)

Here, natural law is a sublime principle of plasticity that obeys its own exigencies rather than the conventional demands of "eminent virtue." Moreover, Emerson is criticizing (two sentences earlier) those who "prate" of self-reliance (clearly already a shibboleth) and exalt it as an "instant of repose" and self-sufficiency. For Emerson, who, as many have observed, anticipates American pragmatism, self-reliance is not a settled condition, but "resides in the moment of transition from a past to a new state." Paradoxically "residing" in transition, self-reliance is thus "power not confident but agent" (142). Perhaps surprisingly, Emerson's point that self-reliance is not an instant of repose is Hobbesian. Recall that for Hobbes, the ineradicability of desire in humans means that there is no "repose of a mind satisfied. For there is no such *Finis ultimus,* (utmost ayme,) nor *Summum Bonum,* (greatest Good)" (160). Emerson would never frame self-reliance as a problem of desire, but like Hobbes's "Felicity," self-reliance is not stasis and completeness but motion—agency. We might better think of it as a re-lying—a relocation, a transportation and reformation of self. (Do not "talk of reliance" but "speak rather of that which relies because it works and is.")

For Emerson, the power that is self-reliance and that is the law of nature does not establish the proprietary rights of labor that the Mississippi settlers invoke. Emerson (like Hobbes) is concerned not with inherent rights but with capacities, not with a priori properties but with agency, with power that is power because it is constantly being effected. In this way Emerson's thought at once perfects and supersedes liberalism's antinomies; he posits a radical individualism (not metaphysical, however, but practical and antinomian) yet refuses to specify the political (or moral) order appropriate to self-reliance. Emerson's antinomianism has some possibly troubling aspects. Conventional duties and notions of virtue must often be transgressed; the self-reliant person is the person who overrides others and in whom "all concentrates" (143); moreover, the American Scholar, who knows that "in yourself is the law of all nature" (62), shall realize in "private life" "a more illustrious monarchy ... than any kingdom in history" (59). Such a vision would seem able to tolerate political arrangements that the Mississippi settlers invoke the law of nature to oppose.

Nature as Negotiation

I have looked briefly at invocations of the law of nature by congressmen, by land claimants, and by Emerson in order to suggest the multifarious and hybrid conceptual and political purposes to which the same semantic unit can be put. Related or even identical formulations have significance only in the context of their production and circulation. I stress this point in hopes of modifying some recent and powerful ideological analyses of what may be called the deep structure of the American imagination. Bercovitch's work is justly admired for probing such deep structures. In doing so, he has examined the "restraining power" of America's millennial imagination, its capacity to accommodate and "absorb the spirit of protest" by furnishing the terms and boundaries of dissent.[12]

Bercovitch's work belongs to a recent resurgence of interest in historicist literary criticism, and specifically in the relation of American self-conceptions to history. As exemplified in Carolyn Porter's discussion of "American ahistoricism" (3–22), numerous scholars are adapting Louis Hartz's thesis that American liberalism has imagined itself as outside, or as the fulfillment of, history. America inherited the ethos of liberalism without having to rebel against *native* political and religious institutions, Hartz argues.[13] Consequently, American liberalism, unlike Continental liberalism, has not appeared to be an achievement. Feeling like an already completed and "natural phenomenon,"[14] American liberalism could define itself only in relation to the present, and sustain a fantasy of universality, with the attendant "tyranny of opinion" noticed by de Tocqueville.

Porter, Bercovitch, and Michael Paul Rogin,[15] among others, have brought to the Hartz thesis somewhat different analytical emphases, inflected by poststructuralist and neo-Marxist premises. On the present topic of the American conception of nature, Myra Jehlen has recently conducted a Hartzian analysis. In self-conception, Jehlen argues, America is the incarnation of "the art of nature and nature's God" (*Incarnation* 41).[16] Consequently, American individualism and society seem the "complete realization" of natural law (80). This "monism of the American incarnation" (226), which Emerson's *Nature* epitomizes and legitimates, enables America to elide conflict and the temporal dimension of the dialectic of change (72).[17] Incarnate America appears dialectic's concluding synthesis, homogeneous and "posthistorical" (131). But America's transcendence of history entails "the suppression of dissent" and of "the capacity to project alternative forms"; dissent and the acknowledgment of change threaten America's self-justification (125, 73). American virtue, then, flourishes at the expense of debate, difference, conflict, and change.

Where Hartz delineates the historical basis for the often conservative or even conformist content of American liberalism, Jehlen, more in the tradition of critical theory, would disclose the unconscious "grammar organizing speech."[18] This "ideological grammar" (21), informing and transcending the content of any particular act or speech act, encompasses and can explain all acts within a paradigm. Jehlen's analysis is an abstract, Hartzian criticism of the conviction that American destiny is immanent in nature, and she shares Bercovitch's sense that American culture effectively "represses alternative or oppositional forms" (Bercovitch, "Afterword" 433). The point of Jehlen's ideological grammar and Bercovitch's ideological analysis is to replace the dream of incarnation with the recognition that notions of nature and hence of institutions and cultural forms are not natural but constructed and thus malleable.[19] However admirable this political objective, it seems to me founded on an inaccurate, overly syncretic account of the liberal invocation of nature.

Exponents of the Hartz thesis generally contend that American liberalism (either because of its historical origins or because of its structuring ideological grammar) has had an essentially univocal bearing that transcends the specific content of acts and utterances and the specific context of their production. The materials I have been presenting suggest a less univocal and more conflictual ethos than Hartz's progressive analysis and more recent ideological and neo-Marxist analyses conclude. As Pocock has argued about Enlightenment possessive individualism, nineteenth-century American formulations of nature and value were "problematic and precarious" (1985, 71).[20] If, as Perry Miller wrote, by the nineteenth century "the identification of virtue with Nature [was] axiomatic" (1956, 212), nevertheless the content and political and cultural effects of instances of that identification varied. As an attempt to assess the relation between phenomenal form and value, and as a way of seeking or claiming virtue, the appeal to nature had no single, unitary force. Naturalist appeal hardly took the natural for granted; indeed, suspicion of contrary naturalist appeals was an indispensable rhetorical strategy. Nor did naturalist appeal always or necessarily elide or preempt conflict;[21] rather, *nature* was the term whose meaning and valence were disputed and being negotiated. Invocation of natural law was a crucial element in, to adopt Stephen Greenblatt's terms, the continuous exchange and renegotiation that make up the process of cultural production.[22]

All acts of representation (all rhetorical figures) are acts of exchange—one sign standing for another—and the justificatory appeal to nature was one of the central exchanges constituting the representation and production of cultural value. In debates over public-land policy, I have tried to indicate, the authority of natural law had a protean and contested charge. Quite self-consciously in Payne's speech and in the exchange between Doolittle and Chestnut, the invocation of nature was the very mark and subject of conflict. Emerson's praise for the Emancipation Proclamation and his later assessment of the nation's progress illustrate this point. Lincoln's act, not entirely popular in the North, "does not promise the redemption of the black race," but through it whites "have recovered ourselves from our false position, and planted ourselves on a law of Nature" (McQuade, 773). The natural law Emerson invokes here is Lockean, averring that all persons have a right to their labor and the franchise. If here Emerson invokes nature against entrenched racism and invidious hierarchies, in 1878, in "The Fortune of the Republic," he makes the kind of remarks for which his later career has been thought embarrassingly conservative and his vocational commitment to nature has more recently been branded repressive. "The revolution [of American fortune] is the work of no man, but the eternal effervescence of nature"; and "the prosperity of this country has been merely the obedience of man to the guiding of nature" (1883, 412). Here the optative mood becomes jingoism. But even this orthodox adulation registers discord and extends a challenge to the exalted social order. Emerson inventories concentrations and corruptions of power—monopolies and Tammany Hall—and systematic injustices against women, children, and the causes of labor and civil rights. The "great sensualism" of the nation (413) and its acquiescence to custom and current institutions make Emerson wonder if nature's "vital force" is not finally vitiated in America. His praise, then, is partial: America's *prosperity* consists (rather tautologically) of achievements of will in accord with natural law, but "'tis certain that our civilization is yet incomplete" (419). He invokes nature as a model of culture, specifically of active public debate, in order to chasten his audience.

Naturalist appeals were often (though of course not always) instruments of conflict, not its denial. The sheer fact that naturalist rhetoric punctuates nineteenth-century discourse and debate suggests that champions of the ethos of nature understood conflict to be its soliciting condition. Legislators, for example, may have tried to *end* conflicts, and may have used teleological rhetoric as an element of persuasion: My proposal is so harmonious with natural law that it will end conflict forever! But these are legislative strategies, rhetoric in the classic sense, often pursuing compromise. Payne, Doolittle, Chestnut, and many of their colleagues and compatriots were well aware that the next debate (and the next appeal to nature) awaited but a few hours or minutes.

The present book's ambition is to examine the ubiquitous, hybrid, and protean invocation of nature in literature and other arenas precisely as a register of conflicts in cultural forms, the conception of value, and liberal premises about the self and its relation to social structures. I have proposed that it is useful to examine these issues as they intersect the at once practical and aesthetic problem of the relation between form and value. To probe the aesthetics of values and the values of aesthetics, this investigation will attend to literary documents and to the economic and legal issues and developments that bear upon them and upon which they bear. In so doing, I hope to contribute to the recent, and revised, historicism we are witnessing, whereby critics consider the way texts (and their authors) are embedded and participate in historical practices and structures.

Influenced by the work of cultural materialists, neo-Marxists, and Continental poststructuralists of various sorts, Anglo-American literary criticism has in recent years been reinvigorated by probing the sociology of literary and other aesthetic acts.[23] I have already mentioned some of its practitioners in the field of American studies. Historicists begin with the insight, as Walter Benn Michaels has phrased it, that aesthetic acts take place "very much within [a] culture" (1987, 27). They are informed by and recirculate cultural paradigms and practices, but also constitute part of culture in dynamic relation with other arenas of it, often not just responding to, but also intervening in it.[24] This sociological attention tries, as Amy Kaplan has nicely formulated a Foucauldian principle, to "situate . . . texts within a wider field of . . . 'discursive practices'" (6). But more, we try to specify historically what cultural work aesthetic and rhetorical practices accomplish.[25] As Leo Marx puts the problem (1988, 215) on what plane does form (literary or otherwise) mediate experience?—a question that we should not interpret to presume the autonomy of any text or form. Our task is to calibrate the pressures informing texts and their production, and also the pressures exerted by the bearers of those pressures. Always of concern is the nature—or rather the constitution—of the self that works and acts, and that is at the same time acted upon and disseminated, within a historical matrix of practices and relations that conditions, though it does not determine, the individual subject.

The main arena of my sociological focus, the economic, is a good place to study this dynamic, since the economic is precisely what individual acts contribute to and constitute collectively, and also what is irreducible to individual acts. The economic exceeds individual control, and moreover helps constitute the field and parameters of action and subjectivity. To adapt a formulation by Foucault, the economic illustrates how agency is both individual and subjective, but also transindividual (*History of Sexuality*, 94–95). The self is at once agent and sublated in the economic, necessary and negligible, and many of the documents I study explore this peculiar relation, which I will term transcendent agency.

But even as I identify and explore this paradigm, I hope to establish its permutations and heterogeneity. Culture is a dissemination, a constellation of events, not a monolithic entity. Consequently, I must juxtapose instances of high and low culture, the most philosophical with the most practical voices; legislators will be compared with, for instance, Kant, the trust with Emerson, and fictional financiers with labor organizers. This clustering will trace changes—in literary works and form, economic or legislative debate, business organization—that both result from and help to transform the content and shape of specific historical and conceptual negotiations.[26] I stress here the specificity of conflict, content, and analysis for empirical and historical reasons, and also for analytical reasons. Although the nineteenth century exhibits a tendency toward incorporation and corporate agency, and finally toward the transcendence of incorporation, I propose no grammar or theory of American culture and action. I refrain from doing so mainly because the formal undertakings and historical conflicts I examine do not suit the syncretic imperative of a grammar. But in addition, I am concerned that much recent historicist analysis, in aiming to disclose a general grammar or ideology of action, risks more than historical inaccuracy. It may also exhibit the impulse to universalization that was the scientistic fantasy of Enlightenment liberal thought.[27]

Jehlen rightly calls the "moral thesis" of incarnation a tautology (85): we express nature, nature expresses us. The logical scandal of tautology is precisely how little it elucidates, how much less it explains than it purports to explain. An ideological grammar risks a similar circularity. Abstracted from context, the content and effects of rhetorically kindred remarks look the same. All acts and utterances will appear contained in the same way within a cultural imagination. One of the most cited phrases in recent historicist work is Fredric Jameson's remark that ideology and cultural forms are at base strategies of containment (1981, 193). But I think this understanding of how cultural forms operate is too neat, even idealistic. It assumes that something—presumably the subject—precedes the culture in which it emerges, and that containment is a secondary phenomenon.[28] But not just Foucault has argued for thinking of paradigms, conventions, and institutions as what Bernard Yack calls "enabling constraints" (1967). That is, the parameters of action and thought—of what Foucault tends to call subjectification—are their possibilities of existence, including the possibility to resist or alter parameters. All experience—including dissatisfaction and dissent—occurs in terms of what Veblen and William James called "habits of thought," and what Pierre Bourdieu calls the "habitus," probably a better term than Kuhn's paradigm because it conveys the animateness as well as the collectiveness of any paradigm.[29] But the ineluctability of habitus doesn't make all action conservative or repressive or preemptive of dissent in any substantial sense. Rather, it means, as Foucault's late work urges, that dissent and change arise and have meaning only in relation to their conditions of emergence. The effects (and hence experience) of acts and speech acts occur and are known only in terms of particular contents, contexts, and intentions.[30] Concomitantly, the diverse content and impact of acts and events are available only through specific acts of analysis.[31]

The present book's commitment to this proposition about meaning and analysis aligns it with pragmatism in the philosophical sense. It thus dissents from and would alter the theoretical urge and premises of much sociological analysis, which foregrounds ideology as an analytical tool or goal. Like Raymond Williams and Foucault, I am suspicious of the usefulness of the term. If *ideology* means anything other than "false consciousness"—the veil of experience[32]—its denotation is so tautological as to be unenlightening. Althusser's technical, Lacanian definition of the term—"the imaginary relation of . . . individuals to the real relations in which they live" (1970, 165)[33]—notoriously retains Marx and Engel's sense of ideology as distortion, along with the romantic implication that one can truly experience the world only if not wholly inside ideology (which is to say, by not inhabiting the world). If ideology is distortion, then everyone is its captive and there is no position from which to effect meaningful critique and change.[34] When critics try to eschew Althusser's vestigial romanticism, however, the term *ideology* is emptied of substantive content. It may mean, as Terry Eagleton writes, "the ways in which what we say and believe connects with the power-structure and power-relations of the society we live in" (14); or it may mean that historical factors are integral to and structure "the way the mind interacts with nature and society" (Jehlen 1987, 10). These definitions are symptomatically unspecific: experience (of self, world, and culture) is mediated; ideas and motives inhabit the material and the social; and physical, corporeal, and linguistic acts are inextricably embedded and also instrumental in historical and political structures and patterns. If so, however, all subjectivity and action, whether creative or analytical, are ideological. Even were one to propose, as Jean Howard has, that rather than a monolithic ideology, there are many ideologies,[35] then the term even more clearly refers simply to the manifold embeddedness of subjectivity and action in historical formations. As a critical tool, then, *ideology* is at best redundant, with little substantive force—unless it denotes false consciousness and therefore suspends the conditions of legitimate change.[36]

This book studies the relation between form and value in nineteenth-century literary works and in economic, social, and legislative debate partly to consider how change was conceivable then. I will attend to resonances and interferences that exist between aesthetic and rhetorical performances and coterminous social formations and practices. The aesthetic partially structures and is also structured by the social field it helps to compose. In other words, particular linguistic, aesthetic, or rhetorical acts and historical events and discourse are isomorphic cultural phenomena. Let me explain this last phrase, in hopes of averting charges that I have simply found homologies and made diverse aspects of a culture equivalent. I use the term *isomorphism* to indicate that different species of acts are coordinated around what Williams calls structures of feeling (128) because they have a related morphology. They have emerged in related environments—historical, geographical, conceptual, or otherwise. Thus, isomorphic cultural phenomena, in my view, are kindred, arising in what Edward Said calls an "affiliated network" (175). Yet if diverse acts may have a related genesis or underpinning, they are nevertheless hybrid, distinctly not identical, nor do they have the same effects.[37] Indeed, I am very interested in the way opposed parties employed related formulations and tried to appropriate key cultural tropes and strategies.

Surely when people as different as George Henry Evans, Emerson, Henry Carey, Orestes Brownson, Willa Cather, John D. Rockefeller, Lawrence Gronlund, Charlotte Perkins Gilman, Eugene Debs, and Mother Jones invoke the law of nature, the effects of these differently positioned and motivated, though structurally affiliated, speech acts differ. Precisely because the law of nature was a contested article of faith, my analysis of isomorphic cultural formations yields no metadiscourse nor perspective to encompass all actors and documents populating this book. All may have believed, as Emerson wrote in "The Method of Nature," that "nature represents the best meaning of the wisest men" (*CW* 1:204), but the content of this wisdom and the identity of its proprietors were regularly disputed. It is the permutations of these disputes, and their conflicted proclivity toward the paradox of action and proprietorship that I will call transcendent agency, that this book probes.

Sublime Possession, American Landscape

An American School

The mixed instrumentality of nature and its representation is evident in the nation's first recognized, indigenous aesthetic movement, the American School of painting, also called the "native" or New York School, and later (after 1879) its familiar name, originally derisive, the Hudson River School.[1] This chapter will extend the remarks of the Introduction in order to probe two primary and related aspects of the aesthetics of the American School, and especially of the work of its progenitor and greatest practitioner, Thomas Cole: the at once liberal and sublime basis of American School landscape art. The American School sought to possess nature by representing it; to do so, it nationalized the Enlightenment and romantic sublime tradition. The sublime, commencing with nature's threat to human understanding, property, and even life, may seem inimical to liberal proprietorship. But appropriating nature was a crucial element of the sublime tradition, and this fact at least partially explains the tendency in the nineteenth century to imagine value as sublime. The sublimity of value was an important, often explicit, tenet of the texts and related historical debates this book will subsequently examine. The American School, however, exhibited more than one account of sublime possession. It therefore illustrates the at once intimate and multilateral relation between aesthetic values and the agency of nature.

The American School was an enterprise in appropriation—of landscape, continent, and cultural heritage; its goal was self-possession, a "native" identity and culture produced through representing America's unique and sublime landscape. As is well known, the early national period saw a litany of calls for a national culture distinct from Continental, especially British, traditions, a culture that would not simply anticipate future prospects—supposedly all that American arts were capable of—but make use of America's own history and resources. As early as 1789, Colonel John Trumbull—entrepreneur, historical and landscape artist (painter of *The Declaration of Independence* and of well-known panoramas of Niagara), long-time president of the American Academy of Fine Arts, and a discoverer of Cole—declined to serve in France as Ambassador Thomas Jefferson's private secretary, so that he could continue to "produc[e] monuments" to America's already glorious history.[2] In 1824 Gulian Verplanck counseled, "Our national existence has been quite long enough" to serve as the "pledge" and "type of the greater future."[3]

The continent's wilderness became the central icon in the typological production of American history and character. The wilderness had gradually been rhetorically transformed from the devilish and inhospitable environment the Puritans represented it to be into a nationalist symbol. Now witnesses attested that "the new world was fresher from the hand of him who made it," as William Cullen Bryant wrote in 1830.[4] Bryant's nationalism both supplements and is abetted by the associationism promulgated by Archibald Alison. The Scotsman's *Essays on the Nature and Principles of Taste* (1790), adapting Kant to moral instruction, glorified the divinity of landscape and its exalting effect on the mind. Since Matter is actually the expression of Mind (the divine mind that created it and the human mind divinely inspired to appreciate it), nature's beauty and sublimity reflect the divinity awakened in perceivers by natural stimuli. Representing nature, then, serves a high moral function, stimulating human divinity.[5]

By establishing the moral mission of landscape painting, associationism helped redeem its reputation, disparaged by such as Joshua Reynolds as mere imitation and therefore as inferior to the imaginative composition of historical painting. Landscape painting seemed the unique opportunity for American artists, because the continent's landscape, as they characterized it, typified the fledgling nation's special mission and capacity. America's cataracts, forests, and mountain ranges were "unequaled," and its "wild beasts and savage men" unaccountable, New York Governor De Witt Clinton told the American Academy of Fine Arts in 1816; therefore, no country is "better calculated than ours to exercise and to exalt the imagination." America's resources and vistas typify its divinely designed destiny, and in celebrating nature, American art would commemorate the nation's culture by producing it. In 1826, artists disaffected from the American Academy's traditionalism—its glorification of European models and its aristocratic organization and pretensions—founded the National Academy of Design to promote America's aesthetic appreciation of itself.[6]

The nation's first art movement was virtually willed into existence, as John K. Howat puts it (1972, 28), not just to forge a national identity but to consecrate political and economic achievements. The aesthetic impulse attended the prosperity of New York and of the "new men" of business who were its art patrons and, sometimes, artists. Cole, one of the secessionists who founded the National Academy, achieved his first major sales in the year the Erie Canal opened (1825).[7] The young nation and its metropolis needed an indigenous art to declare themselves, respectively, a culture and a center of culture. By the 1830s, notices anointed landscape art America's unique contribution to culture, and by 1840, the adventures, travels, and personal fortunes of American School celebrities provoked comment in newspaper gossip columns.

The American School's success constituted as it celebrated American progress, and was said to culminate in the acclaimed 1859 exhibition of Frederic Church's monumental (5½ feet by 10 feet) *Heart of the Andes* (figure 1), perhaps the most publicized and profitable single-picture exhibition of the century.[8] Flanked by portraits of the Founding Fathers, the painting was displayed as an

FIGURE 1. Frederic E. Church, *The Heart of the Andes*, 1859. Oil on canvas, 66⅛ × 119¼ inches (168 × 302.9 cm). The Metropolitan Museum of Art, Bequest of Mrs. David Dows, 1909 (09.95).

allegory of nationalism. *Heart of the Andes* promised international achievement as well. Newspapers noted Church's debt to Alexander Von Humboldt's *Cosmos: Outline of a Description of the Physical World,* where he asked, *The Home Journal* and the New York *Evening Post* reported:

> Are we not justified in hoping that landscape painting will flourish with a new and hitherto unknown brilliancy when artists of merit shall more frequently pass the narrow limits of the Mediterranean, and when they shall be enabled, far in the interior of continents, in the humid mountain valleys of the tropical world, to seize, with the genuine freshness of a pure and youthful spirit, on the true image of the varied forces of nature? (quoted in Roque 44)

Von Humboldt envisions an imperial art, circuiting the earth, penetrating its most exotic climes to seize the true image of nature's forces. In proudly aspiring to Von Humboldt's ideal, Church's painting, and the accolades showered upon it, register not just American progress but the impulse to empire, on which nationalist pride seems to have been predicated.[9]

The American School takes for granted and seeks to realize the appropriative impulse that Locke had called the law of human nature. Consistent with Richard Ray's 1825 advice that artists paint nature in order to "adorn our houses with American prospects" (32), the sublime subject of many American School works— for example, Asher B. Durand's succinctly titled and often reproduced *Progress* (1853) (figure 2)—is the appropriation that founds progress. Representative figures—families, shepherds, farmers, or more obviously artist figures (painters, sketchers, writers)—occupy the right or left foreground, which often overlooks rolling hills, or meadows or lakes, with a crossing diagonal of mountains, and then a blue heaven, in the background (see figures 3, 4, 5, 6). Signs of expanding civilization punctuate the point of vision or the scene viewed. Typically, smoke trails from trains or chimneys, symbolizing productivity, blend into the landscape like wisps of cloud, intimating that productivity is a natural event. The complementarity between environment and human artifact turns the moral principle of associationism into a nationalist economic imperative.

This typology is powerfully realized in Jasper Cropsey's *Starrucca Viaduct, Pennsylvania* (1865; figure 7). Two travelers survey a vista of curves: the curving Susquehanna River, curving foreground foliage, both intersecting the curve of the Starrucca Valley, along the base of which is tucked the viaduct built for the New York and Erie Railroad. Even the trail of smoke from the train's engine, though moving counter to the curve of the valley, finally harmonizes, curling like another creek back between two ranges, angling to the rear and to the right. This harmonious scene has been framed by God, with trees framing the foreground and a proscenium of clouds framing the valley. The blue sky, God's vestibule, eclipses the central mountain, minimalizing its potential obstruction of progress by incorporating it into the composition of progress.[10] The viaduct itself occupies little of the painting's panorama; nevertheless, it is entirely appropriate that this view of primarily a natural scene should bear the title of a human artifact, for as the type of progress, the landscape is de facto an American artifact.

FIGURE 2. Asher B. Durand, *Progress*, 1853. Oil on canvas, signed and dated 1853. 48 × 71¹⁵⁄₁₆ inches (121.9 × 182.7 cm). From The Warner Collection of Gulf States Paper Corporation, Tuscaloosa, Alabama.

FIGURE 3. Asher B. Durand, *Dover Plains, Dutchess County, New York*, 1848. Oil on canvas, 42½ × 60½ inches (107.9 × 153.7 cm). National Museum of American Art, Smithsonian Institution, Gift of Thomas M. Evans (1978.126).

FIGURE 4. Jasper Cropsey, *Harvest Scene*, 1855. Oil on canvas, 21¾ × 35¾ inches (55.2 × 90.8 cm). Signed and dated at lower left: J. F. Cropsey / 1855. Collection of the San Antonio Museum Association, San Antonio, Texas.

FIGURE 5. Frederic E. Church, *New England Scenery*, 1851. Oil on canvas, 53 × 36 inches (134.6 × 91.4 cm). Signed and dated at lower right: F Church '51. George Walter Vincent Smith Art Museum, Springfield, Massachusetts (1.23.24).

FIGURE 6. Jasper Cropsey, *Autumn—On the Hudson River*, 1860. Oil on canvas, 60 × 108 inches (152.5 × 274.3 cm). National Gallery of Art, Washington, D.C. Gift of the Avalon Foundation (1963.9.1).

FIGURE 7. Jasper Cropsey, *Starrucca Viaduct, Pennsylvania*, 1865. Oil on canvas, 22⅜ × 36⅜ inches (56.8 × 92.4 cm). Signed and dated at lower right: J.f. Cropsey 1865. Toledo Museum of Art, Toledo, Ohio. Gift of Florence Scott Libbey, 1947 (47.58).

Cropsey's design exemplifies what Leo Marx, adapting Panofsky, calls the "iconology" of American landscape art (1985, 95–96). *Starrucca Viaduct* presents "wildness" as domestic. One of the travelers holds a gun, and the two men may be hunting with their dog, but effort is supererogatory, this landscape is already appropriated, and they relax, like shepherds or, what they were, tourists.[11] This iconology is illustrated as well by Albert Bierstadt's *The Rocky Mountains, Lander's Peak* (figure 8), whose exhibition in New York in 1863 catapulted Bierstadt (as he had hoped) to the rank of Church's main competitor. The imposing peaks in the distance of this canvas (even larger than *Heart of the Andes*)[12] nurture, through the waterfall they engender, an Indian community. The painting's crossing diagonals move upward, but its color scheme (deeper colors below luminescent clouds and snow-covered peaks) draws attention to the comfortable meadow, itself surveyed by a structure for suspending water that resembles an easel, or camera, or surveying equipment. (Bierstadt traveled in the West with a government survey expedition led by Colonel Frederick W. Lander.) In competing with Church, Bierstadt emulates him in domesticating exotic environments and communities for public, leisure consumption, a consumerism that elevates the leisure public to the status of culture, or at least its connoisseurs.[13]

Sublime Economics

These prominent examples of the American School epitomize one understanding of nature in America, as typological or immanent, an available model of virtue. To the extent that sublime exaltation was in all contemporaneous accounts the goal of landscape art, the sublimity of this work is absolute, exaggerating Enlightenment and romantic thinking about the sublime. Following Richard Ray's 1825 counsel, these works present the source of exaltation to be American prospects, inspired and installed by divine agency. The sublime subject is progress itself. The humans in these paintings are rarely workers, and never industrial workers (although industry is sometimes represented). Unlike Courbet's landscape figures, these human figures are unobtrusive, indeed scarcely different from other figures in the natural landscape. As in Cropsey's *Harvest Scene* (1855) or Church's *New England Scenery* (1851), they are beneficiaries of an ineluctable historical/natural process (figures 4, 5).

Like the 1802 Mississippi settlers' claim to own the river by the law of nature, which I discussed in the Introduction, this vision of sublimity simplifies the appropriative transaction delineated by Locke (not himself a participant in discussion of the sublime, but whose sensationalist psychology helped precipitate this discussion). For Locke, we must appropriate because nature does not of itself supply our needs. In *Starrucca Viaduct,* however, artifice is so consonant with natural setting that the entire composition appears to be at once divine artifice and naturally the property of those—travelers and audience—who observe it. The titles of Samuel Colman's *Storm King on the Hudson* (1866) and of *Lander's Peak* have similar effects, presenting human fabrication and property as natural phenomena. In Bierstadt's canvas, the question of rightful proprietorship of western lands, a vio-

FIGURE 8. Albert Bierstadt, *The Rocky Mountains, Lander's Peak*, 1863. Oil on canvas, 73¾ × 120¾ inches (186.7 × 306.7 cm). Signed and dated at lower right: A. Bierstadt 1863. The Metropolitan Museum of Art, New York, New York, Rogers' Fund, 1907 (07.123).

lent political controversy at the time, is suspended, or rather settled. Though Indians reside beneath it, the peak sustaining their habitat belongs—in name, which is to say, conceptually—to Colonel Lander, leader of the survey expedition that Bierstadt accompanied, and finally to the audience. It belongs not to those who merely enjoy its material benefits, but to those able to infuse the scene with the proper imaginative and moral import.[14]

Bryant's 1830 remarks in *The American Landscape* clarify the assumptions of immanence here. The "far-spread wildness" of "our country," "suggesting the idea of unity and immensity, and abstracting the mind from the associations of human agency, carrie[s the mind] up to the idea of a mightier power, and to the great mystery of the origin of things (quoted in Merritt 17). Bryant rehearses here, with idiosyncrasies to which we will attend, some of the classic topoi of the sublime tradition: verticality and immensity exceeding and overwhelming customary categories of comprehension; devastation of expectations and the imagination; dissonance caused by the inadequacy of comprehension and the sublation of human agency; finally exaltation, as subjectivity, first abstracted from itself, is "carried up" (Kant says "incited" and "aroused") to the idea of a power mightier than mere matter. It is important to note that the sublime, here, is not an inherent, static quality of nature. Rather, as Kant (more than Burke) emphasizes, it is a liminal experience, an experience of transport (*ekstasis,* to use Longinus's term), a dynamic of apprehension. Thomas Weiskel has delineated this dynamic as a narrative of threat, dissonance, and exaltation, an awful transition from fear to empowerment. The sublime experience may not entail the claim, as Weiskel puts it, "that man can . . . transcend the *human*" (3; my emphasis), but it does contemplate the transcendence of current limits.[15]

But Bryant also adapts the tradition whose precepts he has internalized. Bryant readily identifies the landscape as "our country." The anxiety characteristic of sublime experience and of sublimity's characteristic threat to property and human jurisdiction[16] are absent from this easeful assimilation of wildness as a metaphor for mind. Bryant's satisfied sublime, like much work by his American School friends, overlooks Kant's dicta that (1) "no sensible form can contain the sublime properly so-called"; "no adequate presentation [*Darstellung*] is possible" (83–84); and that (2) the sublime "is an object (of nature) whose representation [*Vorstellung*] determines the mind to think the unattainability of nature as a presentation of ideas" (108).[17]

For Kant, feeling the unattainability of objects or ideas *in themselves* marks the brute fact of human finiteness and is awakened by the "chaos" and "irregular disorder" of natural phenomena, and by the "devastation" or "desolation" (*Verwüstung* [167]) such disorder awakens in perceivers (84). "The forms of nature are so manifold" and appear under "so many modifications" that human understanding cannot comprehend a universal principle ordering nature's infinite permutations (16). Excessiveness of magnitude or might in sensible forms (the mathematical or dynamical sublime) appears unmotivated, defying our customary expectation that action is purposive, and thus threatening the conventions for ordering experience.[18]

But because it "strains the imagination to its utmost bounds" (108), sublime purposelessness also incites reason to "intervene." Reason attempts to represent phenomena to consciousness, imputing to the totality of nature a supersensible purposiveness (a *law* of nature) that transcends the irregularity we perceive locally. The effort to represent sensible phenomena and supersensible purposiveness (the art of God) fails, but this effort arouses the mind to feel a moral capacity, by which Kant means "a satisfaction independent of mere sensible enjoyment," a capacity to "exercise dominion over sensibility" (109). Thus, feeling the inadequacy of comprehension and custom ultimately issues in an intimation of moral empowerment.

Kant's analysis of the sublime is the philosophical corollary to the problem of the relation between form and value that in the Introduction we saw Congressman Payne encounter when he seeks the elusive intrinsic value of nature. As Payne sensed about value, sublimity exemplifies the noncorrespondence between forms and both their provenance and affect. In sublimity, Kant writes, "there is nothing at all that leads to particular objective principles and forms of nature corresponding to them" (84). One feels a disjunction between sensible forms, the general principles they fail to make transparent, and the experience they occasion.[19] The disjunction is twofold: phenomenological, between form and the experience of form; and ontological, since particular forms do not obviously point to or stand for the larger principles engendering purposiveness. That is, for purposiveness to be perceptible (and hence perception and action to be legitimate), finite forms need to have a synecdochic relation to higher principles. Kant thinks we make sense of forms only diacritically, by reference to other forms and to encompassing orders. This referencing is finally teleological, as forms signify governing principles. But the perceptible relation between form and principle is finally no more than metonymic, and the contiguity of metonymy expresses contingency rather than necessity. Hence, the purposiveness of forms and the stimulus of experience are not transparent, and the legitimacy of values is not self-evident but in need of justification.

This problem was nicely posed for antebellum artists by Washington Allston, the nation's leading painter before Cole, who read and discussed idealist philosophy with Coleridge. The *Lectures on Art* he began in 1833 probe whether what we might call the sublime structure of experience can be made sensible. Any experience (of aesthetic or natural forms) consists in a relation between the part and the "mass," or between individual "forms" and the principle of "Power" that objectification manifests. The mass depends on the part's being "the very sign of its power" (148),[20] yet the nature of the relation cannot be fixed or finally specified, Allston concludes, and the content of the mass, or value of any form or experience, is indefinite.

The noncorrespondence between forms and their referents and affects has obvious resonances for questions of aesthetics and interpretation, arenas that provoked eighteenth- and nineteenth-century investigations of the sublime. But it underlies nineteenth-century inquiries into value as well, as I suggested in the Introduction through the example of Payne. Economic discourse often probed its subject in the language of the sublime. For Marx, for example, the nonidentity between "outward appearance and the essence of things" is what makes possible and necessary economic "science" (as opposed to vulgar economics, which uncritically presumes correspondence) (*Capital* 3:817). Moreover, the noncorrespondence between the substance of value and its phenomenal form leads Marx to speak of commodities as hieroglyphs of sublime social relations, in need of deciphering but often remaining stubbornly obscure.[21]

The sublime hermeneutics of value were used to figure the economics of art, as well. In the preface to *The Spoils of Poynton*, Henry James characterizes life as "all inclusion and confusion," inundation by sensible stimuli; "life has no direct sense whatever for the subject" of art, and "is capable . . . of nothing but splendid waste." But "the sublime economy of art" makes capital of the inundation of ordinary experience. "Being all discrimination and selection," art searches for "the hard latent *value*" of experience. The "rescue" of value from the confusing inundation of sensation, accomplished by "investing and reinvesting" the effort of discrimination, constitutes art's "sublime economy" (*Spoils* xxxix–xl).[22]

The question motivating sublime discourse—the adequacy of formal representation—is central to the aesthetic and historical problems this book will be addressing. The final question in each arena concerns the *legitimacy* of values. Legitimacy is a matter of dispute precisely because of the noncorrespondence between form and its provenance and affect, the noncorrespondence that provokes Marx and James to speak of the sublimity of value and art, respectively. Similarly, for Kant, as for Burke, it is the noncorrespondence of sublime irregularity to natural or moral principles—paradigmatic of the nontransparency of all forms—that provokes our awe and desolation. In the sublime moment we feel the inadequacy of our faculties to comprehend phenomena, to know their value, which is to say, their relation to natural (universal) and human principles of conduct. The disjunction between sensation and natural law excites fear and awe because it signals our alienation from divine purposiveness.

For Kant, this devastation can be transcended, and indeed inaugurates a redemptive moral capacity. In his narrative of sublime psychology, the desolation precipitated by the irregularity of nature can also arouse "in us a feeling of a purposiveness quite independent of nature." This feeling is, finally, "the sublime merely in ourselves" (84). It begins with "a movement of the mind" seeking the satisfaction of purposiveness, and intuiting that such satisfaction and purposiveness are subjective relations, "referred through the imagination either to the faculty of cognition or of desire" (85). We will see in later chapters that the question of whether the operative faculty is cognitive or appetitive generates problems for economic discourse; but for Kant the crucial feature of sublime satisfaction is that it arises in the feeling that our cognitive and appetitive faculties—and also the principles of conduct we use them to consider—are not wholly determined by sensible experience. Exalted satisfaction is "independent of mere sensible enjoyment" (109). Thus our sublime awe at our alienation from divine purpose comes eventually to signal our freedom from determination, and hence signals the possibility of the moral capacity.

As Weiskel and Frances Ferguson have observed, sublimity is an economy of power. This point is explicit in the German words for the faculties of imagination

(Einbildungskraft) and judgment *(Urteilskraft);* both are *Kräfte,* powers. Exaltation results from the feeling that "merely the ability to think . . . evidences a faculty of mind surpassing every standard of sense" (89; *Kritik* 172). Thus "the might
of nature . . . calls forth our forces" (104) in a distinctly liberal undertaking: our
intuition subordinates "the particular in nature to the universal" and subsumes
"contingent" permutations under necessary law (16); and we overcome our
instinctive fear of overwhelming magnitude or might. Sublimity, finally, becomes
the capacity to feel "that we are superior to nature within and . . . to nature without us" (104). The occasion and site of awe shift, from external threat to internal
capacity.[23] Exaltation is a feeling of interior safety, a superiority to external threat
and internal astonishment,[24] as the spiritual movement aroused by the sublime's
"violence to the imagination" (83) provokes "the extension of the imagination by
itself" (87). Kant figures this territorial clash as a utilitarian economy.[25] Just as
Bentham's career is beginning, Kant conceives sublimity in Benthamite fashion,
as an expenditure and return: the imagination "acquires an extension and a might
greater than it sacrifices" (109).

Kant formulates his "transcendental exposition" of unconditioned sublime
judgment in order to revise Burke's "physiological" and "merely empirical exposition." For Burke, the "satisfying horror" (Burke's own word is "delightful") is a
corporeal, emotional movement produced by the natural object itself, and he
seeks, Kant believes, "the empirical laws of mental changes" (119–20). Sublime
stimulation, if not so excessive as to be "noxious," has for Burke a purgative effect
on the nervous system, unblocking dangerous "encumbrances" of habit and custom (Burke 136). But if sublime feeling is only corporeal gratification, Kant
responds, then agreement in taste is only accidental, a coincidence of egotistical
preference, and no general standards of taste can be valid, nor any "censorship of
taste" legitimate (119). As Ferguson stresses, taste and the sublime engross both
Kant and Burke as principles for gauging and legislating social order. In the title
of one section, Kant indicates his concern with "judgment as a faculty [for] legislating" (15). But if the sublime is a cognitive and aesthetic laboratory for judging
the stability and legitimacy of the social compact, Kant worried that the atomism
of Burke's mechanistic exposition jeopardizes having a principle or "substrate"
(93) for intuitions and hence for orderly conduct and consent.[26] Instead, we
"should appeal to the natural right of subjecting the judgment . . . to our own
sense and not that of any other man" (119).

Here is the paradigmatic gesture of liberalism's Protestant reformation of
Christian virtue. Kant locates the substrate of judgment and consent in an unconditioned faculty capable of knowing the a priori "inner nature" of principles of
judgment. With experience free from sensible determination, self and judgment
precede the social and therefore can "prescribe" standards of taste and conduct.
Only the idea that our criteria are a priori, Kant believes, authorizes us to "pass
sentence on the judgments of others" (119–20). The question of whether the conviction of cognitive freedom is in fact free from *habitual* and *conventional* determination is of course pertinent to Kant, as it is to Locke and Burke, and will be
central to nineteenth-century economic treatises. But the problem of the freedom

of the will is less pertinent, here, than a feature that makes Kant's sublime a fundamentally Lockean political economy: its emphasis on alienation as the condition of value and validity.

The sublime experience is multiply alienated and alienating. It is alienated from natural law because sensations and their representations fail to correspond with phenomena, which themselves depart from their provenance. Hence, satisfaction is not "*immediately* connected with the representation" of phenomena. But this is exactly why judgment can be universally (Kant says "pluralistically") valid. Our alienation from totality (our finitude) is also our capacity to appropriate natural forms, so that all judgment (albeit claiming to be disinterested) is, like Lockean value, an artifact. The sublime appeal to natural right is profoundly Lockean in that it observes a principle of alienability and appropriation. It is altogether appropriate, then, that Kant's crucial word for validity is *"gelten,"* cognate with the word for money, *"Geld."* Validity in judgment is a problem of exchange, valuation, and circulation. Moreover, again in Lockean manner, the self experiencing Kantian sublimity is at once presocial but also fundamentally social, for sublimity arises in a mutually productive transaction with objects and other beings. In the sublime moment, ineluctably subjected to the determinative force of others, you are constantly attempting to subject others to your (autonomous) judgment. Kantian transcendence, like Lockean ownership, is a compensatory economy to protect the self's borders by expanding its territory.

The Manifest Sublime

Kant's transcendental sublime is a narrative of dispossession and possession—of both nature and self.[27] In positing the immanence of nature, then, the American School does not utterly abandon its Enlightenment inheritance. For comprehension of divine immanence is the ideal outcome and very purpose of the transcendental sublime. When Kant writes that "there is in our imagination a striving toward infinite progress and in our reason a claim for absolute totality" (88), he might well be describing the assimilative ethos of much American School work. Allston characterized cognition in the language of assimilation: the forms (natural phenomena) in which ideas "become cognizable to the mind" are "assimilants"; our "intuitive" power that gives meaningful context to forms "assimilates" them (3–4).[28] In his preface to an edition of engravings and natural descriptions, *Picturesque America,* Bryant clearly announces the appropriative creed of sublime art. Bored with European scenes made familiar by reproductions, spectator and artist are impelled by the desire for "new combinations" to explore this continent. Facing the fresh variety of American landscape features—mountains, ravines, prairies, savannahs, its unique canyons—one is "overwhelmed with a sense of sublimity." But the visitor has come here because "Art sighs to carry her conquests into new realms," a desire facilitated by "lately opened" overland railroads and communications lines (iii).

If Bryant's narrative of aesthetic conquest replays the Kantian dynamic of sublime dispossession and possession, often, however, American School work elided the stage of dispossession. When Bryant refers, in *The American Landscape*, to "our country," or, in his funeral oration for Thomas Cole, to "our aerial mountain tops" (14), he is positing a nature already possessed, hailing conquest without struggle, transcendence without anxiety or liminality. Such remarks epitomize what Barbara Novak has called America's Christianized and nationalist sublime.[29] My discussion of the transcendental sublime is meant to indicate its Protestant, legitimationist, and expansionist substrate; that is, sublime discourse was from the start Christian and nationalist. Yet Novak is rightly describing a specific positive content infusing much discussion of the sublime in America. This positive content derives from the collateral picturesque tradition,[30] but it also fits Hegel's category of the "positive sublime," which itself resembles the Burkean or Kantian notion of natural beauty. Beauty comforts and ministers to the will, expectation, and custom, as if it were, in Kant's phrase, "preadapted to our judgment" (83).

Leo Marx has called this aesthetic that infuses the sublime with the picturesque "allegorical" (1985, 104), and Bryan Wolf calls it "metaphorical" (1985, 328). These terms aptly describe the American School's sense of nature as the type of the mind. We might also call this notion of the sublime manifest or typological—Perry Miller called it "missionary"—or the domestic sublime. This notion of the sublime is clearly expressed in the well-known *Letters on Landscape Painting* (1855) by Durand, who after Cole's sudden death in 1848 (apparently of pneumonia) donned the mantle of leading American artist.[31] Nature "minister[s] to our well-being" and its forms are manifestly "types of Divine attributes" (McCoubrey 112). Emerson writes in this mode—he calls it "analogical" (McQuade 19)—when he says that nature "receives the dominion of man as meekly as the ass on which the Saviour rode" because "the whole of nature is a metaphor of the human mind" (22, 18). Elsewhere he baldly states, "I wish to know the laws of this wonderful [universal] power, that I may domesticate it" ("Natural History" 13). Similarly, if in a more popular voice, James Batchelder wrote in 1848 that because the landscape was a "most desirable habitat" it typified "the sublime destiny" of the nation, *"foster mother"* to the *"governing* race of man" (quoted in Miller 1965, 57). The continent's sublimity is synecdoche for divinely nurtured, manifest destiny.

We should recall that what is sublime here is not nature per se but Richard Ray's "American prospects," of which nature is type. As Allston wrote, "nothing in the world of sense [can] so fill us with wonder" as "the present complicated scheme of society" (McCoubrey 63). This glorification of social capacity does not so much forswear the sublime tradition as streamline it. The manifest or domestic sublime abates the anxious dynamic of Kant's proto-utilitarian economy—the painful movement in which self feels its alienation from nature, then recovers itself in a supersensible exaltation that appropriates natural forms (still alien from the mind). Kant is careful to specify that "the ground" of the imagination's extension over fear remains "concealed," and so exaltation is a feeling not just of empowerment but also of "the sacrifice or the deprivation" that occasions it; that is, the imagination remembers "the cause to which it is subjected" (109). For Kant, exaltation retains a residual feeling of the sacrifice and subjection that incite subjectivity. Kant's vision is closer to what I earlier called pragmatic apprehension of nature, in which the natural is a quality constantly being achieved; the effort of appropriation is paramount.

Kant's dynamic of sublime possession—a double movement of subjection and empowerment (never achieving Hegelian synthesis)—will be taken up in Part Three of this book, in discussions of corporate agency and property. But much American School work elides this dynamic and depicts the supersensible as felt immediately in natural forms, and this immanence registers the sublimity of American political and cultural achievement. As in *Starrucca Viaduct,* nature is often a great human artifact. This typological ethos was of course widespread; yet it did not have a univocal application, and could be used to criticize "improvement." Whitman, for example, often critical of middle-class life, was especially fond of a passage in an 1852 essay in *Graham's Magazine* proclaiming that "the elements [of nature] would be only blank conditions of matter, if the mind did not fling its own divinity around them" ("Imagination and Fact" 42; Tichi 224). Even the progenitor of environmentalism, George Perkins Marsh, conceived that, as the author of "Imagination and Fact" put it, "Nature was thus endowed from the beginning" (42) by human divinity. Marsh wrote *Man and Nature,* his 1864 study of the damage progress inflicts on ecosystems, to show "that whereas [some] think that the earth made man, man in fact made the earth" (ix). Marsh criticizes civilization for ruining nature's typological capacity.

The Course(s) of Empire

Not all American School landscape art expressed the domestic, manifest sublime. The movement, precisely because it was evolving, was not uniform in its aesthetics. Another model of the sublime, more like the Kantian dynamical sublime, was also circulating, in, for example, Allston's *Lectures on Art.* In language freely adapted from Kant, Allston rendered the travails and triumph of "subordinat[ing] the senses to the mind" (McCoubrey 63). Although experience begins with the corporeal sensation of objects, "neither the senses nor the object possess, of themselves, any productive power." They are "simply the *occasion,*" not the "agent," of experience. Yet the agency of the mind is felt only when we are subjected to "that which has the power of possessing the mind, to the exclusion, for the time, of all other thought, and which presents no *comprehensible* sense of a whole." In this painful evacuation of thought, the observer becomes "the subject of . . . sublime emotions." Sublimity begins for Allston in a dispossession and subjectification of self, and this condition occasions desire, an "excess of interest" having "an outward tendency." Excess of desire "strains to the utmost" the faculties of understanding "prompt[ing] the mind beyond its limits" (McCoubrey 65–66).[32]

The economy of dispossession and possession Allston imported informs much of Thomas Cole's work. I do not say all of it, because, like American conceptions of the sublime in general, Cole's work is not uniform in its vision. Cole's May 1835 address to the American Lyceum in New York, published the next year as "Essay on American Scenery," expresses well the contrary dynamic of the transcendental sublime. Nature is Americans' "birthright," but "the multitude" displays both apathy toward American scenery and "meager utilitarianism" in its preoccupation with improvement. Both indifference and mere utilitarianism poison the intellect, forfeiting our birthright. Those who love nature, however, may learn to "cultivate" it, a different skill from mere improvement. Improvement, for Cole, is merely, or meagerly, material; cultivation exalts social, cultural, and moral capacities, and is a higher form of appropriation (1–3).[33] Experiencing nature—Cole's example is, typically, a waterfall, specifically Niagara—exposes the mind to the "apparently incongruous idea, of fixedness and motion—a single existence in which we perceive unceasing change and everlasting duration." Overwhelmed by sublime "impetuosity" and "uncontrollable power," "our conceptions expand—we become a part of what we behold!" (8). For Cole, the sublime is simultaneously threat and cultivation. Sublimity and cultivation constitute a double assimilation, by and of nature. Cole imagines nature as a realized will, our "birthright," and as an independent "uncontrollable power." Sublime nature is at once self and other.

Cole's vision of art as cultivation helps explain why he, like his friend Allston, avowed that he practiced "composition" rather than exact mimesis. His patron Robert Gilmor preferred "*real American* scenes to compositions," and complained that Cole's compositions, which combine different views, often from a physically impossible vantage, were unnatural. Cole rejoined that composition was not "a departure from Nature" but a surmounting of the partiality of perception and natural forms. Unlike most of his contemporaries, Cole did not paint directly from nature, but sketched or took notes, painting later.[34] From many sketches of the wild, the artist must "make selections, and combine them," so that the "parts of Nature may be brought together, and combined in a whole that shall surpass in beauty and effect any picture painted from a single view" (Cole, "Correspondence" 45, 47). Like Bryant's quest for "new combinations," Cole's composition is a combinatory art of cultivation, both the expression and achievement of Americans' "birthright" to nature, of their sublime capacity to "surpass" particularity and intuit the intangible principles inspiring disparate forms. As an act of appropriation (Cole seems always to have visited nature with the intention to sketch), composition marks both the mind's combinatory action and its mediated access to natural forms and law.

The paradoxical economy of the transcendental sublime that Cole's essay rehearses and that his technique of composition implies also informs a notebook entry of 1827 known as "The Bewilderment" (Noble 46–51).[35] Bryan Jay Wolf has discussed this entry as a "sublime narrative," an Oedipal drama in three acts: an unsettling act of interruption; autoerotic response; and sublimation, redirecting the threat to a possible moral end. The concluding sublimation exhibits the double movement of sublime possession. While hiking, Cole experiences "those vicissitudes of feeling which result from the change of natural objects." He soon loses his way, mainly since a recent tornado has annihilated the trail, and he suffers "inaction," unable to swim when he falls into a pond, and (a familiar Freudian trauma) fears losing his eyes. Cole's salvation begins when, perched on a shelf in a pool into which he has fallen, he can tolerate inaction no longer. "Desperation now took sudden possession of me, and I determined to rescue myself at every hazard. . . . With complete self-possession, I then commenced a process of exploration . . ." (49). Eventually he arrives at a log cabin with family and dog. Sublimation—the cathexis of anxiety onto external objectives—occurs in Cole's narrative as domestication and self-possession that are also the devastation and possession of self by natural power.

Cole not only sketched this economy of sublimation in his notebooks; he publicly exhibited it in, for example, *View from Mount Holyoke, Northampton, Massachusetts, after a Thunderstorm,* completed in the year "American Scenery" was published and better known as *The Oxbow* (figure 9). The cultivated right portion and the wild left portion are at once in balance and in tension. Following the bend in the river, cultivation presses leftward, yet the thunderstorm, although passing, remains an impenetrable wall of cloud. Scholars have noted that gouges in the hillside in the rear-center spell out the Hebrew words for Noah and, when viewed upside down, the Almighty *(Shaddai).*[36] This invocation of Jahweh's redemptive promise to Noah recapitulates the formal tension of the painting. The promise of ultimate redemption and protection requires devastation by inundation; moreover, this structure of devastation-for-the-sake-of-salvation is never transcended but compulsively repeated in Hebrew Scripture. The painter in the foreground is also divided: he is in the brush, while his equipment leans toward the cultivated area. Finally, this painter, who faces the audience with both an inviting and challenging countenance, is located where he could not have painted this scene. The painting's vantage deprives its putative agent of ultimate authority for it.[37]

This vision of the sublime is generally pragmatic, understanding natural law and the naturalness of culture as continual negotiations. Other landscapes of Cole, however, emphasize the withdrawal of nature and the inadequacy of human faculties to traverse physical and temporal boundaries. Bryan Jay Wolf has examined Cole's sublime landscapes as generally unresolved Oedipal dramas between emblems of human agency (figures for writing and phallic symbols, like series of outcroppings leading to monumental peaks) and natural threats to human agency. Wolf notes the dead trees, exposed roots, and decaying vegetal matter that typically occupy the foreground of a Cole landscape (1982), like *Lake with Dead Trees (Catskill)* (1825), *Falls of the Kaaterskill* (1826) (figure 10), *A Wild Scene* (1831–32), or *A Tornado* (1835) (figure 11). The decaying matter connotes temporality and nature's inhospitability to humans' physical needs. Further, the diagonals of Cole's compositions often intersect to subordinate human agency to wild purposelessness, intensifying the self's estrangement from the universal purposiveness ideally to be imputed to the composition. In *The Notch of the White Mountains (Crawford Notch)* (1839), for example (figure 12), the mountain's verticality and the increasing density of tree growth eclipse the houses; the awfulness of the verticality frustrates penetration and domestication. Finally, Cole's chiaroscuro effects and vertical storm clouds bar the human figures usually in the foreground (and by extension observers of a painting) access to the furthest peak, whose availability would symbolize our capacity to intimate a universal perspective. Compositionally, then, these paintings stress the self's subjection to local and physical

FIGURE 9. Thomas Cole, *View from Mount Holyoke, Northampton, Massachusetts, after a Thunderstorm (The Oxbow)*, 1836. Oil on canvas, 51½ × 76 inches (130.8 × 193 cm). The Metropolitan Museum of Art, New York, New York. Gift of Mrs. Russell Sage, 1908 (08.228).

FIGURE 10. Thomas Cole, *Falls of the Kaaterskill,* 1826. Oil on canvas, 43 × 36 inches (109.2 × 91.4 cm). Signed and dated, 1826. From The Warner Collection of Gulf States Paper Corporation, Tuscaloosa, Alabama.

FIGURE 11. Thomas Cole, *A Tornado*. 1835. Oil on canvas, 46⅜ × 64⅝ inches (118 × 164 cm). In the collection of The Corcoran Gallery of Art, Washington, D.C. Museum Purchase, Gallery Fund (77.12).

FIGURE 12. Thomas Cole, *The Notch of the White Mountains (Crawford Notch)*, 1839. Oil on canvas, 40 × 61½ inches (101.6 × 156 cm). National Gallery of Art, Washington, D.C. Andrew Mellon Fund (1967.8.1).

limits. The desire and recognition of cognitive freedom incited by sublimity issue mainly in more desire and cognitive play rather than the appropriative expansion of Kant's transcendental sublime. Subjection and dispossession remain unredeemed.

A close look at one detail, the lone Indian in *Falls of the Kaaterskill* (figure 10), will elucidate the contrast between this compositional mode and that of the domestic sublime. Included partly to establish the monumental scale of the falls, Cole's Indian does not obviously figure the domestication of the natural and the exotic, as the Indian tribe does in Bierstadt's *Lander's Peak*. Instead, the Indian figure amplifies the inhospitable quality of the scene. Nor does it have the effect Ray imagined in his influential 1825 lecture to the American Academy. Ray rehearses a standard sublime scenario: the artist "lose[s] himself," with

> a wild scene . . . frowning around him; and though, thanks to our equal laws, no bandit is seen issuing from his hiding place, yet there he may plant the brown Indian, with feathered crest and bloody tomahawk, the . . . native offspring of the wilderness. (31)

For Ray, the Indian is not just primitive but a natural growth, a species of vegetation, whose bloody weapon culminates the threat of wildness. Political problems, represented by banditry, have been solved, and the "wild" Indian offers an opportunity for the imagination of the gentleman artist to awaken from the slumber of propertied leisure (6). Durand's *Progress* (figure 2) deploys red men according to Ray's program. Christian progress, illuminated by fingers of God, is surveyed by three Indians standing among Colean dead trees in the left foreground. Indians and trees together betoken a receding topographical and historical threat. Ray's sublime Indian and Durand's primitives are the obverse of Bierstadt's pastoral tribe. Their retreat enhances the triumph of Christian proprietorship. Cole's lone Indian, in contrast, his bowed head gazing over the precipice, bow purposelessly outstretched, "assist[s] the idea of solitude," as Cole's patron Gilmor characterized the advantages of introducing figures into landscape, his example being an Indian hunter ("Correspondence" 44).[38]

These paintings by Cole depict exaltation without the redemption of sacrifice central to Kant's sublime economy. Even were the depicted lands owned, these landscapes are scarcely appropriable. Emphasizing the desolation and subjection of the sublime, rather than its possible renovation, these landscapes empty subjectivity of the reappropriative prospect of Cole's Kantian model. In this sense, they adopt a distinctive departure from Kant by Allston. Allston's rehearsal of the sublime is striking for its negative (more Augustinian than Protestant) account of subjectivity. For Allston, awareness of self as a "power of calling forth . . . emotion," is fundamentally a "sublime condition." Ordinarily, Allston says, we think of the self as the center toward which all energy and resources converge; but in the sublime moment "the center here is not himself." Allston's subjectivity consists not in self-recovery or the possession of self and natural law, but rather in self-displacement. He stresses the finally "unattainable, the *ever-stimulating,* yet *ever-eluding*" aspect of sublimity (67), which works as a repetition compulsion.

Cole's acclaimed series of five paintings, *The Course of Empire,* though his most populated work, effects an Allstonian evacuation of subjectivity (figures 13–17). Exhibited in the same year "Essay on American Scenery" and Emerson's *Nature* were published, the series blends the landscape and historical (specifically epic) traditions, combining the two wings of the nationalist art movement. In the panoramic style,[39] a technique becoming popular in contemporary representations of New York (Bergmann 124), the series' vantage circumambulates a harbor, following a diurnal solar cycle and the historical cycle of an empire.

The first two canvases follow a tribe from *The Savage State* to *The Pastoral State.* The most prominent feature of the first canvas, especially when viewed from a distance, is its fiery cloud movement. The cloud movement to the right reveals the distant (and eschatological) peak. Chiaroscuro effects (like those of *St. John in the Wilderness*) highlight the actions of the tribe. Conflict, social and natural, is evident in the canvas's content and in its form as well, as the tribe's fire repeats (and indicates the direction of) the meteorological turmoil. But prospects for appropriation are suggested in the revealed, distant peak and in the continuity between the line of weather and the line of human action. This composition illustrates what I have been calling the transcendental sublime ethos, or dynamic sublime, with an appropriative and redemptive exaltation fulfilling Cole's ideal of cultivation. As Cole wrote when proposing the series to Luman Reed, the entrepreneur and respected art patron who had commissioned him to "paint pictures to fill one of [his] rooms," the scene is "appropriate for cultivation" (Noble 129; letter of 18 September 1833).[40]

Pastoral State expresses the domestic or manifest sublime—cultivation utterly in concord with the totality of nature. Compared to the first canvas, its most prominent feature is an *absence* of movement. Missing are the swift vertical clouds, chiaroscuro, and Cole's trademark dead trees. The only movement is cultural (organized as intersecting diagonal lines). Rudimentary arts, conventional symbol of cultural development, are evident. Children are playing and being educated. The community farms and shepherds, of course, but it also builds ships for trade. In the left foreground sits perhaps the emblematic figure of social achievement and art: a man writes in the soil with a stick. His representational activity occupies the apex of a triangle linking the painting's other productive activities: farming, shepherding, and shipbuilding. The social evolution Cole depicts has two related effects. First, in moving his vantage back in the harbor, Cole smooths the topography, now more hospitable. Second, the shifted perspective amplifies the promontory where the tribe camped in the first canvas and where a Druidic temple now stands. So not just cultivation but indeed natural transformation seems precipitated by cultural evolution.

Thus nature and human effort appear in equilibrium here. For example, the absence of meteorological agitation is repeated in the demeanor of the laborers, as tranquil as those at leisure. Moreover, the eschatological mountain peak appears readily accessible. The harmony between nature and culture is matched by and signifies an internal cultural harmony. The painting's title, *Pastoral State,* denotes the community's political as well as its technological evolution. And no political discord is evident. All the painting's features work to make evolution

FIGURE 13. Thomas Cole, *The Course of Empire*, 1833–1836. Courtesy of the New-York Historical Society, New York City (1858.1–1858.5). *The Savage State*, 1833–34. Oil on canvas, 39¼ × 63¾ inches (99.7 × 160.7 cm).

FIGURE 14. Thomas Cole, *The Course of Empire*, 1833–1836. *The Arcadian or Pastoral State*, 1833–34. Oil on canvas, 39¼ × 63¼ inches (99.7 × 160.7 cm).

FIGURE 15. Thomas Cole, *The Course of Empire*, 1833–1836. *The Consummation of Empire*, 1835–36. Oil on canvas, 51 × 76 inches (130.2 × 192 cm).

FIGURE 16. Thomas Cole, *The Course of Empire*, 1833–1836. *Destruction*, 1836. Oil on canvas, 39¼ × 63¾ inches (100 × 160.7 cm).

FIGURE 17. Thomas Cole, *The Course of Empire*, 1833–1836. *Desolation*, 1836. Oil on canvas, 39¼ × 63¼ inches (99.7 × 160.7 cm).

appear so harmonious as to involve neither palpable change nor its accompanying potential for upheaval.

Nevertheless, Cole's serial use of panorama finally registers temporality and change, which in this series constitute the essence of nature and sublimity. He had proposed to Reed a series to "illustrate the *history* of a natural scene"; "natural scenery also has its *changes*" (Noble 129; my emphasis). In this formulation, the history of the scene *is* its changes. The question the series keeps raising, and to which it gives varying answers, concerns the agency of those changes. The third canvas, *Consummation of Empire,* indicates that the equilibrium of *Pastoral State* was only apparent, one moment in the appropriative vicissitudes of the natural scene. In *Consummation of Empire,* the transcendental stimulation of the highest peak in the first two canvases has its objective correlative in the imperial expansion wholly occupying the harbor. Like industrial, railroad, and suburban expansion in the Jacksonian era, developments Cole often remarks in his notebooks, this empire stretches to the tops of the framing peaks. The sublime peak (on the right) now houses prime real estate. Here, sublimity consists of the aroused state's subsumption of nature. As Angus Fletcher has observed, in this painting "the sublime in nature is replaced . . . by the sublime human artifact" (376).

Up through this canvas, the series appears to predict the natural culmination of exalted human agency, whose sublime goal, in Kant's schema, is totality. With even less cloud movement and light contrast than *Pastoral, Consummation* presents the empire of exaltation as the completion of the natural history of the scene. Empire consummates nature. Yet this naturalization exhibits its own contradiction. Its very title expresses irony: the absoluteness of "consummation" is qualified by the expansionism (or temporality) of "empire." Moreover, the formal tension structuring the composition of the putative consummation reveals it as but one moment in the ever-stimulated (to adopt Allston's term) sublime economy of empire. This affect is conveyed by the excess of the composition. It is the largest of the series, and population overflows from every cranny, including the roofs of buildings. Crowded, one desires to expand further. Too many scenes and narratives transpire for the eye to comprehend them at once. In this affect, the temporality, finally the mortality, of the historical moment depicted—the inevitability that it will be superseded—becomes the real sublime subject experienced by the viewer. Dwarfed by cataracts of civic stimuli, the imagination strains outward to comprehend it, and the imperial exaltation of this canvas is at once exemplified and transcended in the viewer's incessant and unfulfillable attempts to assimilate it.

The ineluctable failure of consummation implicit here is explicit in the concluding canvases. *Destruction* and *Desolation* make clear that nature has been throughout, as Noble put it, the "terrific actor" of the history (173). If *Pastoral* implies the absence of conflict in the passage of time (and hence that temporality and change are at most negligible), and if *Consummation* invites the exalted assumption that conflict can be fully transcended, the final canvases evoke history as a distinctly unteleological territorial clash. They depict nature not as an amalgam of forms but as the formidable agent of temporality and change reclaiming itself.[41]

In the "pandemonium" style of *Destruction,*[42] temporality appears as explicit social conflict, but it is figured chiefly by the resurgence of meteorological turmoil and fiery light contrast. The surmounting of human agency by nonhuman agency is emphasized by Cole's placement of his signature at the right on the base of the statue of the god of war. Before and below the statue, a woman is being violently abducted, her outstretched hands constituting a gruesome parody of supplication—to the war god? to the painter's signature? in fact, to nothing. No authority will save her. Cole also signed *Pastoral* and *Consummation* on artifacts, on the pastoral state's stone bridge and on the entryway to the imperial public hall at the right. It is such cultural achievements that *Destruction* annihilates. *Destruction*'s bridge, technology for traversing limits, collapses under the weight of frenzy, inadequate to bear the ever-expanding imagination it symbolically conveys. *Destruction*'s signature, sublating the woman's plea for salvation, evokes the futility of the reiterative desire to inscribe the will on nature and time.[43]

In the dusk of *Desolation,* clouds hardly move, and a vine reclaims a phallic column. But the sparseness of movement makes the transcendence of human endeavor all the more inexorable. The temporal agency of nature is conclusively revealed as the one force that can retain its identity through change, for its sublime identity *is* change. In the "Essay on American Scenery," Cole sees in nature the "incongruous" yet simultaneous "fixedness and motion," the "unceasing change and everlasting duration" that is the essence of the sublime (8). Finite human consciousness, however, feels change finally as its desolation. In *Desolation,* only birds (here herons), creatures capable of wider movement than humans, occupy this scene. Most evidence of settlement, like the buildings on the sublime peak, have disappeared. Human artifacts are being reabsorbed as the genesis of cataracts. *Desolation* includes the least precision of detail in the series; specific actions, the only kind of events human faculties can effect or comprehend, are surmounted by the temporality that is also nature's sublime eternity.

This revelation, however, revises the series' overall economy.[44] In the first canvas, the operative economy is Kant's utilitarian, dynamic economy of sacrifice redeemed; *Pastoral*'s economy is the harmonious domestic sublime, typologically completed, apparently, in *Consummation.* But both the transcendental and typological sublime modes are now encompassed by, and shown to be always part of, the last canvas's economy of desolation—human agency's final inefficacy to redeem temporality. As Cole asked his notebook at the turn of the year 1836, "Another year is gone. Who shall redeem it?" The affect of *Desolation* is the evacuated subjectivity in Allston's *Lectures.* As Cole writes, "the mind rushes forward" in sublime arousal, but it cannot contemplate "what is." "The lamp of experience" illuminates nothing (Noble 155).

If *The Course of Empire* ultimately thematizes an economy of desolation, its individual canvases nevertheless exhibit the two other economies of sublimity available to and practiced by the American School: the Kantian transcendental (dynamic) and the manifest, typological, or domestic. *The Course,* that is, deploys the three political economies of nature circulating in antebellum America. If the domestic ethos was predominant in the American School—epitomized in Durand's famous *Kindred Spirits* (1849) (figure 18), rendering Cole and Bryant, neo-

FIGURE 18. Asher B. Durand, *Kindred Spirits,* 1849. Oil on canvas, 46 × 36 inches (116.8 × 91.4 cm). Collection of The New York Public Library, New York City. Astor, Lenox, and Tilden Foundations.

classically framed, standing over a distinctly unthreatening waterfall—American painters were capable as well of the other modes. *Aurora Borealis* (1865), for example, by Cole's student Church, unlike the more domesticated English representations of the arctic, suggests a more classically Kantian dynamic. Some of Church's paintings of South American volcanoes as well, like *Cotopaxi* (1862), evoke primarily the inadequacy of human faculties to comprehend and act upon the sublime object.[45]

Nor do these variant aesthetics bear a consistent or allegorical relation, affirmative or adversarial, to actual political events. As I suggested earlier, the domestic or immanent ethos underlay George Marsh's proto-environmentalist criticism of American progress. And though *The Course of Empire* ultimately dramatizes the evacuated subjectivity that Allston's *Lectures* elucidates more powerfully than any other writing on art in the period, Cole's political motives differed substantially from Allston's aristocratic predilection. Though widely acclaimed in reviews and in public lectures and letters,[46] the series' criticism of America's progressive, imperial designs was both disturbing and overlooked. The series was generally taken to represent a pagan contrast to America's Christian destiny.[47] But if many needed to ground manifest destiny in the immanence of nature, Allston found it in his alternative, evacuative account of sublimity and subjectivity. The sublime dispossession of self he envisioned would issue in *"good without self."* The sublation of ego reveals "legitimate" and "permanent, universal principles" (McCoubrey 68, 64). His legitimationist politics are evident in his fairly common response to the wars in the Southwest against Mexico and the Indians: though he himself could not undertake the commission, he favored a request for a monumental commemoration to the sublime annexation of Texas (Flagg 265–67).

The efforts of the American School and the differentiable accounts of sublimity circulating in the period illustrate what I have been calling, after Emerson, the mixed instrumentality of nature and appeals to natural law. The rhetorical, ethical, and political instrumentality of the appeal to nature is multiple, often intermixed in single works, at times confused. The tradition the rest of the present book will trace is for the most part transcendental, and often also progressive. It is progressive because this was the dominant nationalist ethos of the period with which I am concerned; transcendental because this is the intellectual tradition important to even the domestic impulse of landscape art, to Emerson, and to the specific developments of corporate trusts and literary naturalism with which I conclude.

The desire for "new combinations" that we have seen both Bryant and Cole express, and that is expressed also by Emerson among many others, is a central, if controversial, element of the transcendental urge. Combinatory desire culminates historically in the corporation's consummation, the trust, and fictionally in Dreiser's Frank Cowperwood, but also in the radical unionism of Eugene Debs. Ultimately, I will examine the way a transcendental logic of agency and value was both the substrate of and potential threat to the liberal exposition of (aesthetic and economic) value. If figures like Emerson, Dreiser, Debs, Cather, and the Standard Oil Trust generally navigate the antiliberal vector of the logic of transcendent agency and value, others like Twain and Norris repress its implications for their aesthetic ideals. Howells provides an important example of ambivalence toward the structure of valuation prevalent in the period. To continue examining the appeal to natural law as the ground of value, I will turn to a movement both progressive and transcendental: protectionist economics. Antebellum protectionism will introduce us to the logic and rhetoric underlying both the definition of value in the period and also the series of protectionist economies we will encounter, beginning with Emerson's, and which Cather is most bold in relinquishing.

CHAPTER 2

Transcendentalism
and Protectionism

Protecting Markets, Perfecting Character

"The divinity of man moves him to works . . . which have their origin in no individual preference, but in that same Principle, by which the idea of the divinity is conceived as a creative power." It was by this espousal of the divine source of human agency that the transcendentalists distinguished themselves from the Unitarians and the simple individualism and empiricism underlying their liberalism. The "Principle" inspiriting individual agency transcends the singularity and mere preference of individual agents. Moreover, divine power "recognizes no property" and "is unwritten, for the same reason that the movements of the heart are unwritten." Thus this transcendent principle contests two central means (or idolatries) for defining and regimenting individuals and the instinct of the heart: property and doctrine. The fact that these transcendentalist sentiments appear in an article advocating a high protective tariff in the conservative *American Whig Review* might seem to, but does not in itself, disqualify their anonymous author from membership in the club of transcendentalism.[1] Indeed, accepting many of the assumptions fueling the quarrel with Unitarianism, he may readily have felt, as ardent protectionist Horace Greeley declared to Emerson in 1842, that he was a transcendentalist (*Letters of Emerson* 3:19).

Protectionism and transcendentalism would seem unlikely conceptual confederates. Protectionism—advocacy of high tariffs to protect the home market— was then a conservative plank, and even today, though advanced by political liberals, tariffs connote the protection of entrenched interests. Shouldn't a transcendentalist side with the free traders, who wanted social intercourse to be independent of imposed law?[2] "The philosopher," Emerson wrote in his *Journal* and later in "Politics," "will of course wish to cast his vote with the democrat, for free trade." Of course, Emerson also observes here that if the democrats had the better cause the conservatives had "the best men,"[3] and character always counted for more in Emerson's thought than particular creeds or policies. If transcendentalists were concerned foremost with the relation between character and action, and insisted on the priority of spirit over conventional form, then protectionist premises and goals were indeed those of the transcendentalists: both transcendentalist challenges to Unitarian practice and protectionist assaults on free-trade policy sought to reform internal economy (individual and national) in order to perfect character so that property could be employed as an instrument to higher ends. The attack on free trade was mounted in the rhetorical arenas of custom, nature, and universal law, delineated by Kenneth Burke as the three stages of transcendental ascent (191). My aim in remarking affinities between protectionist and transcendentalist tenets is not to expose the transcendentalists as Whigs fighting for higher tariffs, but rather to suggest that transcendentalists were adapting a logic of critique circulating widely in a culture anxious about the effect of market volatility upon character, the fundamental and distinctive property of individuals.

Henry Clay's "American System" of tariffs was instituted five years after the young nation's first major tribulation of market volatility, the crash of 1819. Supported for five more years by many Jacksonians, the American System combined a domestic laissez-faire policy with high tariffs, which were calculated by an excruciatingly complex formula. This tariff system was intended less to attract revenue than to protect "infant" industries endangered by Europe's superior technology and cheaper labor. The Nullification Act of 1832 dismantled the minimum tariff system and the Compromise Tariff of 1833 gradually reduced tariffs till a low uniform rate took effect in July 1842. Two months later, amid much hyperbole blaming the reduced tariff for the panics of 1837 and 1839, and bolstered by lobbyists like Greeley, who founded *The American Laborer* in 1842 to lobby for home protection, another high tariff was passed. The tempering of this tariff in 1846 revitalized protectionist outcry, now led by Henry C. Carey, progenitor of the American School of political economy, America's antidote to the notorious pessimism of Ricardo and Malthus (see Griswold).

The exact influence of the tariff system upon the nation's economic health was then, as it is now, debatable. Tariff policies alone probably do not determine or fully explain the dynamics of economic history. For my purposes, it is irrelevant whether the protectionists were right to claim that the "unsteadiness" of trade, production, and currency was caused by low tariffs.[4] What is significant is that protectionists imagined their cause as the way to protect self-dependence and the communion of private interest with the universal will.

Protectionists began their quest with an attack on attachments to custom and prejudice that is identical to Thoreau's vilification of custom in his search for a *point d'appui*. Clay noted the "obduracy of fixed habits" (1824, 294); and because "ties of custom" (Taussig 60) endure amid rapid alterations in conditions, practices often fail to address new exigencies. Thus, "once formed," wrote John Rae, habits "make slaves of their former masters" (2:123). By smoothing fluctuations in conditions, the tariff, protectionists argued, would improve man's capacity to anticipate and adapt to change. This practical liberation from enslavement to habit was enabled by a commensurate analytical liberation. In his 1834 treatise on the principles of economic inquiry, Rae derided free trade as merely the canonization of "prejudices" (vi). He compared the difference between his inquiry and that of the free traders to the difference between the inquiry of the philosopher and that of the "mere sailor." The sailor has only "practical knowledge" of the wind, its direction and velocity. And the wind is "simply . . . [a] particular aspect of the heavens at [any] time." The protected person, in contrast, can be the philosopher, "gathering together all that consciousness makes known to us of what is within" (iv) in order "to ascertain the nature of the wind itself" and "to inquire into the general causes producing all . . . phenomena" (2). (Rae's distinction between empiricist and philosopher resembles the one Emerson draws in "The Transcendentalist" between the materialist and idealist.)

Rae's metaphor of protectionism-as-philosophy promotes protectionism as not merely a trade policy but a mode of consciousness that comprehends natural law. His general conceit, grounding economic principles in natural tendencies, was typical of economic discourse since Aristotle, and his protectionist application of this conceit had been common since Clay[5] and the work of Daniel Raymond. A decade earlier, in the first (and for a generation the most popular) treatise on economic principles written by an American, Raymond hailed protectionism as inquiry into "the principles of nature in their utmost purity" (400).

Clay's, Raymond's, and Rae's goal of revealing natural law was shared by their disciples and culminated in the influential work of Henry Carey, son of the Philadelphia publisher and activist, Matthew Carey. Carey regularly sought to discover and invigorate "natural inclination," so often "thwarted" by "human institutions" ("What Constitutes Real Freedom of Trade?" 134). In the year Thoreau left Walden, Carey published *The Past, the Present, and the Future,* "designed," he proclaims in Thoreauvian manner, "to demonstrate the existence of a simple and beautiful law of nature . . . , a law so powerful and universal that escape from it is impossible, but which, nevertheless, has heretofore remained unnoticed" (5). Figuring his philosophical and sanctifying project, like Thoreau, as a sojourn in the wilderness, Carey sets after the universal law of nature by detailing the efforts of "the first cultivator," without family or tools, to discover and furnish his absolute necessities and thus found a civilization. Carey rehearsed this primitivist scenario in much of his work, foremost in his popular and frequently reprinted *The Harmony of Interests* (1848), wherein he argues that "the object of protection . . . is to restore the natural [tendency]" of discrete interests to harmonize with the universal will (53).

The protectionist movement's primitivist goal of universal harmony is readily understood as the justificatory strategy of a young expansionist nation anxious to secure its territories against foreign powers and non-white domestic forces. But the expansionist policy of protectionism, shoring up what are now called windows of vulnerability, was also—indeed was erected upon—a profoundly romantic vision of the divinity of the individual and consequent harmony of individual relations. Free-trade doctrines were charged with vulgar individualism, a result of absolutizing Adam Smith's notion of the Invisible Hand. Smith argued that competing private interests coalesce to serve the general welfare; hence, competition is not intrinsically a social evil. But free traders, Greeley complained in 1843, codified Smith's notion as doctrine and declared "Free or unlimited Competition"— which Greeley called "the idol of Free Trade worship"—the only social good. Likewise, Clay mocked the "favorite maxim" of free traders, one familiar today among many economists and legislators, "Let things alone!"[6] Since Smith's selfish agents need not be aware of their ultimate altruism, selfishness may remain unregenerate. Free-trade idolators effectively deny labor its just reward and exacerbate social inequity because they forget that the social compact is a complex network of relations and obligations. They imagine that production and consumption are discrete rather than inseparable activities, and that value results from a *"single act"* rather than from a series of interdependent acts.[7] Thus, just as Emerson observed that contemporary politics tended "to insulate the individual" (62), Carey wrote that free trade was the "school of discords," keeping individual action only individual (*Harmony* 77).

Protectionism reconceived and reformed the nature and scope of individualism. By "enlarging the sphere of industry," wrote Willard Phillips, protectionism would enlarge citizens' consciousness of the relation of their individual acts to an entire network of capital, labor, merchant, farmer (133). From such consciousness issues harmony in the vast matrix of social relations. The protected citizen was like Emerson's American Scholar who "raises himself from private considerations" to "comprehendeth the particular natures of all men" (McQuade 56, 59).

Protectionism's revelation of the universal intention of individual acts would foster "self-dependence" ("Paradoxes" 7). Free traders were accused, finally, of having a false conception of freedom, of equating the absence of tariffs with the "self-emancipation" Thoreau seeks at Walden. As Thoreau writes in "Civil Disobedience," free traders "postpone the question of freedom to the question of free-trade" (228), to a set of laws called the "doctrine of instructions" ("Paradoxes" 10). In his widely read "Junius tracts" of 1842, Calvin Colton echoes Emerson's scorn for the Unitarian doctrine of miracles and pronounces the doctrine of instructions "monstrous."[8] The monstrosity of free trade, like that of the doctrine of miracles, is that it renders an individual, specifically here the laborer, a mere *"Agent of Power."* Free-trade instructionism "enslaves" individual labor to cheaper foreign labor and proliferating commercial intermediaries.[9] Free-trade doctrine, goes Carey's Thoreauvian charge, "degrades" men into machines, into slaves to machines and laws; even the rich man simply grovels to maintain the burden of his property, and "is far less free than he who is sold to a master." Free trade is "precisely the sort of freedom we [do] NOT want," Carey declared in a public letter to President Buchanan after the panic of 1857, attributing the panic in part to reduced tariffs.[10]

Protectionism, proponents declared, was a higher form of individualism than the mere freedom from laws that free traders propounded. The protected self, they argued, was the truly free self. By stabilizing the market, Carey writes, protectionism "would free [laborers] from all *dependence* on the movements of distant markets, making them independent" (*Harmony* 48). Carey seeks the same "self-emancipation" Thoreau sought when he exhorted New Englanders "not [to] depend on distant and fluctuating markets" (*Walden* 4, 43). Carey's protectionism is a condition of being in which men "directly" "exercise power" over their arenas of action and experience (*Harmony* 69; "Two Letters to a Cotton Planter" 9). Virtually reproducing Orestes Brownson's summary praise for Emerson's Divinity

School "Address," Carey declares that protectionism would help "the people of this country learn to think for themselves . . . instead of borrowing all their ideas," in this case from English free traders and their doctrine of instructions.[11] The main charge against the "doctrine of miracles" was that it located authority for experience in external phenomena and in Congregational dogma, thus mediating between a person and God. Invoking Jefferson's vision of a small-farm culture, Greeley complains that free-trade legislation made trade even among neighbors "intermediate" rather than direct ("Grounds of Protection" 542). The argument was finally ontological. As would the radical ministers, the protectionists argued that evaluation should occur, wrote Alexander Everett (brother of the cultural and political figure Edward), only in "reference to our own experience" (24).

It is not incongruous, then, that Carey, originally a free trader, declared even after his conversion to protectionism that "we are *all free trade men.*"[12] Clay had insisted the same (1820, 221). By licensing the internalization of authority, protectionist legislation would commission *true* freedom. Free traders "put no faith in . . . human nature" ("Paradoxes" 2). Protectionism, on the other hand, promotes the conditions in which citizens can "distinguish between a true and a false economy," Alonzo Potter wrote (22); and just as American exponents of Swedenborg extolled intuition as the source of discrimination, protectionist discrimination would proceed, Greeley wrote, "instinctively" (*American Laborer* 194). Like Thoreau's "philosophical" economy, Potter's true or "ethical" economy concerned "conduct and feelings" (52). Improving "modes of thought" (Carey, *Harmony* 78), and elevating "the moral and intellectual culture of the laborer" (Potter 20), protectionism would secure the internal economy of the nation and its citizens, achieving what Emerson called "the domestication of the idea of Culture" (59) by enabling citizens to engage in "the highest possible activity" (Greeley, *American Laborer* 273). "A man is free," Potter pronounced, "only when . . . he has his moral faculties perfect" (12).

It is in the hope of such perfection that protectionism speaks, its goal "*legitimate* free trade," freedom to trade in the market one chooses—and because one chooses to—rather than in just "one market" (Carey, "Two Letters to a Cotton Planter" 9). One is reminded here of the lad from Vermont or New Hampshire Emerson describes in "Self-Reliance," modeled on Thoreau and rehearsed in the history of Holgrave in *The House of Seven Gables*. The lad assays many professions and trades rather than being limited to just one. The transcendentalist ideal of plasticity in occupations is a typically American vision of social mobility, and so is the protectionist ideal of flexibility in choosing markets. Because protectionism, an expansionist policy, would finally globalize the intuitive choice of domestic markets, the ultimate goal of protection's perfection of character is, Carey argued, "the abolition of custom-houses" (*Harmony* 228). Thus protectionism enacts, one advocate wrote, the paradoxical metaphor "catachresis" (Phillips 154): the figure that signifies by obliterating its own referent. By protecting the home market, and thus perfecting individual desire to make it no longer merely individual, protectionism would perfect the workings of world markets to make protection superfluous.

Unprotecting Transcendentalism:
Antinomianism in the Market

I could proliferate evidence of the logical and rhetorical affinities between the work of protectionists and of those who were called transcendentalists. The point of this comparison is not to suggest a secret or unconscious wish on the part of Emerson, Parker, Thoreau, or others of the group to join the Whig Party (although Emerson was suspicious of Andrew Jackson and his popularity);[13] or that they, like many Whigs, comfortably tolerated slavery,[14] an institution notably condemned by Carey, who joined the Republican Party at its inception. But like much contemporary conservative criticism of American values, antebellum protectionist critique of American culture was as fundamental as that by critics more familiarly branded radicals.[15] Both spiritual reformers and economic conservatives were distressed at developments that seemed to threaten self-dependence, and both formulated the problem within an available restorative discourse about nature, universal law, and character.

In the Introduction and chapter 1 we have observed the classical liberal tradition legitimating itself through a complex discourse of nature, combining Enlightenment premises (identifying reason as the essence of human nature), classical premises (appealing to nature as a set of laws for guidance), and sublime and religious, specifically Protestant, premises (exalting nature as an ideal that exceeds formalization). It was, shall we say, natural that nervousness about expanding commercialism and industrialism should be expressed by invoking natural law or a lost (probably nonexistent) natural moment. That is, the physical removal to nature we associate with Thoreau, and the spiritual or linguistic appeal to nature we associate with Emerson and Whitman, were common currency in American culture, part of the primitivist tradition mythologized in figures like Paul Bunyan, Daniel Boone, and Davey Crockett. No less than *Nature, Walden,* or Brook Farm, Carey's *The Past, the Present, and the Future* typifies antebellum "Robinsonades," as Marx, thinking of Defoe, called investigations into the origins of economic organisms.

In discussing transcendentalism's reformism, critics have often overlooked the embeddedness of transcendentalists' ideas and strategies in the culture they sought to reform. Instead, scholars have emphasized transcendentalists' antinomianism and idealism—their celebration of intuition, self-reliance, and originality—as a radical alternative to commercial and other cultural developments. When sociological criticism, best exemplified in the work of Carolyn Porter, Anne C. Rose, Michael T. Gilmore, Amy Schrager Lang, or Cornel West, discovers affinities between transcendentalist thought (mainly Emerson's) and the ambient middle-class culture, it views the affiliation as a scandalous compromise of transcendental ideals. Although, as Theodor Adorno observed, in *Prisms* for example, no cultural criticism can be absolutely transcendental—proceeding from a perspective external to culture—criticism has generally adjudged transcendentalism's immanence in its culture to discredit or vitiate its reformism. In the vocabulary of the specific topic I am investigating here—the relation of transcendentalist thought to protectionism and to the general issue of economic exchange—criticism has practiced its own protectionism, protecting transcendentalism, or wishing transcendentalism had better protected itself, from the cultural imagination it often challenged. We may say that criticism has retained the purity of classic idealism—thought's transcendence of materiality and contingency—as the necessary condition of change and virtue.[16]

Rather, as Richard Poirier has recently urged, Emerson believed that change can be measured only as an "inflection" upon prior conditions and conventions (1987, 167), to which all actors and language users are heir (and partly bound), and in which (and against which) action emerges (see Poirier 1987, 74–79). As decisively illustrated by Emerson, the relation of transcendentalist thought to market culture is best understood in this way, with reform an inflection on forms of practice available to an agent. While not identical, the relation between form and market practice is, in Emerson's view, symbiotic: for him, market practice was emblematic of the relation between spirit and form that transcendentalists were attempting to perfect. This relation is neither artificial in the strong sense nor, as many would conclude, particularly conservative.[17] Indeed, this view of practice and market forms has no inherent political impact, but is, instead, emblematic of the structure of change informing and even constituting nature.

Despite the rhetorical affinities I have assembled, protectionist desire to perfect the market would seem to breach transcendentalist aims, often formulated as critiques of the curse of trade, of material acquisition, or of what Emerson called in a late essay the nation's "great sensualism, a headlong devotion to trade" ("Fortune" 413). Due to such sentiments, one of the seminal tenets of American criticism has been the categorical opposition of intuitionism—the fulcrum of self-reliance—to a market economy. "The reliance on Property," after all, "is the want of self-reliance" (McQuade 152), and the inappropriateness of trade and property as measures of the self, value, and virtue has been central to critical understanding of transcendentalism. The postulated hostility of self-reliance to trade seems compromised by the distressing optimism of Emerson's late essays, when he could write, in "Wealth," that money is "representative of value, and, at last, of moral values," and that "property is an intellectual production" (707, 705). The ease with which nineteenth-century business executives appropriated Emersonian slogans, often framed and hung over desks, adds practical urgency to acquitting Emersonian principles from complicity with market mechanisms. This enterprise has engaged generations of critics since the 1930s—classic formalist, deconstructive, and Marxist critics alike.

The project tends to yield unsatisfactory results. Gilmore and Porter, for example, have identified even in *Nature* an incipient capitalism in Emerson's desire to appropriate nature as symbol.[18] Emerson directly exhorts appropriation in lectures he delivered in the early 1840s to groups like the Mechanics' Apprentices' Library Association and the Mercantile Library Association ("The Present Age," "Reforms," "The Young American," "Man the Reformer"). Moreover, Emerson undertook these lectures largely because during the recession of 1840–1841, the City Bank of Boston suspended dividend payments on investments from the estate of his late wife, Ellen Tucker. Thus self-reliance seems compromised from the start, promulgated as a means of income and affirming the Lockean principle that "it is the nature of the soul to appropriate all things" (McQuade 171). The arguments of Gilmore and Porter continue a tradition formulated perhaps most brilliantly by Richard Poirier a quarter century ago in *A World Elsewhere.* The poet's "aesthetic contemplation" of nature, in the opening of *Nature,* parodies and would finally "relinquish" "more palpable and profitable claims to ownership," "so that he may be possessed by and come into possession of the cosmos." But these "revolutionary sentiments" remain captive to the rhetoric of sentimental gentility; Emerson's recourse to "the social forms [he] wants us to disown" bars him from nature's alternative economy and authoritative voice (61, 63, 67, 68).

Holding revolutionary self-reliance to be absolutely antipathetic to the appropriative behavior Emerson would reform misconceives the nature of his radicalism;[19] it confuses his insistence on the appropriation and useful conversion of nature with his attitudes toward market practice in his time. About actual practice he was often indignant, and he charges the conductors of commerce with abuses and venality even when he most seems their apologist, in the closing pages of "Wealth," for example, or in some parts of "Man the Reformer." In the often approbative "The Fortune of the Republic," Emerson calls the very "spirit of our political economy . . . low and degrading" ("Fortune" 402). Nevertheless, the principle of appropriation underpinned his vision of the office of the soul, and even in early lectures commerce figures as the manifestation of the impulse to appropriate, and thus as emblem for spirit. "Every trade," reads "Trades and Professions," "if nearly examined proves to be through all its processes a study of nature," a "symbol . . . of all doctrines and all duties" (*Early Lectures* 2:115, 126).[20] "Trades are the learning the soul in nature by labor," he wrote when planning his "Philosophy of History" lecture series of late 1836 (*JMN* 12:163).[21] Commerce and property are not in themselves lapses from virtue, for Emerson; virtue or sin lies in one's relation (of mastery or dependence) to property. The more radical and activist Theodore Parker expressed a similar view, when he criticized merchants for their materialism, venality, and treatment of labor but nevertheless honored "the fierce energy that animates your yet unconscious hearts" and exhorted them to realize the "wonderful revolution" created by trade against custom and class (455).[22] Neither Parker nor Emerson doubts that persons should trade. Rather, Parker worries the ethics of commerce, and Emerson the ontology of value and how to discern that ontology. Around this cognitive problem revolve questions of identity and of the morality of action.

A dichotomy between trade and intuition is inviting, since it would ensure the apprehension of values and a self independent of the mediation of convention and material forms. "The Transcendentalist" expounds the analogy between commerce and the mediated knowledge of Lockean sensationalism. The "sturdy capitalist" is the type of the materialist, accepting at face value "the data of the senses," insisting "on the force of circumstance and the animal wants of man." Insisting instead "on the power of Thought and of the Will," Emerson's idealist "affirm[s] facts not affected by the illusions of sense, facts which are of the same nature as the faculty which reports them [the spirit], and not liable to doubt." The idealist's method mounts a challenge to materialism, enjoying a cognitive superiority to the materialist like that which John Rae's protectionist-philosopher enjoys to the empiricist "mere sailor." Emerson's idealist "admits the impressions of sense, admits their coherency, their use and beauty, and then asks the materialist for his grounds of assurance that things are as his senses represent them." The crudely empiricist capitalist suffers a headache because he cannot explain "on what grounds he founds his faith in his figures." "The senses give us representations of things, but what are the things themselves, [the materialist] cannot tell." Mistaking representations of things for the things themselves, he is victim to "illusions of sense." But the idealist, intuiting the true grounding of sensual representations in consciousness, can discern in phenomena "the laws of being," which themselves "reiterate the law of his mind"; he thus sounds the "Unknown Centre" in himself that is the presence of God. Idealist spiritualization of sensual representations is the method of self-reliant action; the idealist overcomes the materialist's liability to determination by circumstance: "I make my circumstances," he declares (87–90).

Note that the difference between idealist and materialist is not finally epistemological, nor, hence, absolute. The idealist, after all "concedes all that the other affirms" (87). He "does not deny the sensuous fact: by no means; but he will not see that alone." Rather he looks to "the other end" of phenomena, to the spiritual fact they complete (88). He does not, that is, see differently from the materialist; he uses conventional means to see further. The difference consists in a modulation of shared capacities. The idealist's "respect" for phenomena and "all products of labor, namely property, . . . as a manifold symbol [of] . . . the laws of being" (89) is justified because, unlike the sturdy capitalist, he recognizes the mediated nature of faith. This is why, to survive, "every capitalist," Emerson recommends, "will be an idealist" (87).

If he is a consummate beholder of phenomena and their mediations—if he is consummate symbologist—the capitalist may join the club of transcendentalism. This point tells us not about Emerson's entire attitude toward capitalism but about his faith in the spiritual capacity of property. As Richard Grusin has shown, Emerson himself fretted habitually about the ability, or rather the inability, of his bank to pay dividends on the late Ellen's inheritance (1988, 39–42). He lectured to recommence and enhance commerce with both the public and God. He might not advocate business school as a course of study; nevertheless, material property can be the emanation of spirit, though always the authenticity of values must be verified with respect to other spheres of action.

This interpretive faculty constitutes the appropriative authority of *Nature*'s poet. The goal of "Self-Reliance" is to command "the moment of transition from a past to a new state," to enact one's thought and will by converting them and phenomena to usable form (142). Emerson goes to nature to learn "the lesson of power": how to make nature "useful," how to "subordinate [nature] to the mind" by "the exercise of the Will" (22, 26, 33, 22). This "Art," "the mixture of [man's] will with [Nature]," seeks "the perpetual presence of the sublime" not merely to expose the limits of bourgeois property but to perfect ownership by purifying the Lockean account of the validity of property it closely paraphrases (4, 5).

For Locke, property arises when one "mixe[s] his labour with [nature]," where labor is the expression of the divine gift of reason (134). Emerson's poetic faculty also mixes an inspired self with nature, the Me with the Not Me, to generate not property as such, but the very principle of entitlement. Emerson remarks that Miller, Locke, and Manning own portions of the "charming landscape." "But none of them owns the landscape. There is a property in the horizon which no man has but he whose eye can integrate all the parts, that is, the poet. This is the best part of these men's farms, yet to this their warranty-deeds give no title" (5–6). The problem with the Lockean principle of property, with Mr. Locke's title to his field, is its discreteness and privateness, its apparent unrelatedness to other titles and to the landscape it has appropriated. The poet's comprehensive practice beholds the "original relation" of self to not-self that makes property possible. Emersonian spontaneity is a particular experience of mediation; more specifically, originality is the palpability of relations, not their absolute generation.[23] Originality is assimilative, capturing "the integrity of impression made by manifold natural objects" (5) in which discrete selves and phenomena, "private purposes" and ends, ineluctably serve their "public and universal function" (23).[24] The poet's power to achieve "perfectness and harmony" of the private and universal (13), earns him aboriginal entitlement.

The Transactions of Value

The poet's interpretive economy authenticates the value of the smallest material unit of measure, the commodity. Emerson's consignment of the commodity to the lowest class of the uses of nature is famous, and contributes significantly to critics' insistence on the antipathy between self-reliance and the marketplace. But Emerson consigns the commodity to the lowest order of the uses of nature for the way it is often used, not for its essential nature; he seeks its perfection, not its obsolescence. The problem lies in its objective and exhaustible form. When a commodity is used it is often used up, and its finite form limits the range of its physical uses. Therefore it tends to be "regarded by itself," treated as an end in itself rather than as an instrument to spiritual ends; "this use of the commodity . . . is mean and squalid." When spiritually regarded, however, the commodity "is to the mind an education in the doctrine of Use" (23). When the commodity, like nature, is regarded within this spiritual poetics and used according to the needs of the will,

> . . . the real price of labor is knowledge and virtue, whereof wealth and credit are signs. These signs, like paper money, may be counterfeited or stolen, but that which they represent, namely, knowledge and virtue, cannot be counterfeited or stolen . . . [and] cannot be answered but by real exertions of the mind, and in obedience to pure motives. (166)

This passage from "Compensation" (which Emerson had used earlier in "Trades and Professions" [*Early Lectures* 2:127]) anticipates the insight of later analysts of value (Georg Simmel and John Dewey come to mind, as do some utilitarians), an insight that, according to Marx, eluded the mystified classical economists: value is an imaginative production, a double transaction or transformation, inspiriting objects and objectifying spirit. The materialist tradition has led literary and cultural critics to focus on the latter aspect of the valuative transaction, which Marx analyzed as the fetishizing tendency of exchange-value and Georg Lukács made notorious as reification. Faced with a world of goods labeled by price, persons may experience value as a thing, a quantity, a commodity's price, or even the commodity itself. Process and experience seem to be reduced to things.

Commodification arises because of the need to represent interminable process and intangible, perhaps ineffable qualities in some determinate form. In all production and exchange, human labor, qualitatively different for different products, is abstracted into "objective form" in commodities and their price. We see not labor and skills being exchanged but television sets and cameras, or money. Consequently, Marx writes, "the social character of labour appears to us to be an objective character of the products themselves." The abstraction of labor into a quantity enabling comparison for exchange confers a "mystical character" upon commodities, which now step forth as "something transcendent," as autonomous "beings endowed with life." In the theater of exchange, men's "own labour is presented to them as a social relation, existing not between themselves, but between the products of their labour."[25] That is, "social relations between individuals" "assume [the] fantastic form" of social relations between things. As a result, price and social relations become naturalized: "proportions" of exchange "appear to result from the nature of the products" themselves rather than from social relations of production; concomitantly, "all the accidental and ever fluctuating exchange-relations" of a social organization seem the result of "an over-riding law of nature" (*Capital* 89, 86). Citizens, just like, Marx claimed, the classical economists (epigones of Ricardo), accept (or worship) price and value as natural facts.

Marx and Lukács delineate what they take to be the social and finally moral repercussions of the cognitive structure of valuation, of the fact that human agency is apprehended only in objectified form. Lukács felt that reification was the "essence of the commodity-structure,"[26] turning men, as Thoreau might have said, into tools of their tools, or rather tools of their products. Because commodities seem to behave "independently of the will, foresight and action of the producers," Marx writes, "social action takes the form of the action of objects, which rule the producers instead of being ruled by them" (*Capital* 86). Vivacious commodities seduce citizens to labor simply to acquire them, rather than to improve life. In this view, the fetishism saturating market life dispossesses moral capacity.

Thoreau's well-known charge that the "curse of trade" enslaves citizens to their labor deploys a consonant premise. The citizens he describes, their heads bowed in labor, have forgotten even labor's material objective. Concerned with things rather than with their uses, his neighbors appear "so occupied with the factitious cares and superfluous coarse labors of life that its finer fruits cannot be plucked by them." A man's "labor [is] . . . depreciated in a market," and he "has no time to be anything but a machine," not so much because his labor is exploited by capitalists as because men do not labor to live but simply to acquire new and maintain old possessions. Because it is so obsessively directed at acquisition that it becomes an end in itself, labor enslaves laborers. Labor produces only more labor and "the laboring man cannot afford to sustain the manliest relations to men" (*Walden* 3).

The materialists Marx and Lukács and the idealists Thoreau and Emerson describe similar effects of market culture. Yet unlike the materialists, the idealists do not denounce exchange as a mode of organizing economic experience or as a metaphor for experience in general. In this difference, we can detect the finally essentialist basis of classic materialist analysis, despite its antiessentialist claims, and recognize a fundamental materialism informing intuitionism. In finding fetishism the necessary consequence and condition of commodification, the Marxian analysis, as Jean Baudrillard writes, fetishizes labor and productivity themselves as transcendent values (88). Marx's analysis reduces the symbolic dynamics that by his own account constitute value.[27]

Marx understood exchange-value itself to be a phenomenon, the necessary manifestation of the processes of production. Like Saussurian linguistic value, the value of any object is representational and dynamic, arising in differential relation to objects with which it is compared. Thus money is an emblematic instance of symbolization, valuable, Marc Shell has written, because it expresses some "relationship between the substantial thing and its sign" (6). Antebellum economic writers, conservative and critical alike, fairly uniformly held this view of money as symbolic: money "is the representative of all other values," wrote the protectionist Edward Everett, Massachusetts governor and senator, later secretary of state, and once a professor of theology at Harvard, whose oratory inspired the likes of Emerson and Margaret Fuller, even if he opposed their views on religion (22). The very same language was used by John Whipple, whose 1836 pamphlet was regularly reprinted (especially during economic panics) to inspire the working classes against the greed and larceny of bankers and investors: money is "an instrument to represent the value of all other articles" (19).

More precisely than most, Marx recognized that the symbolic function of money was also active. Exchange-value, for Marx, is a late avatar of barter, the representation of one commodity by the quantity of another for which it is exchanged. When barter is no longer convenient, he writes, money arises "spontaneously in the course of exchange" as an agent of "metabolism": in "representing the exchange-value of all commodities," money transforms a commodity into money and then into another commodity.[28] Exchange-value is no corporeal or essential thing, then, but a doubly metabolic "social relation" (*Contribution* 49, 95).

Marx understood exchange-value as a specifically linguistic relation, a "communication" among commodities and kinds of labor. Money is "the language of commodities" by which they "inform us" of their "sublime reality as value." Marx terms the reality—value—that money communicates "sublime," presumably because it comprises the whole expanse of the relations—geographical, cultural, and psychological—whose individual expressions or transactions are at once insignificant and elemental. Since, in the Kantian sense, the sublime totality of social relations exceeds human comprehension, value can "betray its thoughts" only in finite formalization. Consequently, there "exists neither value, nor magnitude of value, anywhere except in its expression by means of the exchange relation of commodities, that is, in the daily list of prices current." By intimating the total social reality animating individual productive acts, "value . . . converts every product into a social hieroglyphic," which "we try to decipher . . . to get behind the secret of our own social products" (*Capital* 60, 70, 85).

Exchange-value, then, does not exactly supplement or objectify (and degrade) labor-value and use-value; rather, it is the mediation that is the condition of their emergence and expression. Labor-value and use-value don't come first in any absolute sense, but themselves have value (if, in certain cultures and periods, primary value) sublimely, only within immense networks of relations. Value emerges and can be apprehended only in transaction, only in differential relation with other objects. Only formalized in commodities and the "hieroglyph" of price, then, do the "sublime" secrets of social products exist and become comprehensible. So commodification or price is not a fall from pure labor, not the disfiguring or betrayal of value, but its very production.

Emerson, too, recognized the necessity of formalization. An object, or symbol of man's life, serves as a "hieroglyphic" of the relations constituting experience (22). In Emerson's account, humankind itself resembles Marx's money. "A man is a bundle of relations," he writes in "History."[29] A *Journal* entry of 1837 elaborates this point: "Man is but a relation;—at least all his knowledge & all his thought are relations. He subsists not from himself." Emerson compares human existence to caoutchouc, the rubber tree. "Wholly a relation is Caoutchouc," given "proportion and form" by molds, and "like man . . . an educated being," "never absolute, always conditioned." Caoutchouc is foremost "plastic," both a register of and the principle of metamorphosis (*JMN* 6:277). Emerson's idealism, then, performs a variation on materialism. Like money and value for Marx, the unconditioned self-reliance of human divinity is conditioned and metamorphic. "We are symbols and inhabit symbols," reads "The Poet"; "workmen, work, and tools, words and things, . . . all are emblems" (McQuade 312). Emerson's intuitionism specifies not a different *conception* of cognition or of the experience of meaning, but a different *content* to symbolization. For him, the referent of the formalization of agency is not social relations but divine intentions, spiritual or higher laws—or rather, the higher laws (of using) embodied in the social relations expressed in commodities. A symbol serves as a hieroglyphic of the divine "order of things" (22), and hence Emerson, like Marx, calls price "sublime" (166).

In both the materialist and the idealist visions, the category of value is paradigmatic of interpretation: a price (or another commodity, in the case of barter) represents the value of a thing; and in general value and the qualities of things make sense (and therefore exist) only in cognitive relation to other qualities with which we are already familiar. The apprehension of objects and values is a phenomenal or cognitive transaction that exemplifies the transaction that is cognition, the construal of substance, intention, and vast relations from surfaces and forms. But this means that the risk of misconstruing value is a structural necessity, for no law exists to link visible price or form unmistakably to the invisible values or relations it incarnates. Because value arises and is known only in the multiple transactions of representation, the sublime relations at once transacted and expressed in exchange will not always be, in Marx's phrase, "perfectly intelligible," that is, transparent. Both the result and sign of the nontransparency of values, commodification exemplifies the condition of cognition and interpretation. Objects of cognition (like products of labor) are perceived as external to the cognizing ego. But our phenomenological difference and distance from phenomena and from linguistic and economic tokens means that character and phenomena may not be perfectly intelligible, because we cannot comprehend them in themselves.

Commodification is one (cultural) form of the objectification necessary to any social or cognitive act. The materialist tendency (especially in postmodernist analysis of consumer culture) to decry commodification as reification, as morally disabling, misconstrues in two ways the action of commerce (economic or otherwise). First, it reprises the demand for transparency that materialists explicitly deplore in most economic, theological, or idealist discourse. Second, it arrests the dynamic quality of objectification. As Baudrillard writes, excoriating commodification is a form of sanctification (92), hypostatizing a symbolic relation of differential meanings into a semiological code of fixed signification. The relationality of value is codified as entirely systematic in order to secure psychological and political effects; the work of interpretation, or the possible "opacity" of forms (65), is obviated in the immediate transparency of the semiological code. Thus the obsession with commodity fetishism is itself fetishistic, Baudrillard argues, seeking to arrest the dynamics and risks that by Marx's own account constitute value (92–95, 98).[30]

We must remember that exchange relations entail a continual transference and retransference of energy, form, value. The smallest-scale, self-sufficient production works this way, no less than credit cards or computer brokerage accounts. In every economic instance, the form encountered is apprehended as different from itself, as the medium to other forms and actions. Indeed, this is the decisive and defining characteristic of the economic. In a structure that is identifiably sublime, and which we now associate with the Lacanian account of self, an economic form or transaction always points to something not itself: to past action or production, to future action or need. In any transaction, identity emerges in a moment of cognitive and symbolic loss or sacrifice of present form and magnitude or measure.[31]

The metabolic agent money, then, works as Donald Pease says "metaphorizing power" works, when he discusses Emerson's transparent eye-ball figure (1987, 226). Mòney is itself by being something other than itself; and something has value because it *could* be something else—exchanged, used, or consumed in some way. Metaphors, of course, compare or associate different qualities. But money and metaphors also illustrate the irreducible disparateness as well as the affiliations of things compared: goods, values, words, signs. That is, comparisons inevitably effect the figure that, we saw earlier, Willard Phillips used to describe the protectionist project, catachresis; value signifies and transpires by superseding its referent.[32] Money epitomizes the catachrestic quality of comparisons (they are never exact and always motivated; the inequity that may result is the real scandal of exchange). In this sense money is not merely, in Marx's term, an agent of metabolism; it is metabolism itself, valuable only in its capacity to become something else. As Emerson observes in "The Young American": in itself "money is of no value: it cannot spend itself" (*CW* 1:236).[33]

This idea was central to classical economic thought, which, since Hume and Adam Smith, conceived of money as the "instrument" or "great wheel of circulation" (*Wealth of Nations* 266). As Carey, quoting Mill, declared: "Money, as money, satisfies no want, answers no purpose" (*Money* 7). Adapting Hume, and at least since his 1837 *Principles of Political Economy,* Carey defined money as the "machinery" for "effecting changes of place and of form," and hence for facilitating "combination," "association," and circulation in society.[34] It is because he so insisted, quoting Hume, that money is the "oil" lubricating the social circulation by which "every thing takes a new face" (Hume 1985, 286), that Carey, in some quarters scandalously, advocated expanding paper issue and credit to ease market fluctuation and, especially, contraction after the panic of 1837.[35] For money and value are nothing in themselves but, rather, the capacity for social action, change, and combination.

Materialist analysis has tended to indict commodification as necessarily fetishistic precisely because labor-value is never directly represented in commodities or price. But in seeking to ground valuation and social forms in a first cause—in the primal moment of labor—such analysis, firstly, maintains a simplistic, correspondence notion of representation (to which classical economists like Smith, Hume, Ricardo, and Carey did not subscribe); and, secondly, as Baudrillard writes, it fetishizes labor as a transcendent "moral agency" (88). Marx, of course, appeals to labor power as a standard to measure the deviation of price from value, that is, to gauge exploitation. But defining labor power as the "uniform standard of value," Georg Simmel will later argue, begs "the question of how labour power itself became a value." "Even if labour power is the content of every value," it, "too, enters the category of value only through the possibility and reality of exchange." Labor power has value only because "the activity of labour in producing all kinds of goods [gives] rise to the possibility of exchange." Moreover, labor power is already a transaction; it "receives its form as value only" because it is expended, like Kant's sublime economy, in "a relation of sacrifice and gain" (Simmel 96). Labor power is a form of exchange both in the market and in relation to (natural) resources. A so-called "natural or self-sufficient economy" involves "the same basic form" as all exchanges (84): it involves a difference and resistance between subject and object, and a translation of one form of energy to another and from one site to another.[36] Not only are labor power and preexchange production already embedded in networks of transaction, they achieve iconic value precisely by virtue of this imbrication.

Emerson's biography hardly suggests he would care to contest this analysis. He was a consummate urbanite (or perhaps suburbanite), and famously declined, for example, to join the experiment at Brook Farm. If Emerson's appeal to nature as a model of action is not equivalent to advocating a subsistence economy, Thoreau's sojourn at Walden is a more likely candidate. Yet even as he searches for necessity and a "simple economy" to surmount habit and fashion, Thoreau recognizes the contingency of necessity.

Thoreau explicitly removes himself to his cabin at Walden and away from "all trade and barter" (1966, 43) so that he might "transact some private business with the fewest obstacles" (13) and achieve a transparent economy, wherein he hears "the language which all things and events speak without metaphor" (75). Nature ideally offers an ontological vantage from which to discern true necessity,

> to learn what are the gross necessaries of life. . . . For the improvements of ages have had but little influence on the essential laws of man's existence; as our skeletons, probably, are not to be distinguished from those of our ancestors. (7)

The essential necessaries and laws of existence are ideally ahistorical, skeletal. Yet if Thoreau's interest in naturalist studies seems ignorant of basic principles of evolution already then being promulgated, his very definition of what is *"necessary to life"* includes change and contingency as fundamental components: necessity is "whatever . . . has been from the first, or from long use has become, so important to human life that few, if any, whether from savageness, or poverty, or philosophy, ever attempt to do without it" (5).

Necessity itself is a product of long habit, and changes with changing historical conditions. If necessity is different at different times, there is no point (or *point d'appui*) before or outside transaction. The relative failures of Thoreau's attempts to cultivate seeds of virtue in the bean field, to predict the itinerary of the mocking loon, or even to measure the depth of and find the source of Walden Pond,[37] suggest that the moment of nature, simplicity, and necessity he sets out to identify is known only in a difficult transaction with volatile phenomena. We always have an account of necessity as a point of appeal, Thoreau suggests, but its substance is a function of contingent, cultural relations, which will never wither away.

If foundations, necessity, and value are known only in transaction—in spending money (Carey and Emerson), or in an unpredictable encounter with nature (Thoreau)—then, as Emerson writes of the American enterprising spirit, "All depends on the skill of the spender," "for, obviously, the whole value of [a] dime is in knowing what to do with it" (*CW* 1:236–37). For Emerson at least, exchange or commerce is not intrinsically a degradation of self-reliance or virtue but the condition in which they are achieved. Indeed, he warns in "Compensation," "Beware of too much good staying in your hand"—virtue removed from social circulation "will fast corrupt" (McQuade 165). Virtue is a function of specific actions and consequences, not the result of the (impossible) retreat from exchange. Some "adventurers" cheat Indians and others, some buy "corn enough to feed the world" (*CW* 1:236). On this view, the phenomenon of money is not itself a political question; the political question concerns its circulation, the distribution of wealth, and the nature and administration of governing institutions.

This is not to say that mistakes cannot be made, nor inequities enforced. Since value is not transparent, "abundant mistakes will be made" (*CW* 1:236). In "Compensation," Emerson decries the "infractions" committed when "the senses . . . make things of all persons" (McQuade 164). Emerson, like many of his contemporaries, called such a mistake "fetich" (*JMN* 8:281). In a late essay, "The Philosophy of Fetichism," Frederic Henry Hedge, participant in Transcendental

Club discussions, succinctly explained fetishism as well as its difference from symbolization: "Both [fetichism and symbolism] are homage paid to things, but in one case it is the thing itself, for its own sake; in the other it is the thing in its representative capacity, as sign of something else" (345).[38] But if persons can mistake signs for things and means for ends, this is all these confounded economic/symbolic acts are: mistakes, infractions, and political inequities rather than the fall into alienation implied by the notion of reification. Valorizing labor-value and use-value as Archimedean points, the materialist account of the genesis of value and its perversion in capitalist exchange is a typically normative (and formalist) vision of primitive production and consequent social relations. This very hierarchy, however, reveals and reprises the desire for and dependence on an original point of moral grounding, the natural moment Marx explicitly repudiates in classical economics but then tries to realize in the base of labor and use and need.

Though as a vocational and (even after resigning his ministry) habitual cleric, Emerson is deeply concerned with the morality of conduct, his transcendental account of the transactive nature of value holds in abeyance the moral problem implicit in his project in appropriation. One axiom of "Self-Reliance" asserts that conventional morality is not sufficient to judge of virtue. On what basis, then, do we determine which acts of appropriation are manifestations of the universal will, acts of the self-reliant poet, the true Napoleon, and which are merely acts of the tyrant exploiting vicissitudes of custom?

The Sacramental Poetics of Commodities

The transactive and symbolic nature of commodities makes their agency, in transcendentalist terminology, sacramental. "Let us learn the meaning of economy," Emerson urges in "Man the Reformer." When properly conducted, "economy is a high, humane office, a sacrament" (*CW* 1:154). Thoreau expresses a similar view in *A Week on the Concord and Merrimack Rivers*. Reading one night the "list of prices current in New York and Boston," he observes that "commerce is really as interesting as nature. The very names of commodities were poetic," and the list "seemed a divine invention, by which not mere shining coin, but shining and current thoughts, could be brought up" (229).

Thinking of economic forms as sacramental emphasizes the hieroglyphic and transactive function of forms. This function has for Emerson a structure resembling Jacques Derrida's notion of the trace, where signifying forms always point to another referent, which itself, since it can be apprehended only as another form, points to yet another referent. In Emerson's view, symbolization has this infinitely differential structure because nature does. He writes in "The Method of Nature": "Every natural fact is an emanation, and that from which it emanates is an emanation also, and from every emanation is a new emanation" (*CW* 1:124). Forms are endlessly emanative, and because symbolization is interminably transactive, forms incarnate agency, and the self-reliant person has a duty to invest their animate quality. The distinction of Emerson's poet (so reads "The Poet") is the capacity to reanimate symbolic values we ordinarily forget how to "originally use"

(McQuade 312). The poet makes the "circuit of things through forms . . . trans-
lucid to others" (316). Recognizing and repairing "the accidency and fugacity of
the symbol," the poet is poet because "he uses forms according to the life, and
not according to the form" (313). The poet invigorates the instrumentality of
forms.

For Emerson, a commodity, like any form or natural fact, is "a phenomenon,
not a substance" (27). The dynamic character of nature and commodities and of
their proper, poetic perception constitutes the prime and primal virtue of sacra-
ments, the capacity for ecstasy, a perennial aspiration of Emerson's work. "Life is
an ecstasy," Emerson avows in "Fate" (691). In this notorious essay, he opposes
life and ecstasy to fate, which "is known to us as limitation" (680). Fate, or determination by historical necessity, is, like a "mean and squalid" economy, "an
expense of ends to means." It is "organization tyrannizing over character" (674).
Ecstasy is a different economy, traversing limits, as its etymological root suggests.
Ecstasy is both an effective and cognitive triumph of individual genius, effective
because it engenders events, cognitive because we now experience our fate as truly
necessary—related to, rather than the arbitrary determinant of, our character. If
history or fate (now redefined) is finally the economy between character and context, ecstasy makes sensible "the copula" linking the two. Ecstasy does not simply
invert the tyrannical economy of fate. Rather, in ecstasy we are lifted out of a
present fate to another and, more crucially, we experience that movement.

Ecstasy, then, is the catachrestic experience of the constant transaction that
is "life," and registers and expresses the dynamic or emanative structure of commodities or any natural fact. The particular content of ecstatic movement is irrelevant here; only the *fact* of ecstasy (which may be delusion as well as love or inspiration) interests Emerson. He is interested, he writes in "The Method of Nature,"
in beholding "the true order of nature," which is the fact of "the visible . . . proceeding from the invisible," the fact of copulation, to adapt Emerson's term. In
this essay, "ecstasy is the law and cause of nature" (*CW* 1:124, 132).[39]

Ecstasy imbues all particular forms. As he does elsewhere (in "The Transcendentalist" or *Nature*), Emerson here describes nature as an "excess of life" (*CW*
1:127), a "rushing stream" whose "permanence is a perpetual inchoation"; nature
is a constant flux of and violence against particular forms. "If anything could
stand still, it would be crushed and dissipated by the torrent it resisted, and if it
were a mind, would be crazed" (124). Even nature itself is subject to its torrent,
since nature constantly "contraven[es] her own laws" (*JMN* 5:214–15).[40] Emerson's description of nature calls to mind Thoreau's ecstatic description of the
excremental events of "Spring," when nature seems to turn itself inside out, sacrificing old forms and sculpting new ones. Nature seems most itself in this sacri-
ficial destruction, where every act of destruction is also an act of creation. "What
we call nature is a certain self-regulated motion or change," a constant "metamorphosis" (McQuade 314). Emerson acknowledges that knowledge and events
are manifest only in particular forms: "Each individual soul is such in virtue of
its being a power to translate the world into some particular language of its own"
(*CW* 1:128). But experience has the power of life only in transition from old to
new states, as we have seen him write in "Self-Reliance," in the metamorphic and

ecstatic transgression of particular forms; more specifically, in one's sense that the particular forms of knowledge and actions are subtended by nature's immense and fluid matrix of relations: "the spirit and peculiarity of that impression nature makes on us . . . does not exist to any one or to any number of particular ends, but to numberless ends and endless benefit" (*CW* 1:126–27).[41]

Individual power, then, is not absolutely or discretely individual. Poirier has written that Emerson's notion of the individual is at odds "with the vulgar meanings assigned the term" (1987, 143), and Emerson explicitly contrasts his understanding of self-reliance with the common notion of which others "prate." Self-reliance is not self-sufficiency: rather, in a phrase that closely adapts Hobbes,[42] "Power ceases in the instant of repose." Self-reliance is capacity, not essence; it is "power not confident but agent." Much to popular dismay, Emerson writes, the soul *"becomes"* rather than *is*,[43] and the "self-relying soul" is such because it "works and is" (McQuade 142–43). "Do your work," Emerson exhorts. Self-reliance is action and transition, always liminal, or "vehicular," Emerson wrote in 1835, its "worth" stemming from the "transgression . . . of routine" (*JMN* 5:70). We may, then, take the term *self-reliance* to mean self-*re-lying*—the capacity to effect the transport of the self from one condition to another. And this transport is never merely individual. Emerson scorns the belief in "strict *individuals*" (*JMN* 5:344–45); rather, the self-reliant person or poet or scholar is one through whom "heaven & earth traverse freely" (*JMN* 5:351).

Thinking of the traversed self as ecstatic helps to elucidate its workings. As in its pagan origins, as well as in its Protestant legacy, ecstasy accompanies enthusiasm, the ingesting of or "inundation" (*CW* 1:132) by the divine. And enthusiasm entails the traversal and therefore sublime expansion of individual form and the individual will. Adapting the *Symposium,* Emerson employs the idea of love to exemplify "overpowering enthusiasm." Love for Plato is an intermediate tending between knowledge and ignorance, between having and not having the object of desire. Emerson's love is a similar transitional state, "never self-possessed," "all abandonment" (*CW* 1:133). The ecstasy of Jesus exemplifies the way in which the self may be "ravished" by an object of desire, specifically by the "severe harmony" of spirit (McQuade 72). In the condition of ravishment, ordinary egotistical limits are transgressed, as the self consists in its intending toward forms of life different from itself. Hence, the moment of ecstasy or enthusiasm is "so much death," for the present form of identity is relinquished. But self-relinquishment is a greater redemption, for individual agency becomes transcendent: in ravishment, "our action is overmastered and characterized above our will by the law of nature" (163). Thus, just as any particular manifestation of value does not signify in itself but points elsewhere, action and valuation express "no private will" but "the surrender of [the poet's will] to the Universal Power, which will not be seen face to face" (*CW* 1:134, 127, 132). The individual is at once willful actor and emanation of the universal will. Even the agency of Napoleon, Emerson's frequent model of the strong individual, is subtended by universal forces. "Napoleon like all men of genius," Emerson wrote in 1838, "is greatly impersonal in his habits of thought" (*JMN* 5:508–9). Agency signals simultaneously the investment and divestment of self.

Commerce, as a metaphor for cognition, works within the same sublime structure of transcendent agency; it is a "double event," to use Simmel's term (83). Simmel means by this phrase that in any transaction one thing is gained, another sacrificed. He means also a psychological, rather utilitarian description that resembles Kant's sublime economy: value arises in a sacrifice of the agent's energy, or at least in the willingness to sacrifice energy to acquire the object of value. He expresses also a fact of cognition, that it is relational: the subject feels that the content of perception or belief is compelled by the object even as these qualities are known only in comparison with habitual experiences of other objects.

Emerson's sense of the phenomenality of value is consonant with Simmel's, but even more radically sacramental, toward both the object of value (money, commodity) and the transacting agent. The object is spiritualized in that its objectivity is an event, an emanation of immaterial relations. It is both object and not-object, essence and the erasure of essence, form and the transcendence of form. In this view, objectification is already spiritualization, a point that squarely defines the central Christian sacrament, the Eucharist. The value of nature arises in precisely such a transaction, which Emerson, consistent with his notions of ecstasy and love, describes as a copulation: "All the facts in natural history taken by themselves, have no value, but are barren, like a single sex" (McQuade 16).

The myriad and ecstatic transactions of value and identity constitute a poetics of commodities. This insight is what enables Thoreau to consider commerce "really as interesting as nature," and the list of prices current "a divine invention," "the very names of commodities . . . poetic." Similarly, Emerson's poet reanimates the instrumentality of property and commodities, forms we have forgotten how to use, by using them "according to the life, and not according to the form." This is why the use of the commodity, when "regarded by itself," "is mean and squalid," but when apprehended for its service to other ends, offers "an education in the doctrine of Use." Commodities have a poetics, or rather the *use* of commodities does. In itself, the commodity, like persons or other phenomena, does not exist. Thinking of commodity-qua-commodity is legitimately a "fetich," for then "the end is lost sight of in attention to the means" (38–39). When regarded as emanation, however—as simultaneously the objectification of spirit and the perfection and subjection of the self—the commodity, like nature, is a hieroglyph intimating the divine.

Virtue as Reform and Reinvestment

The danger of fetishism, of mistaking means for ends, survives, however, since the intelligibility of the symbol (in this case, the commodity or value) cannot be guaranteed. Although "Political Economy is as good a book . . . as any Bible" in which to read "the ascendancy of laws over all private . . . influences" (McQuade 706), the proper reading of this book cannot be guaranteed. Because all practice employs sense impressions, values are in principle falsifiable; "signs, like paper money, may be counterfeited or stolen" (166). The ineluctably phenomenal character of experience and valuation means that formalism, or the idolatry of forms—the sin with which Emerson charges Unitarianism in the Divinity School "Address"—is always possible.

Yet commodification or objectification in itself is neutral with respect to ethics and epistemology. In "the transactions of the street" we may discover "the present action of the soul of this world," reads "Compensation" (154). When we "act partially," squalidly taking the formal representations of things for spiritual facts themselves, "we sever the pleasure of the senses from the needs of the character" (160). But in principle, we must use—though we must not rely on—property and material values as transitional phenomena or events in spiritual production. The mistake, he insists in "The Method of Nature," is to "attach the value of virtue to some particular practices, as the denial of certain appetites in certain specified indulgences." Sin lies in loving particularity in itself, not in the particular enterprise one undertakes. "You shall love rectitude, and not the disuse of money or the avoidance of trade" (*CW* 1:132–33).[44] The virtuous world elsewhere Emerson would inhabit already inhabits the material and conventional world; rather than disowning, it purifies market practice and the use of social forms to incorporate the personal and particular into the universal.

This attitude toward valuation and commodification is entirely consonant with Emerson's theological radicalism, which would purify rather than displace religious forms. Transcendentalists famously criticized conformity to forms and institutions. Yet it was not the principle of formal worship, mediation itself, that Emerson decried, but the formalism of established religion, the idolatry of particular forms of worship, what he called the "doctrine of forms" (McQuade 303). He preached in his resignation sermon (1832) that Jesus sought to "redeem us from formal religion." Yet "I am not so foolish as to declaim against forms. Forms are as essential as bodies; but to exalt particular forms . . . is alien to the spirit of Christ" ("Lord's Supper" 27). In "Self-reliance" he specifies that his radical exhortation against blind obedience to institutions and ritual is not "mere antinomianism" (McQuade 161), and Unitarian fears that the Divinity School "Address" denied the need for preachers and churches were unnecessary.[45] Emerson objected to the preacher as "formalist" (76), prescriber of doctrine without regard for context. The true preacher, Jesus, does not instruct but provokes. Emerson rebuked only particular "evils of the church," the fixation on doctrine. As Richard Grusin argues, Emerson's antinomianism does not reject the principle of institutionality but, rather, scorns codification of spirit in particular form.[46] Indeed, in the Divinity School "Address" Emerson specifies that retaining current rituals is superior to zealously contriving "new rites and forms" and establishing a new "Cultus": "Rather let the breath of new life be breathed by you through the forms already existing." True preaching and worship make forms and institutions "plastic and new" by spirit's ravishment and inspiration of particular forms and selves (McQuade 83–84).

The goal of Emerson's idealism was to transcend not form but forms, or rather to exhaust the particularity of forms by inspiring experience of the "infinite relations" comprising particular forms and individuals (67). The poet teaches us that "we are far from having exhausted the significance of the few symbols we use" (311). In the moment of using forms according to the life rather than the form, the particularity of a form is exhausted, used as a bridge to the sublime. As Kenneth Burke puts it, transcendence is really "a further step," when "particulars [are] treated in terms that transcend their particularity" (1968, 192). For Emerson, this liminal effect is what defines "fine genius":

> A man should know himself for a necessary actor. A link was wanting between two craving parts of nature, and he was hurled into being as the bridge over that yawning need, the mediator betwixt two else unmarriageable facts. (*CW* 1:128)

Genius epitomizes the operation of transaction and love. Its traversing affect, like the ecstasy and value of nature, is a conjugal consummation, the generative "union of foreign constitutions." Transcendentalist genius performs a copulative bridging, "whereby," Burke writes, "one realm is transcended by being viewed in terms of a realm 'beyond' it." Burke suggests that Emerson, as preacher, wants to make literal the act of "pontificating" (187, 193).

There is a profound practicality or worldliness to this pontificatory sense of transcendence; it is always in the world. The poet/genius/transcendentalist "cannot live without a world. Put Napoleon in an island prison, let his faculties find no men to act on, no Alps to climb, no stake to play for, and he would beat the air and appear stupid" (McQuade 125). Spirit, though not reducible to any particular manifestation, nevertheless requires materiality through which and on which to work. This is why Emerson distinguished his transcendentalism (a word he often expressed discomfort with)[47] from the Continental idealism being imported by his intellectual and ministerial colleagues, and which was being attacked as dreaming and fancy by Unitarians like Andrews Norton and Francis Bowen.

The worldliness of Emersonian transcendentalism is evident both historically and philosophically. What we now call the Transcendental Club he preferred calling Hedge's Club or the Monday Club, after Frederic Hedge or after the day on which it often met. Hedge ministered in Maine, and the group tended to meet on the day of the week he could visit Boston. Emerson's preferred term suggests that he associated philosophy with its occasion and community. Philosophy here is preeminently mundane. Nor should it be elevated to doctrine; even naming the group of dissenters was excessive codification. While Hedge, for example, eagerly proposed attaining "a transcendental point of view" ("Coleridge" 69), Emerson complained in the spring of 1838 (as the miracles controversy unfolded) that transcendentalism was "misnamed," and that indeed "the disciples . . . do already dogmatise" (*JMN* 5:481). When he did accept the term, in "The Transcendentalist," his definition was historically specific: "What is popularly called Transcendentalism" is "Idealism as it appears in 1842" (McQuade 87)—that is, idealism at a particular moment in America. If the term *transcendentalist* tempts its adopters to doctrine, we might better call Emerson what he called himself while writing *Nature*, "the practical Idealist" (*JMN* 5:135). He found the "completeness of system which metaphysicians . . . affect" "dangerous" ("Natural History" 11–12). So unlike his brother Charles, who when considering any issue looked "steadfastly to the First Cause," Emerson asked different questions of experience: "I take the phenomenon as I find it, & let it have its effect on me" (*JMN* 5:135). This effect is transcendence, which "means," as he cheerfully reported his friend Almira Barlow's characterization, "a little beyond" (*JMN* 5:218).

Transcendence is liminal effects, the reform of current forms of expression, practice, experience, the self. This notion of transcendence, as numerous scholars have noted, anticipates the philosophical pragmatism of William James.[48] For my purposes, the point of noting this affiliation is to emphasize first the materiality of Emerson's thought, but finally the nondeterminateness and plasticity of virtue, which in his view is the *manner* rather than the *content* of action. In *Nature* Emerson defines ethics simply as "the practice of ideas";[49] the idealist practice that is virtue is distinctly Lockean: "it subordinates [phenomena] to the mind" (32–33). More specifically, virtue is the bridging that is transcendence; it is poetry, ecstasy, provocation, nature, genius, the various terms Emerson applies to the transactive metamorphosis he seeks to experience and transact. Virtue is the act of reform, or re-form, or remaking, "the effort," he writes in "Man the Reformer," "to re-attach the deeds of every day to the holy and mysterious recesses of life" (*CW* 1:156).

Emerson does not distinguish the pontification of virtue from social and commercial activity; indeed, he uses the example of banking and capital investment (his major means of income) as an analogy for spiritual virtue.[50] Another word he uses to describe the condition of action to which he aspires is "prudence," a concept he began to evolve soon after the Panic of 1837. Prudence, which combines judgment and action, has a long philosophical history, dating back at least to Aristotle's *Nicomachean Ethics,* in which it connotes not just good judgment or practical reason but proportion and moderation. Emerson uses the term to designate proportionate action (one of his persistent topics) and even temperance, but not moderation in the familiar sense. Prudence is "a name for wisdom and virtue conversing with the body and its wants" (McQuade 222), that is—a point that should be familiar by now—the attaching of sensual stimulation to ineffable spiritual cause.

Because the prudent person can bridge sensual form, "the laws of the world are written out for him on every piece of money in his hand." It is not just money, however, but a specific investment policy that is emblematic of prudence.

> Iron, if kept at the ironmonger's will rust; . . . money, if kept by us, yields no rent and is liable to loss; if invested is liable to depreciation of the particular kind of stock. . . . Our Yankee trade is reputed to be very much on the extreme of this prudence. It takes bank-notes, good, bad, clean, ragged, and saves itself by the speed with which it passes them off. Iron cannot rust, nor beer sour . . . nor calicos go out of fashion, nor money stocks depreciate, in the few swift moments in which the Yankee suffers any one of them to remain in his possession. In skating over thin ice our safety is in our speed. (228)

The last sentence here will remind us of (and perhaps inspired) Thoreau's ironical admonishment, "Let us not play at kittlybenders. There is a solid bottom everywhere" (218–19). Thoreau's point, examined in detail by Walter Benn Michaels in "*Walden*'s False Bottoms," concerns the difficulty of locating the bottom of the pond and the foolhardiness of incessantly testing the ice; eventually it will crack.

When Emerson first drafted this sentence in October 1837 (*JMN* 5:412), he may well have had in mind that his initial response to the panic of that spring was that it rent and exposed "loud cracks in the social edifice," a remark soon followed by "There is a crack in every thing God has made" (*JMN* 5:305). He habitually imagines investment as a perilous skating. But the sentiment that when skating, safety lies in speed is the opposite of a warning, and is a prudent bit of political economy, one resounded after the recent stock market panic of October 1987, though few, preeminently Carey, sounded it after antebellum panics: if expenditure, transaction, and investment are sufficiently rapid and constant, the economic ice won't crack. For Emerson, prudence and investment are structurally equivalent, transplanting exhausted forms to new contexts. Like investment, "every intellection is prospective"; that is its greatest "value," he wrote in October (and later published in "Intellect") (*JMN* 5:389; McQuade 280).

But of course the economic edifice can and did crack, and Emerson's *Journal* entries register his recognition that irresponsible speculation was at least a partial cause.[51] Yet his sense that the panic cracked the world was not entirely remorseful. "The Crisis of Trade . . . teaches political economy" since it "constrains every man to explore the process involving the labors of so many by which a loaf of bread comes from the seed wheat to his table." The panic is thus "instructive," "show[ing] us something of ethics & . . . practicks" (*JMN* 5:441) because it exposes citizens to the "bundle of relations" that they inhabit and that compose them (McQuade 125).

Thus for Emerson the Panic of 1837 threatened the simplistic notion of (discrete) individual identity and agency that it was his vocation to criticize. The panic offered a like opportunity to Emerson's colleague, Orestes Brownson. "The Laboring Classes" (1840), written in response to the long depression following the panic, was one of the period's most thoroughgoing and controversial attacks on what Brownson calls the "individual system" of property. To relieve the national crisis, we must convert citizens from the notion that "each man is his own centre," "a whole in himself" (379). A single act of labor has value only in exchange, Brownson writes the next year in "Our Future Policy," and hence all labor is already *"associated labor"* (109), and policy must better manifest our "relations to . . . invisible powers" yoking and influencing individuals, and of which we "stand in awe" ("Laboring Classes" 379).

Like Brownson (and many other commentators), Emerson also saw the panic as a stimulus to discern transindividual relations and salve the social fabric. When the panic broke, Emerson, like others, was shocked, both by the novelty of panic (albeit widely anticipated) and by the prospect of the social conflict it threatened to unleash. Yet in May Emerson struck, perhaps in compensation, a nearly enthusiastic tone at the instruction the panic offered. "The black times have a great scientific value. It is an epoch so critical a philosopher would not miss," for as the social process "yawns apart," it "discloses its composition and genesis." The "causal bankruptcy" that put "Prudence . . . at her wits' end" was complacency with social forms and patterns of circulation. Citizens were "pleased with themselves" (*JMN* 5:332–33)—content with acquisitions. The panic painfully demands the revitalization of prudence, the reyoking of "yawning needs," which, like consummation, will generate new events. Because it occasions instruction in virtue—in poetic or intuitive vision; there is no important difference—Emerson calls the panic itself a transaction (*JMN* 5:441). In essence, the panic marked for Emerson the need to revitalize the nation's circulatory system, an outcome Carey sought (especially after later panics) by controversially calling for increased paper issue and expanded credit.

In his *Journal* Emerson sketches the difficulties and anxieties the panic engendered for laborers, farmers, merchants, and capitalists. The panic exemplified, as he translated from Goethe, how "metamorphosis is . . . inestimable, but also perilous," destroying and dissolving knowledge and forms (*JMN* 6:301). Yet Emerson's prudential analysis also partly anticipates the Social Darwinism of the postbellum period, and too quickly forgets some of the human costs of the panic. Dross is abraded in the panic, Emerson opines, a sentiment consonant with remarks even in *Nature* that property is a right measure of virtue. His analysis of social action starts to approach Pope's notorious "Whatever is, is right." "Power is, in nature," he writes in "Self-Reliance," "the essential measure of right" (McQuade 143).

Yet Emerson's sentiment is not justificatory but expository—whatever is, *is,* and is so for a reason. Truth and virtue may appear in any form, but this nonpredictability is precisely the difficulty intrinsic to his understanding of virtue, a pragmatic practice without a priori content and which differs with different exigencies and agents. Emersonian virtue intuits and enacts the subordination of the material (corporeal, commercial) to the needs of the ineffable. Virtue both treats and results in material form: its method is "to spend on the higher plane," "to invest," "that is to say, to take up particulars into generals" (720); its goal is the "power to execute design, power to give legs and feet, form and actuality to . . . thought" (702).

But the inextricability of the spiritual from the material and personal means that Emerson, though he sets out in *Nature* to discover "a true theory" to "explain all phenomena" (3), will not devise a theoretical—universal or systematic—formula for distinguishing degraded from virtuous practice, for distinguishing, that is, merely private interest from inspiration, from investment on a higher plane. Because sordid and virtuous conduct may take the same form, the morality of action—what acts are undoubtedly acts of virtue—cannot be prescribed. Indeed, to stipulate such a method for distinguishing would commit the sin of formalism: thinking that the content or meaning of action or a sign follows from its form. In one *Journal* entry Emerson typically refuses to specify what is moral: "Yesterday I was asked what I mean by Morals. I reply that I cannot define & care not to define" (*JMN* 4:86).[52] Two weeks after the publication of *Nature* (in an entry he included in "Spiritual Laws" [McQuade 190]), Emerson does define virtue: "And what is Virtue? It is adherence of actions to the nature of things" (*JMN* 5:204). Emerson's definition of virtue is radically formal: virtuous actions adhere to the nature of things. But deliberately abstaining from specifying "the nature of things," it is so formal as to be merely tautological and therefore unhelpful as a guide to action or to identifying virtuous acts. The definition must be merely for-

mal because virtue is, he wrote in "Spiritual Laws," simply "great actions," irrespective of their particular content (McQuade 192).[53] Moreover, reads "Circles,"
"there is no virtue which is final; all are initial" (272). Were virtue content-specific
or doctrinal, it would be final, an absolute formalism, and in his view monstrous.

Thus Emerson's admiration for the "great proprietor" who "represent[s] the
law in his person" does not abate though this man's acts may cause suffering
(McQuade 139); virtue or sin cannot be measured against the familiar. "Circles"
elaborates on the consequences of enthusiasm: "The terror of reform is the discovery that we must cast away our virtues, or what we have always esteemed such"
(272). Virtue abdicates its reformative and sublime authority when it attends to
conventional morality. In its use and reformation of conventions, self-reliance is
radically amoral; its relation to moral standards is irrelevant to whether "the law
of consciousness abides" (145), to whether an act is an exercise of virtue or of
enslavement to the senses and circumstance. To continue the amorous vocabulary
Emerson regularly employs, virtue is promiscuous, its affiliations infinite and protean. Still, however, virtue's indifference to standards marks the risk as well as the
power of self-reliance. "The bold sensualist will use the name of philosophy to gild
his crimes," and Emerson himself is uncertain whether he is exercising aboriginal
self-reliance or mere "Whim" when he "denies the name of duty to many offices
that are called duties"; "but we cannot spend the day in explanation" (133).

Protective and Promiscuous Acts

We may read Emersonian transcendentalism as a controlled importation of Continental idealism (indeed, imported through Carlyle and Coleridge, since his German was imperfect). Like defenses of the American tariff system, the objective of
transcendentalism was the protection and perfection of American borders and
American character against formalism and crude liberalism, against doctrine, and
against mere individualism of person and interests. As well, American transcendentalists sought the transcendence of vulgar individuality—the kind now
described as the bourgeois or liberal or autonomous self, the notion of self-reliance
currently invoked by conservatives trying to abolish social spending programs. In
trying to universalize interests and consciousness, supposedly atomized in liberal
Unitarianism, transcendentalism perfects the ideal of liberalism. As George Ripley wrote, the "Reformers" sought the "purification" of theology as then practiced
(130). The goal was to emancipate and empower oneself and all citizens to subordinate the world, to perfect the instrumentality of symbols and natural facts.

Protectionists, too, were careful to warn that property is not necessarily
instrumental. Protectionism, in Carey's view, sought to perfect "the power to
command the use of the machinery of exchange." The power to "fashion the great
instrument," to exercise the will upon the world, was the true "art" of man (*Harmony* 126), Carey said in a passage that virtually reprises Emerson's protoliberal
definition of art in *Nature* (an essay he could hardly not have known, especially
since his political allies like Francis Bowen and Edward Everett were Unitarian

opponents of Emerson). Moreover, as Emerson's poet has true "property in the horizon" by the ability to "integrate all the parts" of nature (5), Carey's goal is not simply to accumulate property, but to found and secure property truly by the genuine exercise of the moral will (1840, 78). Finally, this perfection of character also registered character's supersession by larger forces. While proposing policies to reverse the Panic of 1837, for example, Edward Everett marveled at how every transaction moves an individual beyond his proper sphere, and at how an individual act "produces an effect" in the vast network of social and organic (specifically planetary) relations, yet at the same time its act is "insensible" (22). Everett's proposals (including a stiff tariff) were meant to realize a quite transcendental notion of agency: distinct and valuable only when sublimely insensible, any individual act is at once necessary and negligible, discrete and authorized by divine, natural law. In brief, protectionists sought "true genius," wrote John Rae in 1839, which "raises the soul above itself" ("Genius" 321).

Thoreau had little truck with conservatives, with those who wanted to fix the configuration of social relations. Yet he deemed "it is a great evil to" decry the levying of duties upon foreign commodities ("Civil Disobedience" 227). Particular policy is less significant than the exercise and transformation of character, and he declares himself "but too ready to conform to" an "expedient" government that serves rather than enslaves the "conscience" (224–25). With different local aims, the protectionists were Transcendentalists: against the enslaving and liberal idolatry of free trade, they sought the conditions of self-culture.

In their interventions into liberal cultural policy, both protectionists and transcendentalists exploited the rhetoric of liberalism in order to criticize and purify its practice. The idiosyncratic, in real ways radical, transcendentalists and the conservative protectionists criticized not the principle of property relations but the impurity of the property relations (the relation between persons and property) that obtained in the liberal state. Through their protectionist economies of self, both hoped to restore the sacramental operation of the market in order to surmount the antinomy between the will and the world, the private and social, and the individual and the universal, that informed the liberal doctrines that they thought degraded mankind.

The logical and rhetorical intersections between transcendentalist critiques of the market and protectionist attempts to stabilize it show that the claims of natural law and divine intuition were neither the exclusive invention nor exclusive property of those ministers who challenged and sought to purify their liberal Unitarian heritage. Part of the American Jeremiad tradition, and of a larger philosophical and religious (specifically Protestant) tradition, appeals to nature and the universal were standard tools of critique available to a wide range of political interests. Like Emersonian virtue, appeal to the natural or transcendental was promiscuous in its availability and plasticity. In Part II, I will examine several subsequent engagements with the plasticity of nature, and three other protectionist economies. First we will examine Twain surveying the cult of the Mississippi River, and then Howells considering the agency of character in economic action. Both examples could be called realist and liberal, and both exhibit the conflict structuring the liberal and realist mind, between independence and aristocracy on the one hand (Twain), and between fair and exploitive representation on the other (Howells). What was sublime paradox to transcendentalist inquirers into the constitution of individual experience and the state, feels to Twain and Howells like discord. A related problem—the relation between individual interests and larger, speculative networks—is nearly unfathomable to Frank Norris.

PART TWO

REALISM AND
ROMANCES OF FREEDOM

"Ours by the Law of Nature": Reading, Romance, and Independents on Mark Twain's River

Independence Betrayed

In 1866 Samuel Clemens declared in a letter to his mother and sister, "Verily, all is vanity and little worth—save piloting." His literary career was not yet established, his future projects uncertain. "I do not know what to write," he complains, and he belittles the acclaim being accorded his recently published sketch about a jumping frog.[1] It is understandable that someone feeling such self-doubt should invoke a former employment of conspicuous authority. If this letter's bathos prefigures Mark Twain's later pessimism, which held all aspirations vain, in 1866 piloting signified the possibility of valuable human endeavor. In *Life on the Mississippi* piloting continues to elicit the language of value. To know the shape of the river bed, the cub must learn to read the surface of the water, and he receives from Horace Bixby the famous "lesson in water reading" (63). "Now when I had mastered the language of this water," Twain writes, "I had made a valuable acquisition" (67). De Soto, Twain explains in an early chapter, "did not value" the Mississippi "or even take any particular notice of it" (18); but "I," Twain later affirms, "am to this day profiting" from learning to read the river (125).

I begin by pointing out that Mark Twain viewed piloting as a skill of value and source of profit in order to illustrate that for Twain, economic and aesthetic values are densely interwoven on the Mississippi. Criticism of *Life on the Mississippi,* however, has too narrowly conceived the romance of piloting and the independence Twain thought it epitomized. Many have discussed piloting as an aesthetic activity. James Cox, for example, regards the water-reading lesson as perhaps Twain's most revealing discussion of art (112–13). Others, like Edgar Burde, have examined piloting as a metaphor for writing. The aesthetic values of Twain's romance are clearly economic and professional; Twain repeatedly reminds us that "a quarter of a million dollars' worth of steamboat and cargo" depends on the pilot's skill (57). Nevertheless, and even though both Cox and

Henry Nash Smith have noted that the water-reading passage, as Smith writes, "bears . . . directly on Mark Twain's thinking about the problem of values,"[2] neither explores the economics of the aesthetics of piloting.

The central value of piloting, for both Mark Twain and his critics, is the independence it signifies. Twain nominated the steamboat pilot "the only unfettered and entirely independent being that lived in the earth." "His movements were entirely free; he consulted no one. . . . Indeed, the law of the United States forbade him to listen to commands or suggestions" from anyone (93–94). This independence underwrites Twain's romance of piloting, but Twain gives two conflicting accounts of its basis. Although in this passage independence is legally sanctioned, elsewhere it appears to be absolute and natural, since Bixby claims to read the river not by means of professional techniques but by "instinct" (66), by communion with the river. Bixby's instinct is clearly a professional achievement, but his description omits its professional character as work and appropriation. With the shape of the independent river dwelling naturally in the pilot's mind, he seems not to use nature but merely to attend to it. Because it imagines independence as a communion with nature prior to and independent of institutional practice, the romance of piloting exemplifies the theme of independence from custom so prominent in Mark Twain's writings, especially his early work. Most famously, Huckleberry Finn's experience on the river seems to afford him pilotlike, instinctive independence, with the potential for democratic insurgence. Feeling at times "mighty free and easy and comfortable on a raft," as it "float[s] wherever the current want[s] her to go" (96–97), Huck learns to respect a slave as a friend and implements this feeling by risking damnation and determining to liberate Jim from incarceration on the Phelps farm. The apparent invitation to such natural freedom in Twain's early works, as opposed to the pessimism of his later ones, very much underwrites their wide mythological appeal.

But the independence gained from association with the river is betrayed, many have protested, when the river is forsaken or accommodated to the market. When Huck leaves the river, he sacrifices his moral freedom—his concern for Jim—to follow Tom and Tom's "authorities," the adventure books that pattern the cruelly protracted freeing of, as Tom imagines it, not Jim but Louis XVI, the disenfranchised Dauphin (or Dolphin). *Life on the Mississippi* has been judged similarly, through an invidious contrast regularly drawn between Part 1 (comprising substantially unchanged the "Old Times on the Mississippi" articles published in the *Atlantic* in 1874–1875) and the material (beginning with chapter 21) composed after Twain's 1882 journey on the river. In Part 2, postbellum government engineering "has knocked the romance out of piloting" and the "intruding" railroad supplants steamboating (171, 109). Moreover, the formal coherence of "Old Times," with its heroic story of piloting, gives way to episodic anecdotes and unfocused lists of population, trade, and manufacturing statistics. Thus Part 2, Van Wyck Brooks writes, is just "dull notation," resembling "the annual commercial supplement of a metropolitan newspaper."[3] For most critics, the romance of piloting evokes creativity, imagination, and independence, while Part 2 glorifies industrial progress, philistine materialism, and "the tyranny of literalism and direct observation" (Burde 884–85). As a result, Edgar Burde writes, the "individual talent" once exemplified by the pilot is "replaced by economic institutions as the chief source of authority." In this view, Twain's fascination with the facts of progress constitutes both imaginative failure and political malfeasance, his inability to transform resources into art and so to oppose the myth of progress. Both books display a fall—moral as well as formal—from the heroism they promise. In such lapsarian accounts of the river books—which indeed encapsulate most assessments of Twain's entire career—critics argue that individual authority surrenders to institutional authority, just as Huck surrenders to Tom's "authorities." Part 2 and *Huckleberry Finn*'s closing chapters both betray the democratic aesthetics of the man whom Howells called "the Lincoln of our literature" (*My Mark Twain* 84).

Now, it is true that the freeing of Jim is protracted and redundant (since Jim is already free). Moreover, Twain does note the passing of the romance of piloting, and one can scarcely argue for the narrative coherence of Part 2. Richard Bridgman has recently demonstrated the arbitrariness of the associative psychology at work there (105–20). In viewing the two books as allegories of Mark Twain's political dereliction, however, critics have misconstrued Twain's romance and its cornerstone, "independence," the attribute that elevates piloting from a professional skill to romance. Critics have too readily accepted Twain's invitation to think of independence as a natural emancipation from the social and conventional; they wholly follow Bixby, whose description of his piloting instinct omits its professional character as work and forgets that it is acquired through long training. The lapsarian account assumes that the sublime river and its readers must be out of the market to be independent. This romantic assumption, however, hypostatizes what is only one impulse in Twain's account of piloting. As we have already seen, Mark Twain often celebrates piloting as economic mastery through acquired, professional training. This instrumentalist aesthetic, as we will see by way of an excursion into the river's history, was central to the nineteenth-century imagination: appreciating the Mississippi's beauty and power was thoroughly compatible with a desire to profit by it.

Why does Mark Twain present two competing accounts of the economics of aesthetic values, one instrumental and the other ideal? Twain's romance of the free and independent pilot epitomizes at once the liberal premises central to the nineteenth-century imagination and a contradiction in those premises. Twain's romance both exemplifies and attempts to resolve a tension in the liberal conception of independence. The independence necessary to domesticate nature was no transcendental condition but rather a freedom of property rights, a function of market relations. Because Twain, like many of his contemporaries, idealized freedom of property rights as an absolute freedom, and because absolute freedom is, of course, unattainable, he aestheticized piloting as "entirely free" and the pilot's work as instinct. Both points are inaccurate, but Twain exaggerates aesthetic authority in order to justify and insure social authority. Twain's romance of independence on the river figures the idealized security of property rights that a free-market economy seemed at once to promise and occlude. He tries to secure the absolute proprietorship he desires by imagining the pilot as a king, and feudal authority as the perfection of market independence.

A River Worth Reading About

Piloting is valuable because it is "a way to make [the Mississippi] useful" (18). Piloting is thus one instance of the appropriative transaction with nature that constitutes the classical, post-Enlightenment, liberal definition of economic activity. This tradition descends from Aristotle, who defined economy as the "management" or "use of the resources" that "Nature intends and provides" for man (*Politics* 18–19). To justify private property in a world God has given "to mankind in common," Locke revised Aristotle's notion of economy as stewardship of nature and formulated the definition of property underwriting the liberal tradition.[4] Because divinely bestowed on all men, the earth belongs to no particular man, and Locke therefore pondered "how any one should ever come to have a property in anything." But "God, who hath given the world to men in common, hath also given [men] reason to make use of it." When man applies his reason, in the form of labor, to a thing in its "natural state," he removes it "out of the hands of nature," "excludes the common right of other men," and "appropriate[s] it to himself." Property results from "workmanship" that "put[s] the difference of value on everything," which untreated has none (123, 133–35, 141).

Thus for Locke, property arises in a two-step process of differentiation. First, nature is given to mankind; then, individuals appropriate parts of nature from nature and other persons, thereby making it different from itself and from others' property. Economics textbooks published during Twain's days on the river echo Locke's two-step creation of property and value. As the Reverend John Bascom wrote, no "recognized possession and value attach to" natural objects that occur in consumable form. Values are created only when man can "appropriate these natural gifts," when he "fit[s] for his use and enjoyment" "the materials about him" (24–26, 30).[5]

As one form of appropriating nature, piloting exemplifies the liberal tradition's conception of economic activity. But piloting is not the only treatment of the Mississippi described in *Life on the Mississippi.* Narrating a succession of attempts to create value on the river, the book is in a real sense an economic history of the river and by extension, since the river is the "Body of the Nation," of America. Especially because this history is anecdotal and personal, it illustrates changing attitudes toward appropriating nature and toward the economic forms this appropriation took. If we view the book, as Mark Twain did, as "a standard work" on the river, we may distinguish its motive from that of the "Old Times" articles. Critics have generally conflated the nostalgia of "Old Times" with the historical concerns of the later book, but the piloting and Mississippi works occupied distinct places in Twain's mind.

On 24 October 1874, he mailed two letters to his friend and editor, William Dean Howells. Both concern Howells's request for contributions to the *Atlantic* that would capitalize on the success of the recently published "A True Story." In the first letter, Twain abandoned this project for lack of a subject.[6] Only hours later, he wrote to announce his discovery of that subject. While strolling with a friend, the Reverend Joseph Twichell,

> I got to telling him about old Mississippi days of steamboating glory & grandeur as I saw them (during 5 years) *from the pilot house.* He said "What a virgin subject to hurl into a magazine!" I hadn't thought of that before. Would you like a series of papers to run through 3 months or 6 or 9?—or about 4 months, say? (*MTHL* 1:34)

The resultant *Atlantic* articles on piloting differ in inception from a three-hundred-page history of the river he first announced in the January 1866 letter that I cited at the outset.[7] Five years later he wrote his wife, Livy: "when I come to write the Mississippi book, *then* look out! I will spend 2 months on the river & take notes, & I bet you I will make a standard work."[8] Twain always referred to "Old Times" as his articles or "papers" about piloting, and to the longer volume as "my Mississippi book."

While composing "Old Times," he wrote Howells that "the piloting material has been uncovering itself" rapidly. Noting that the articles concerned piloting and not the river per se, Twain previously had suggested "get[ting] the word Piloting into the heading." "I suggest it because the present heading is too pretentious, too broad & general. . . . I have spoken of nothing but Piloting. . . ." Piloting was only a "portion" of the greater river subject, and as soon as he began projecting a subscription volume, in June 1875, Twain resumed speaking of "the Mississippi book." Howells shared his friend's sense that the two works were distinct. Though he never published any revision, he agreed that the "papers" on "Western steamboating" merited a different title: "Piloting it shall be," he answered. Howells referred to *Life on the Mississippi* by the same sobriquet as Twain, and when he relayed Thomas Hardy's estimate of it in 1883, he wrote that Hardy "praises your Mississippi."[9]

That "Old Times" concerns piloting while *Life on the Mississippi* examines economy on the river may be too obvious to need emphasizing. Yet this distinction is crucial to understanding Twain's attempt in the later work to describe and reform postbellum culture. If Part 1 describes piloting, antebellum economic activity on the river proper, Part 2's commercial statistics constitute Twain's attempt to apprehend postwar activity along its shore, with the "on" of the book's title now signifying a different location. The opening three chapters of the book, not part of the original *Atlantic* papers, set the stage for the accounts of appropriation to follow. Because critics have equated the intention of the piloting papers with that of "the Mississippi book," these chapters have been neglected and disparaged as mere padding to achieve subscription length. Rather, these chapters explain why, as their first sentence has it, "The Mississippi is well worth reading about" (13). They introduce the relation between man and river, or man and nature, that is central to the book.

This relation revolves around the Mississippi's status as property, and Twain presents two contrasting accounts of this, distinguishable as the river's "physical history" and its "historical history" (15). Physically, the river is "remarkable" (13) because it resists human management. By its cutoffs and flooding it "plays havoc with boundary lines and jurisdictions" (14), enabling a man to go to bed in one state and wake up in another. Island 74 epitomizes this effectively sublime quality of the river: after a cutoff, this "freak of the river" belongs to neither Arkansas nor Mississippi, and so "has sorely perplexed the laws of men and made them a vanity and a jest" (208). To those like the "old-time" mate, Uncle Mumford, who feel the muddy Mississippi, unlike clear and hard-bottomed European rivers, "ain't [the] kind of a river" (174) humans can manage, the river's mockery of human endeavor demands what appears to be a purely aesthetic appreciation of the river. One "old gentleman" recommends "unbounded admiration" for "the primeval wildness and awful loneliness of nature and nature's God." "What grander river scenery can be conceived," the gentleman declares, than the "enchanting land-scape" of the river's bluffs and valleys (332–33). Twain employs similar language to lament the effects of learning to pilot. When the cub Sam Clemens "had mastered the language" of the river, "I had made a valuable acquisition. But I had lost something, too" (67); "the romance and the beauty were all gone from the river. All the value any feature of it had for me now was the amount of usefulness it could furnish toward compassing the safe piloting of a steamboat" (68). Beauty, here, seems sacrificed to professional apprehension.

Contrasting with this primitivist aesthetic is a more appropriate mode. If Mark Twain at times employs what Smith calls the "conventional aesthetic vocabulary" of the "sentimental tradition of art" (1972, 79), he also tends "to suspect" the "lurid eloquence" of the old gentleman's encomium (332). Sure enough, the gentleman is a tour guide whose admirations are his wares. Twain is intrigued by the river's "historical history" (15), the story of human attempts to harness the river's power. In this view, the river's sublimity is not transgressed by human appropriation but is its very possibility and stimulus. This is a more Kantian vision of sublimity, though more instrumentalist than the Kantian sublime, approaching the American School's typological or domestic sublime. In the "historical" mode of appreciation, the river is always at least incipient property.

Twain's discussion of La Salle illustrates this point. He recounts how La Salle wooed the Indians with gifts, then raised a "confiscation cross" to take "possession of the whole country for the [French] king . . . while the priest piously consecrated the robbery with a hymn." Twain disdains La Salle's pious robbery not because the French appropriated "the mighty river" from nature or God, but because they "stole" it from its true "owners," the Indians (22). The river's status undergoes no transition from inviolate nature to property, because to be an object of appreciation it must be, at least potentially, property, in this case the Indians', though they themselves may not view the river as property. In the whites' view, the river is preordained to be made "ready for business" (23). As the Mississippi River Improvement Convention held in St. Louis in 1881 maintained, the river was "prepared by the Creator for the use of the people" (*Official Report* 77).

Twain's point is yet more radical. To admire the river's sublimity, even to notice the river at all, means already to have understood its economic potential. Because De Soto "was not hunting for a river," when he "found" the Mississippi "he did not value it or even take any particular notice of it" (18). In Twain's formulation, when De Soto has no interest in the river, not only does he fail to remark its majesty, but he hardly perceives it.[10] The "mere mysteriousness" of the river's course and size "ought to have fired curiosity," but "apparently nobody happened to want such a river, nobody needed it . . . ; so, for a century and a half the Mississippi remained out of the market and undisturbed." The river's power alone does not stimulate curiosity. Instead, the mystery, resistance, and "awful solitude" (21) of the river are remarked and admired by Europeans only after they have remarked its economic potential. The Mississippi is referred to as "the mighty river" (22) only once there is a dispute about who owns it, which race or which nation.

Thus the opening chapters introduce two aesthetics. One, akin to the dominant critical view of Twain's river books, values the river for its sublime independence of human jurisdiction. The other, closer to the Kantian sublime, values the Mississippi as a corollary to human desire; the river's power is not so immediately absorbed to human convention as in the American School's typological sublime, but its resistance to labor has meaning only in the context of human desire to exploit its resources. The first view rehearses a conventional view of American romanticism, which hypostatizes remarks like Thoreau's that Walden is "too pure to have a market value" (134). The assumption here is that we can purify politics and social practice by stepping outside human perspective and practice to an Archimedean ground that might authorize practice. This goal is not just counterfactual but impossible, and if achieved could not issue in any practice, because practice—embeddedness in human conceptual and institutional structures—is its antithesis and contamination.[11]

In the previous chapter I argued that Emerson's emblematic transcendentalism does not entertain such a primitivist ideal.[12] Indeed, the moral (or rather amoral) force of transcendentalism derives from its particular deployment of the Lockean justification of property that structured the most basic premises of American consciousness. Nevertheless, nature was, for Transcendentalists, the ground of value. Appeal to nature as the ground of value was a strategy available as well to the citizen of the 1880s, like Howells's Silas Lapham. Silas defends his controversial advertising practice—apparently a common one, using hillsides and large rocks as billboards—by declaring that nature "was made for any man that knows how to use it." "I never saw anything so very sacred about a big rock, along a river or in a pasture, that it wouldn't do to put mineral paint on it in three colours. . . . I say the landscape was made for man, and not man for the landscape" (14). Howells means for us to laugh at Silas's unselfconscious confidence, and many who held primitivist ideals, like John Muir, scornfully cited this commonplace as the epitome of vulgarity (Watkins 18). Yet Silas's treatment of natural resources in his paint business reveals him actually to represent the voice of nature: he is solicitous of nature's gifts, a propriety most obvious in his Persis-brand paint. Others conduct business by manipulating stocks, caring little for the integrity of any product. Howells's novel, and the realist aesthetic it propounds, husbands a physiocratic ethos, urging that while nature is an occasion for enterprise, we must treat nature properly for our values to be justified.

Transcendentalism and Howellsian realism suggest the profound concord between romantic and post-Enlightenment concerns with the duties of appropriation. But Mark Twain, we will see, finally pursues a radically primitivist notion of unconstrained independence and power. Though no less an act of appropriation than commerce and engineering, piloting possesses for Twain a natural aura that he does not accord postbellum industry, and which he in fact has fabricated. His account of the pilot's absolute independence and natural interpretive method is, at best, exaggerated; and as the following overview of attitudes regarding acceptable forms of appropriating the river reveals, postwar engineering rhetoric in fact exhibits a firmer commitment than did antebellum piloting to obeying natural law. Twain's fantasy about the powers of piloting illustrates a confusion informing his and much popular imagination of the natural law of appropriation.

"Ours by the Law of Nature": Mastering the River's Caprices

The liberal impulse was baldly expressed when Spain closed the Mississippi to trade in 1802. Recent American settlers protested on grounds that they had improved the river and its surrounding lands. "The Mississippi is ours by the law of nature; it belongs to us . . . by the labor which we have bestowed on those spots which before our arrival were desert and barren."[13] In so affirming (probably paraphrasing) the Lockean principle that "the law man was under was rather for appropriating," the settlers' protest rather crudely reduces Locke's two-step notion of the creation of property, virtually eradicating any difference between labor and nature. Labor here not only establishes proprietary rights but constitutes the law of nature itself; and nature is essentially the application of human effort. Later histories of the river and petitions for federal aid for river improvement proudly invoked the settlers' protest.[14] And the settlers' liberal impulse to appropriate was deemed not just a right, but an aesthetic principle.

Timothy Flint, popular novelist and author of travelogues and natural histories, articulates an exemplary early-century aesthetic of the river. In *Recollections of the Last Ten Years in the Valley of the Mississippi* (1826), Flint describes his feelings upon first glimpsing the river. Everything seemed "different from the same things elsewhere." The river's uniqueness relieves Flint's ordinary sense of "sickness and sorrow": "this novel and fresh scene revived . . . delightful images of youth, the springtime of existence . . ." (87–88). Ironically, this primordial spring is populated with tombstones, the "imperishable traces" of generations that have perished by the river's power. Therefore, Flint writes in 1832, "no thinking mind can contemplate this mighty and resistless wave . . . without a feeling of sublimity" (*History and Geography* 87).

The correlative to nature's sublimity is progress. Having "contemplated nature . . . in the original," Flint regards himself an acolyte and "earnest lover of nature." What he admires most is the "magic" "transformation" of "wilderness to fields and orchards" that the river has inspired. The "improvement" of nature, despite "sickness, solitude, mountains, the war-hoop," and other obstacles, is no less sublime than the river's threat to human generation, as unparalleled in history as the river is unique in nature (*History and Geography* x, 130). Flint celebrates the aesthetics of a Jeffersionian political economy of improvement. In contrast to the "dishonest arts" of the speculator, with his "barbarous," "zigzag," "ugly farms," the backwoodsman, working the family plot, creates "virtuous" freehold-ing farms of "beautiful simplicity," "surveyed in exact squares" (*Recollections* 171, 180, 194). Flint's aesthetic logic, in which beauty and sublime power inspire regulated and regulating art, is reproduced in abbreviated form in Samuel Cum-ings's *The Western Pilot*, a river guide first published in 1822 and regularly updated. The islands on the Mississippi and Ohio rivers, Cumings writes, "are of exquisite beauty, covered with trees of the most delicate foliage, and afford the most lovely situations for a retired residence."[15]

Conceiving the river as a gift to commerce and domesticity, Americans acclaimed its capacity to "add vastly to the wealth of the nation" (Ellet 223). Americans sought to improve the river because it was "the property of the nation" (Hayes 619), fashioned by nature for the nation's benefit: "The Mississippi River system [was] constructed and presented by nature at no cost to the people," read one river history (Gould 333).[16] The purpose of this gift was human gratification. "The great valley of the Mississippi," explained an 1843 report to the Senate, was "destined to afford, in the most lavish abundance, nearly everything that human wants can ask." Here, the valley's greatness results from its capacity to oblige human desire and inspire "the natural progress" of American civilization, "des-tined" to surpass "the historical wonder of Egypt."[17] This paean to natural prog-ress contains no hint of primitivism. Nor do the many reverent references to the enormous trade conveyed on the river, said to equal the nation's entire foreign trade or all of Europe's commerce.

But the river's natural state impeded natural progress. Turbulent currents eroded the soft riparian soil, and various obstructions—called shags, planters, and sawyers—threatened cargoes. Like any piece of property, the river must be main-tained. "Like the old woman's house floor," Cumings reminded pilots (3), the riverbed must be swept clean. Men lacked the technology for this household main-tenance, but in language that may remind us of Foucault's analysis of the modern disciplinary society, the authors of an 1823 report to the House of Representatives hope that "the Mississippi will one day be confined, by stable limits, to its bed," with "the empire of its caprices" "mastered by artificial embankments."[18] The sublime aesthetic requires the techne or "'physics' . . . of power" that Foucault has identified as the disciplinary mechanism in the protomodern state (1979, 215).

In the meantime, the steamboat was the most effective means of mastering the river's caprices. An old (and probably unironic) story recounts an elderly slave's response upon seeing a steamboat for the first time. Though like many people skeptical about the newly invented vehicle, he was awed by its ease of movement. "By golly," he shouted, waving his cap, "the Mississippi's got her Massa now" (Hulbert 176). Without this legend's racist overtones, Henry Clay praised the steamboat in similar terms: "Nature herself seems to survey [the steamboat], with astonishment" and "in silent submission" (1824, 293). In gen-eral, Americans saw in the steamboat "sublime power and self-moving majesty" to match the river's own; it seemed to achieve a century's progress in five years, matching the "art of printing" in its benefits.[19]

However remarkable, the steamboat alone could not achieve sublime mastery. Various techniques of river improvement were available: clearing snags and sawyers, constructing levees, evening channel depth. Debate about which to employ weighed financial rather than technical questions, assessing especially the propriety of government funding. Before the war little federal money funded improvement, and this money was devoted mostly to clearing obstructions. Levees were built and maintained by individual proprietors of the river front. This state of affairs had both practical and theoretical causes. Practically, landowners could little afford appeals to Congress, whereas navigation interests (boatowners and merchants) had capital to invest in the fight against taxing their profits.[20] As a matter of principle, however, nearly everyone resisted the idea that the federal government should fund improvement. Clearing obstructions perhaps deserved tax appropriations because the river was, after all, public property, but appropriating funds, especially federal funds, for levee construction and upkeep seemed "a reclamation project for the benefit of private property."[21] Not until 1852 did an engineer, Charles Ellet, Jr., advocate levees as necessary to any comprehensive program of river improvement.[22] Such a sensible recommendation was so long in coming more because of laissez-faire anxiety about government intervention in private affairs than because of limitations in scientific understanding.

The war ravaged the private levee system; uncontrolled flooding and erosion reduced 1870 land values in Arkansas, Louisiana, and Mississippi to one-fourth their 1860 levels (Frank 31). Individuals could not finance renovation, but the federal subsidization now undeniably necessary would no less deniably subsidize private citizens. Rhetorical intensification allayed (as well as signaled) nervousness about government intervention in private affairs. The river grew in geographical and commercial size: in 1874, one congressman likened its improvement to "conquer[ing] the Mediterranean"; another proclaimed the river to be "worth all the canals in the world," an estimate that ups the ante from antebellum days, when the river's commerce was said merely to equal the trade of Europe.[23]

As the river grew rhetorically, its character transformed. Proponents of federal funding effectively elided the fact of government intervention by imagining a virtual merger between natural desires and actions and those of man. "Nature seems to have been mindful of the wants of the interior," declares a Louisiana congressman on 4 June 1874. But, adds Barbour Lewis of Tennessee, "nature's lavish gifts" have been "in vain," because obstructions "neutralize" the river's "free and unrestricted commerce"; thus the valley itself, "the richest and oldest portion of our nation, is robbed of one-half or more" of its commerce.[24] This logic completes the abridgment of Locke's two-step justification of property. Commerce and the desire for commerce belong not to man but to nature itself—to the interior, the valley, or the river. Whereas Americans once sought to master the river's caprices, they now want to fulfill them; more exactly, the river no longer has caprices but productive desires. With partisan interests attributed to nature (and thereby no longer partisan), nature itself (rather than any person or interest group) becomes the beneficiary of labor, which now seems not just a Lockean right but a moral imperative. Mankind seems to have been given nature to protect it from itself, and government intervention is justified because improvement helps nature gratify its desires.

With improvement no longer regulating nature but, rather, liberating and ful-filling it, river policy transcends partisan interest and becomes the test of human sublimity. The first Mississippi River Improvement Convention, held in St. Louis in 1877, proclaimed that the river hosted a commerce of "transcendental magni-tude" (*Memorial* 5). In the same year, President Hayes called improvement "a matter of transcendent importance" (619). With transcendence at stake, improve-ment comes to be promoted as "the grandest of human enterprises," man's most "sublime undertaking." The debaters' hyperbole registers their discomfort at allo-cating funds for private reclamation. At this point, as part of partisan hyperbole by which philosophical liberals justify intervention in the private sphere, we even begin to encounter rhetoric about obeying nature's laws in the transcendentalist tradition of Emerson or Thoreau or Henry C. Carey. We may master nature only by apprehending the "obscure" laws that rivers "obey just as much as the planets in their orbits." Proper improvement, Barbour Lewis declares, "is only carrying out the system of nature; it is only following the law of the great Mississippi River exhibited by itself."[25]

These arguments climaxed on 21 June 1879 with a speech by James Garfield, then a House leader from Ohio. Garfield spoke in response to a joke recounted on the House floor, which nicely encapsulates the primitivist position that the Mississippi is not subject to human management.

> It was said . . . by a great and eminent politician of Mississippi . . . that there were some things which were subject to the laws of science; that there were some things which could be controlled by man's ingenuity and man's devices; but that the Mississippi was not one of those things. He said that God Almighty, when he made [it] and bade its great floods flow from the mountains to the sea, said, "Let her rip; there is no law to govern it." (Laughter)

Garfield insisted to the contrary that this "most gigantic natural feature of our continent, far transcending the glory of the ancient Nile," must be brought under "permanent" "management." It is the duty of government to "devise a wise and comprehensive system" for "perfecting this great natural and material bond of national union" (*CR,* 43d Cong., 1st sess., 21 June 1879, 2283–84). Garfield's speech about systematically and permanently perfecting nature was received with applause and cries of "Vote! Vote!" A week later the Mississippi River Commis-sion was established. The commission's duties rehearse a naturalist version of the regulatory function Foucault discovers in Bentham's panopticon: to devise "a sys-tem of observations" to perfect the river by designing, in Garfield's term, a "com-prehensive system of control."[26]

Miraculous Readers; Absolute Owners

Five decades of debate over the Mississippi conceptually transformed nature's sublimity: as the need for federally funded improvement increased, what once resisted human desire and endeavor became an emblem of human desire and value, and, as such, a transcendent power, admired through systematic regulation. Debate over policy gradually exalted nature to justify intervention in the natural workings of the state and in the business of independent individuals. Mark Twain's standard work on the river reverses this narrative, and radicalizes its premises; Part 1 of *Life on the Mississippi* imagines "the absolute fact of the mysterious river," as James Cox has written (126), in order to create a theater for the pilot's "marvelous precision" (56). In this transformation, Twain's romance of piloting dramatizes the fantasy of economic power potential in the liberal vision of independence.

The extreme character of Mark Twain's declaration of the pilot's independence is clear when read in its entire context. The reason Twain "loved the profession far better than any I have followed since" (93) is that the pilot

> was the only unfettered and entirely independent human being that lived in the earth. Kings are but the hampered servants of parliament and people . . . ; no clergyman is a free man and may speak the whole truth . . . ; writers of all kinds are manacled servants of the public. . . . In truth every man and woman and child has a master, and worries and frets in servitude; but in the day I write of, the Mississippi pilot has *none*. (93–94)

Twain was fond of this notion of the independent pilot in a world of slavery. He once wrote Will Bowen, a boyhood friend who preceded him into piloting, "that *all* men—kings & serfs alike—are *slaves* to other men & to circumstances—save, alone, the pilot." Therefore pilots are "the only real, independent & genuine gentlemen in the world" (*Letters to Bowen,* 25 August 1866, 13–14).

Mark Twain greatly exaggerates the pilot's absolute independence and monarchial power. In his 1889 history of Mississippi navigation, Emerson Gould, himself a former pilot, disputes Twain's claim that the pilot was "entirely free," legally forbidden "to listen to commands or suggestions" from others (94). No such law existed, says Gould; the pilot harkened to the captain, who determined the steamboat's destination and oversaw its course (489–90). Twain's characterization of the pilot's reading, for which his river book is most famous, is no less an exaltation.

The mechanism by which Twain exalts piloting is the pilot's famous instinctive and natural interpretive capacity. Instinct becomes an issue when Horace Bixby admonishes his inattentive cub to learn the A-B-C of the river. Bixby identifies the points the *John Paul Jones* passes along the river. The cub doesn't realize these are items to study, and when examined can remember none. Admonished "to know the shape of the river perfectly" (58), "to know it just like A B C" (49), the cub memorizes "a long string of these names" (57). Cataloguing alone does not reveal the river's shape, however, for its surface features are protean and even illusory. As the boat approaches any point Twain has selected for study, its shape changes:

> the exasperating thing would begin to melt away and fold back into the bank. . . . No prominent hill would stick to its shape long enough for me to make up my mind what its form really was. . . . Nothing ever had the same shape. . . . (61)

Lines, fringes, or ripples on the water can signify reefs, bluff reefs, branches or logs, or nothing at all, mere wind reefs. Conversely, treacherous obstructions can lurk below smooth water. Nighttime is worse, for then remembered points will be either invisible or misleading, with straight lines on the shore appearing curved and curved ones straight.

The semiotics of the river, then, amount to a critique of empiricism, one Twain inaugurated earlier when discussing the deficiency of "a remark which states a fact without interpreting it." If one simply lists facts, like the "astronomical measurements" of a sunset or a catalogue of its colors and their wavelengths, "you get the bald fact of the sunset, but you don't see the sunset." Similarly, "standing by itself" the calendar date of De Soto's discovery has no meaning (15–16). For Twain, as for Emerson, formal features should not be taken out of context, as values in themselves.

Because the river's physical features don't in themselves reveal "the *shape* of the river," Bixby insists that "you only learn *the* shape of the river," the one "that's *in your head,* and never mind the one that's before your eyes" (59). In Bixby's antiempiricist advice, piloting safeguards against perception, converting untrustworthy visual data into knowledge by inverting the ordinary visual process. The pilot masters the form before his eyes by willfully ignoring it. In Bixby's account, piloting is an epistemological romance, apprehending the real only by bypassing visible phenomena.

But if visible features provide no reliable clues to the river's shape, how does that shape ever become available? Instinct succeeds where empiricism fails. After encountering a "wind reef" that looks "exactly like a bluff reef," the cub moans, "How am I ever going to tell them apart?" "I can't tell you," Bixby replies, "It is an instinct. By and by you will just naturally *know* one from the other, but never will be able to explain why or how you know them apart" (66). It does not matter that empirical procedures cannot with any certainty ascertain the shape of the river, because after sufficient training—"by and by"—one naturally and inexplicably "just knows" its shape. Bixby's formulation implies that one must never learn the shape: the knowledge is natural, untransmissible, even unlearned, although admittedly a product of laborious training (lasting two to four years).

Bixby's advice dispels anxiety about knowledge[27] by simplifying Locke's declaration of the compatibility of nature and labor. Bixby's simplification is less obvious but even more radical than the 1802 settlers' equation of nature with labor. In piloting, labor disappears in natural instinct; piloting is so natural it is not labor. Piloting exchanges the "exquisite misery of uncertainty" (77) for the authority of effortless knowledge: "how easily and comfortably the pilot's memory does its work," "how placidly effortless is its way" (87). This ease is why piloting is a "wonderful science . . . and very worthy of [the reader's] attention" (93). The pilot's instinct is worthy because its natural operation relieves anguish about the character of nature and about the expenditure of labor necessary to discover it. Piloting interpretation is work that is not work, mastery that is ease, power that is effortless.

This assuagement is the real basis of the romance of piloting. Although instinct (the pilot's professional art) is acquired through a lengthy and difficult apprenticeship, its quality as labor is erased. The pilot's memory works *"uncon-sciously,"* noting bearings and depth changes "without requiring any assistance from *him* in the matter" (87). Twain's need to eliminate individual labor and agency from the art of piloting—a principle he often applied to any writing or art[28]—leads him to omit mentioning the work of loading and unloading cargo when docked. He glorifies the divine authority and leisure of the pilot, when in fact the pilot had to supervise loading.

There are obvious reasons that reading the river must be as automatic as possible, but this exigency does not require that piloting be effortless. Mark Twain's aestheticization of piloting achieves the romance of piloting by disguising the fact of labor. When he early on discovers that he must sacrifice sleep and a warm bed to steer the boat at night, the cub begins "to fear that piloting was not quite so romantic as I had imagined it was." It now seems "very real and worklike" (47). The cub's sentimentalization of piloting dissolves in the reality of work. The whole point of laboriously learning instinct is to forget having to labor. When he has acquired instinctive reading skill, then the pilot can perform his job, a former pilot wrote, "without realizing that he is making any mental effort" (Merrick 93). With the labor of reading indescribable and invisible, purportedly nonexistent, piloting seems not an act of interpretation but rather a moment of aesthetic, nearly erotic, intimacy: "the face of the water . . . became a wonderful book—a book that was a dead language to the uneducated passenger, but which told its mind to me without reserve, delivering its most cherished secrets as clearly as if it uttered them with a voice" (66–67). When the labor of reading disappears, the water confides in the pilot, and his knowledge enjoys the immediacy of the spoken word that Plato in the *Phaedrus* charged the written word with lacking.

This erotic aestheticization of piloting intensifies (or perhaps constitutes) the gratification aroused by its economic rewards. Twain figures instinct's placid storage of information as a laying-up of treasure that is apparently exempt from Jesus's caveat about the transience of earthly reward: "how *unconsciously* [instinct] lays up its vast stores . . . and never loses or mislays a single valuable package" (87). The utter security Twain associates with instinctive knowledge inspired from the outset Twain as well as the cub. Albert Bigelow Paine tells us that Twain, like the cub, was prompted to explore the Amazon by reading "an account of the riches of the newly explored [upper] regions" of the river (*MTL* 1:33). Unable to sail south, the cub "must contrive a new career," and he enters "upon the small enterprise of 'learning' the Mississippi" (45), where he discovers more familiar riches. Once a licensed pilot with a regular berth, Sam Clemens bragged to his brother Orion in 1859: "I can get a reputation . . . [and] can 'bank' in the neighborhood of $100 a month. . . . Bless me! . . . what vast respect Prosperity commands!" This wealth is not measured just monetarily. A trick Clemens played on those pilots "who used to tell me, patronizingly, that I could never learn the river" illustrates both the respect he sought from piloting and the integral relation between that respect and the art of piloting. When paying dues to the Pilots' Association, Clemens made sure "to let the d—d rascals get a glimpse of a hundred dollar bill peeping out from amongst notes of smaller dimensions, whose face I do not exhibit" (*MTL* 1:43–44). The respect due prosperity is not merely a sham exploiting the inability of others to read the shape of Clemens's money roll. Twain's ruse exemplifies the art of piloting, its expertise in what visible evidence disguises.

The romance of piloting begins with a (cognitive and appropriative) transaction with nature, naturalizes that transaction to disguise the labor of reading, and culminates in the claim of a free self, independent of circumstances and human institutions. This romance of the unencumbered self is not an Emersonian, liminal transcendence of institutions but rather a fantasy of complete transcendence. The fantastic quality of Twain's romance is evident in his exaggeration of the pilot's skill. Another former pilot disputed Twain's characterization of the pilot's unconscious steering techniques.

> While the pilot was running a bend "out of his head" in darkness that might be felt, there were always well-known landmarks to be seen—shapes of bluffs so indistinct as to seem but parts of the universal blackness. But these indistinct outlines were enough to confirm the judgment of the man at the wheel in the course he was steering. (Merrick 88)

It seems that Twain "elevated matters of professional skill to the level of the miraculous," as Louis C. Hunter writes in his standard history of western steamboating.[29] Most pilots were familiar with only a limited portion of the river system, and they liberally referred to river guides, called navigators. I referred earlier to the most popular of these guides, Samuel Cumings's *The Western Navigator,* first published in 1822 and updated with increasing frequency through 1850s. Besides consulting navigators, pilots referred to newspaper reports of river conditions and routinely shared information with other pilots (Hunter 245–46).

Neither self-contained oracles of command nor miraculous readers of nature, pilots belonged to a professional network and operated empirically with the aid of books, charts, newspapers, and colleagues. Mark Twain's undoubted familiarity with this fact helps explain why he valued the Pilots' Association as "perhaps the compactest, the completest, and the strongest commercial organization ever formed among men" (99). Edgar Burde (following Van Wyck Brooks) considers the Pilots' Association the death of heroic reading and the compromise of the self because it subordinated the pilot's innate and individual talent to institutional authority (881–84). But Twain thought the association "a beautiful system— beautiful," he wrote Will Bowen (13), precisely because it consolidates the pilot's independence and power.

Since it organized informal cooperation among pilots into an efficient network of communication, and thus streamlined communication between water and pilot, the Association insured individual control over property, thereby enabling Twain to imagine the pilot an absolute monarch, or even super-regal authority—"here was the novelty of a king without a keeper" (94). "A cramped treasury overmasters" a monarchy, but not "the old business of piloting" (*Letters to Will Bowen* 14). Twain's hyperbole transmutes institutional power into a freedom from all economic constraint, a strategy that both idealizes the freedom offered by a market economy and reprises the idealization of freedom embedded in the liberal vision. The essayist W. S. Lilly expressed the characteristic view that the "essential condition" of liberty was unconditional control of property. Liberty was not a Rousseauian birthright antedating and compromised by convention, law, and the social contract, Lilly wrote. Rather, the history of civilization is "the history of the gradual growth of personal liberty and of private property. The two things arose together, they developed together. . . . Property is nothing else than liberty realized" and liberty is "the power of doing what one likes with one's own" ("Shibboleth" 509–11).

As we have seen, this laissez-faire control structured debates over funding river improvement. This precept was at issue, too, in debates surrounding the Steamboat Act of 1852. Although he relates the story of his brother Henry's death in a steamboat explosion, Mark Twain omits that accidents and explosions were frequent, the result of pilot incompetence and owner negligence. The 1852 Steamboat Act regulated maintenance and for the first time required formal licensing of pilots (Hunter 243). Pilots and boatowners opposed such regulation as "an unjustifiable violation of human liberty" (Gouge 423). Regulation may reduce accidents and save lives, they conceded, but true life and freedom require complete control of property. "Consider the value of a man's life compared with his happiness and his liberty, with the freedom and happiness of our race," insisted one senator. "Life is transient and evanescent, but liberty and equal rights, I hope, will endure as long as man shall endure." Freedom of property, here, transcends and gives value to life. "Let me ask you what will be left of human liberty" if government will "control [man] in the investment and management of his property. . . . Can a man's property be his own, when you take it out of his own control . . . ?"[30] Regulation threatens ownership, and since the self seems scarcely distinguishable from its property, when control of property is constrained, so is the self.

The understanding here—that life is an aspect of one's property—revises as it inherits the Lockean tradition. For Locke, we saw in the Introduction, economic transaction with nature axiomatically expresses the integrity of the self. Each person owns himself, owns an "unquestionable" and exclusive "property in his own person." It is this sense of self as a divinely granted principle of self-possession that makes "the labour of his body and the work of his hands . . . properly his." Proprietorship is imagined as incorporation, the mixing of self with objects. In the ideal, then, property is the pure expression, or vessel, of self. The argument against steamboat regulation focuses on Locke's declaration that in "a state of perfect freedom" man is "master of himself," able to act "without asking leave or depending on the will of any other man" (Locke, *Second Treatise* 122, 134, 210). Hence, government intervention undermines freedom of life on the market.[31]

Just as both the 1802 settlers and Bixby elide what Locke understood to be the original disparity between man and nature, this argument reduces the metaphysical dynamic that is the signal achievement of Locke's account of the relation between the self and its property. It conflates or even reverses Locke's ontology of property. Locke's self owns its property because it owns itself; as a principle of ownership, the self is identified and realized through its property.[32] Lockean selfhood is a double relation, at once source and function of its alienable attributes. But then a tension exists between the self's inalienable authority over and identification with its material properties and expression, because self is at once distinct from and commensurate with its contingent attributes. The relation of authority is reciprocal rather than unilateral. Those who opposed the Steamboat Act, unlike its proponents, acknowledged the imbrication of selfhood with the property it putatively generates, and they attempted to resolve the tension by paradoxically absolutizing both terms of the problematic equation. It is *because* freedom and personhood are known only through property that persons must have complete authority over it. In this view, property is not merely produced and authorized by the self; instead, self is authorized as absolutely independent by unregulated rights to property. Where Locke's notion of the free self served to describe and justify a market economy, opponents of regulation believe it is the free market that generates independence.

The antebellum Mississippi steamboating industry seemed to embody the liberal ideal of the self. While eastern steamboating already tended toward corporate fleet lines, capitalization of western steamboating was private and local, often individually run. With the river considered national property, its use was free. This free passage encouraged private ownership because capital was needed only for the transport vehicle and its maintenance, and not for the right-of-way costs, roadbed maintenance, or terminal facilities demanded by railroad operation (Taylor 1962, 69–70; Fishlow 156). Moreover, the peculiar irregularity of the Mississippi made the fixed schedules of eastern fleets impractical. The "tramp steamboat," locally operated, able to await cargoes or high water, and available to go wherever cargoes awaited, "had a great advantage over packets tied to a scheduled run."[33] Thus, the river's capriciousness seemed to necessitate a pure form of small-scale free enterprise. This conjunction between nature's intractability and laissez-faire independence is emphatically expressed when, during his 1882 visit, Twain glimpses island 74, the one remarkably left by flooding no longer part of Arkansas or Mississippi. The absence of government jurisdiction ensures absolute freedom of ownership: on the island, "the owner is monarch of all he sees" (*Notebooks and Journals* 2:532), his entitling vision equal in scope to yet more tangible than that of *Nature*'s visionary poet.

Enacting the premises of those opposed to the Steamboat Act for interfering with individual action, Mark Twain's romance of piloting is a romance of property rights: a radical idealization of laissez-faire ownership underwrites the pilot's unconstrained self; the pilot's interpretive prowess is a metaphor for unconstrained ownership. This is why Twain unequivocally applauds the Pilots' Association as the climax of pilots' power and independence. The association secures high wages and transfers licensing authority from the government to the pilots themselves; by standardizing the dissemination of river information, it improves water-reading, reduces accidents, safeguards property and life, and thereby forestalls calls for government regulation. Pilots read best when they are free from outside authority and supervision.

The romance of piloting, then, squares with debates about river improvement and regulation of the steamboat industry. Its aesthetic values idealize free enterprise and laissez-faire property rights. The pilot, I've argued, did not enjoy the independence Twain ascribes to him. Nor could he have, for laissez-faire can never amount to the ontological independence Twain claims for the pilot. Freedom always requires institutional sanction: legal protection, or the insurance of a pilots' association, or of the insurance underwriters of the vessels, whose leverage the association exploited to gain power. So Twain's romance denies the labor of piloting; because when labor is visible, the self's involvement with institutions is all too evident, and its independence, the inalienability of its labor and property, too obviously contingent.

It All Belongs to Me: *Huck* and the Feudal Fantasy

How does Mark Twain surmount contingency in imagining authority on the river? As may already be evident, Twain suspends dependence and contingency by formulating his desire for independence in the language of feudal power. To ensure that the pilot is no mere laborer or wage earner, Twain imagines him not just a king but, indeed, "a king without a keeper, an absolute monarch" (94), never, unlike actual kings, "overmastered by a cramped treasury," the only person not a "manacled servant" (93). Thus, the pilot's "pride in his occupation surpasses the pride of kings" (52). The Pilots' Association institutionalized and transcended monarchial power. It possessed, Twain wrote Will Bowen, "more than regal power," "for no king ever wielded so absolute a sway over a subject & domain as did that old Association" (*Bowen* 13).

The language of slavery and feudalism had long figured in political disputes on the Mississippi. When the Spanish closed the river's mouth to trade in 1802, American settlers forecast that inaction would "make us vassals to the merciless Spaniards": "Shall we be their bondmen as the children of Israel were to the Egyptians? Shall one part of the United States be slaves, while the other is free?" (quoted in Ogg 435, 437). This outcry heralds free property relations (at least for landed whites) supplanting a system of slavery. But recall that in Mark Twain's exaltation of piloting, "*all* men . . . are *slaves* to other men & circumstances—save, alone, the pilot" (*Bowen* 13). For Twain, slavery is not superseded by free exchange, but persists, with pilots as master-kings, putting "all their wishes in the form of commands" (94), mastering others as they master the river. Here is independence as primary narcissism. Twain's account of piloting is not a critique of capitalism as the transformation of feudal service into service to capital. In his ahistorical vision, slavery transcends all social and economic forms, and gratification springs from enslaving rather than serving. Acknowledging that the pilot is a salaried worker would shatter Twain's fantasy of power. Consequently, his aestheticization denies the economic in order to guarantee economic rewards for the self he is trying to protect. Aestheticized autonomy is protected by a feudalism where only the enslaving aesthete/pilot is spared enslavement.

Adventures of Huckleberry Finn addresses perhaps more directly the same basic problem—the general security of property rights in the self—and it thus clarifies why Mark Twain must feel more kingly than a king to feel independent. Twain characterizes the journey downriver in *Huck Finn,* widely celebrated as a quest for freedom, in the same monarchial terms and with the same idealization of ownership that inform *Life on the Mississippi.* When Huck arrives on Jackson's Island, he feels "ruther comfortable and satisfied" (34) essentially because "I was boss of it; it all belonged to me" (36). Such unqualified possession constitutes Huck's definition of kingship during his debate with Jim regarding King Sollermun's wisdom: kings "can have just as much as they want; everything belongs to them" (64). Owning everything, a king can have all he wants; like pilots, he need only formulate his desires as commands to gratify them. Feeling absolute proprietorship is essential to why "you feel mighty free and easy and comfortable on a raft" (96). Traveling at night, naked, the world ashore asleep, "we'd have that whole river all to ourselves for the longest time" (97).

Despite the ironic criticism directed at the King and Duke and at southern society's captivity by chivalry, the runaways conceive freedom as the consummation of monarchy. Twain imagines freedom in language contravening many of his actual political commitments because he senses that free-market property rights only precariously secure the self in the way reductions of Locke imagined they could. The contingency of property, and hence of the self that realizes itself through property, is poignantly expressed after Huck and Jim meet on Jackson's Island. Jim explains that he ran away from Miss Watson because he was to be sold for $800; as property, a slave is, obviously, insecure, deprived of (in Locke's phrase) his "property in his own person." Jim then describes his various failed speculations, recounts the "signs" that predict his future wealth, and forswears lending any "mo' money 'dout I see the security." He suddenly realizes he is no longer Miss Watson's property: "Yes—en I's rich now, come to look at it. I owns mysef, en I's wuth eight hund'd dollars. I wisht I had de money, I wouldn't want no mo'" (42).

Of course, when Jim owns himself he is no longer worth $800 and is far from rich. Laurence Holland has demonstrated the historical irony of this novel that goes about "setting a free nigger free" (227) two decades after the Emancipation Proclamation. Miss Watson's will frees Jim "to stand jobless and alone"; this only nominal freedom exposes "the failure of [Twain's] world to flesh [emancipation] out with the family, the opportunities, and the community which would give it meaning" (Holland 80). But Jim's exclamation signifies more than a historical failure of will. It exemplifies the liberal principle that one must sell one's labor to survive. Freedom to own property is the freedom to turn yourself into property. Jim's joy eludicates to us, though not to him, the notion that to be free necessitates in a real sense becoming partially like a slave.

Historians have noted certain abolitionists' sense of the similarity between the market and slave economies. Harriet Beecher Stowe, for example, in essays like "A Family Talk on Reconstruction," felt that the market system was not so much the antidote to slavery as slavery was the epitome of the market. The term "free labor," as Walter Benn Michaels has observed of *Uncle Tom's Cabin,* seemed dangerously like "shorthand for a free market in labor" (1987, 111).[34] Although slaves' alienation from their property in their own persons is a legal fact, while that of free laborers is only a matter of institutional practice, what Stowe called "the condition of service" is common to both economic systems ("A Family Talk" 285). Southern defenders of slavery as a fairer form of exploitation than capitalism adduced the same assessment of free labor. "Property in man" is the goal of all men, writes George Fitzhugh, a leading southern apologist. But in selling themselves, free laborers have less property in themselves and thus "less liberty than slaves"; they "are worse paid and provided for, and have no valuable rights"; free laborers are "miscalled freemen" (20, 32, 221).[35]

Despite the isomorphism between the service of slavery and that of wage labor, the property rights of slaves and freemen, of course, differed vastly. Yet in Stowe's nervousness about and Fitzhugh's pleasure in the structural similarity, we see a possible result of inverting the etiology of the Lockean self and identifying the self wholly in terms of its property. As C. B. Macpherson's analysis of Locke's "possessive individualism" suggests (194–203, 255–62), and as we have seen in earlier chapters, the problem is intrinsic to Locke's conception and is underscored by American laissez-faire attempts to resolve it. If the self is a principle of ownership, expressed and realized in its property, then one's identity may be equated with the materials that represent it. Ideally, one is the inalienable master of oneself, and thus of one's labor and material property; but if identity is available only in terms of material property, which by definition is exchangeable, self and selfmastery are redefined—and hence revealed to be contingent—in every transaction.

Opponents of regulation typically sensed how viewing identity as proprietary, and therefore imbricated with the alienable properties that express it, undermines an ideal self-mastery. To compensate, they radicalized Locke's notion of the market self in two ways that seem incompatible but in fact tacitly support each other. Twain and his compatriots simultaneously defined the self utterly in terms of its property yet insisted on its transcendence of material goods and convention. In this way one's rights to oneself may seem inalienable though realized through property. Without this idealization, the self's identification with and through its property may feel like subjection to a political order, or, in historically specific terms, like slavery. The conflation of self with its property implies and is provoked by a terrifying alternative: stripped of property, the self may seem little more than the site of enslavement as labor. Twain and his compatriots mistake form for substance in idealizing the independence promised by free-market property rights as an absolute autonomy, presocial in its independence of the consent of others, finally atemporal in its transcendence of change, an ontological condition rather than a political fact or disposition of power.

The historical and class specificity of this vision of freedom is illustrated by comparison with that of freed slaves or the "new negro," for whom freedom could *only* be a temporal and political achievement. Twain's pilot, typically, could not afford to imagine freedom as something that moves or results from political proclamations. But for Booker T. Washington, W.E.B. Du Bois, and Frances E. W. Harper, for example, freedom "came" with the Union army and the Freedman's Bureau. It is a "possession," Washington writes, like a garment to "try on" (22, 24). The concreteness, and hence contingency, of freedom is evident when a male protagonist of Harper's *Iola Leroy* ponders the Union army's advance: "Freedom was almost in his grasp. . . . All the ties which bound him to his home were as ropes of sand, now that freedom had come so near" (35).[36] Freedom, here, is not an absolute condition but a transformative phenomenon to be grasped, breaking former bonds upon arrival. In these authors' works, freedom is explicitly realized through property, which verifies the otherwise mere abstraction of freedom by substantiating it.

If this conditional notion of freedom is the condition of black emancipation and what Du Bois called "self-realization" (49), Mark Twain's romance of piloting registers what many (whites) felt were its limits. The rhetoric of independence, initially employed to justify the market system, later came to imagine the market as the very source of independence, and hence absolutized independence in order to protect it from market contingencies. But since property has value and is property only by virtue of its exchangeability, the market itself occludes the inalienability of property it may seem to promise. No one can own in this way.

Unless, of course, you are king and own everything and everybody. If so, your material property is inalienable because exchange is superfluous or negligible, and your property in yourself is secure because you need never labor nor compromise. It is therefore thematically appropriate that Tom sees Jim's incarceration as an opportunity to rewrite the escape and restoration of the wronged Dolphin, the "natural son of Louis XIV" (204). Mark Twain's response to the hazards of the market is distinctive. He does not yearn for an agrarian, premarket world of self-sufficient domestic production; nor does he espouse the southern claim that by extending the family slavery reduces the alienation of free labor; nor does he oppose slavery as either the subversion of market freedom or the institutionalization of capitalism's worst features. Instead, Twain invokes absolute monarchy as the perfection of free-market property rights for the one independent master. "To become independent," Fitzhugh writes, "is to be able to make other people support you, without being obliged to labor for *them*" (18). Fitzhugh intends this statement as a critique of the free labor system; Twain would view it as what freemen implicitly want from property rights but what only enslaving others can achieve.

Jackson's Island and later the raft have often been thought to symbolize the possibility of a refuge from and alternative to social constraint, where natural comfort dissolves the relations of master and slave. Thus the island and raft have seemed, like the newly made island 74, to belong to no state. The assumption here, to which Twain subscribes when he characterizes ownership on island 74, is the one W. S. Lilly criticized: the self is truly free only in the absence of property relations. But ownership does not vanish on island 74; rather it is transferred, from the state to the inhabitant. Huck and Jim feel comfortable not because ownership has ceased but because they imagine they have achieved the monarchial self-ownership—being bosses of everything—that is an idealization potential in the liberal vision. In fact, however, Huck and Jim's freedom, and the reform of master-slave relations into friendship, is a property transfer, no more absolute and no less social than the actual ones that resolve the novel's plot, Huck's arrangement with Judge Thatcher and Miss Watson's manumission of Jim.[37]

The hyperbole of Twain's language suggests the impossibility, and hence urgency, of his fantasy. How could pilots be more kingly than kings? If kings "sit in chains" (93), do pilots sit tethered only by rope? The pilot's freedom, like

Huck's, Jim's, or any free person's, is a freedom of contract, replete with opportunities, constraints, and duties. But because Twain associates any contract with manacled service, no one can enjoy the kind of freedom he and many other Americans labored in service to. If it is a historical and conceptual fantasy, Twain's labor-eliding romance of piloting interpretation and independence is a sensible strategy for repairing and redeeming the contradiction in the liberal vision of independence and self. The more-than-regal independence Twain imagines for the pilot is impossible, and can obtain only outside social relations, in intimacy with the map of one's instinct.

Changing Occupations

It is the romance of property rights that postbellum federal river improvement and the intruding railroad in Part 2 of *Life on the Mississippi* destroy. The epistemological romance of water-reading is not demolished but, rather, changes occupations, transferred from the pilot to the engineer and to the travel writer. After the war, the pilot's reading has had the romance knocked out of it because elaborate charts, lamps, levees, and local channel improvement divest water-reading of the exquisite misery of uncertainty: "one may run in a fog now . . . with a confidence unknown in the old days" (171). The army engineers are the new priests of water-reading. They read the river, the pilot reads their maps.

According to the revolutionary study of the river on which the Mississippi River Commission based its efforts, the engineer divines the shape of the bed below the opaque water. The study's authors, A. A. Humphreys and H. L. Abbott, note "how difficult" it is to make "perfect observations." The bed cannot be inferred from the appearance of the surface. So like pilots, engineers disregard visible features. Measuring the volume of discharge and the velocity of the current at all widths and depths, they collect "fact upon fact, until the assemblage of all reveal[s] . . . the true conditions of the river." By this account, engineering observation is no simple empirical act. Engineers do not just measure; any assistant, like the depth-caller on a steamboat, can do that. Like the pilot, the engineer assembles diverse data to envision the unseen bed and "the laws governing the flow of water in natural channels" (18–20, 28–29). Deducing hidden laws and the hidden channel from data that in themselves do not represent them, the engineer's empiricism, like the pilot's, is an anti-empiricist communion with natural law.

Mark Twain is skeptical about the engineers' "prophecy" to control the river permanently by discovering its eternal laws.

> One who knows the Mississippi will promptly aver—not aloud, but to himself—that ten thousand River Commissions . . . cannot tame that lawless stream, cannot curb it or confine it cannot say to it, Go here, or Go there, and make it obey. (172–73)

Twain's skepticism is shared by Uncle Mumford, who believes the Mississippi, unlike "one of those little European rivers," "ain't [the] kind of river" you can "boss . . . around" (173–74).

Despite his skepticism of the engineers' claims of comprehensive control, Mark Twain admires their heroic enterprise.

> The military engineers of the Commission have taken upon their shoulders the job of making the Mississippi over again—a job transcended in size by only the original job of creating it. . . . [T]he West Point engineers have not their superiors anywhere; they know all that can be known of their abstruse science. (172–73)

Their abstruse science godlike in its ambit, the engineers are the new experts in the mysteries of the river. They have accomplished things that "seemed clearly impossible." Engineers are the latest avatars of pilots, with no superior anywhere.[38]

However much he admires the commission's project, Twain nevertheless doubts the engineers' claims to perfect nature systematically. Part of the nine-teenth-century cult of science, the rhetoric of perfection is, as we have seen, what propelled the 1879 River Commission Act through Congress. Twain doubts the possibility of perfecting the protean river. With change the one constant of the river, no handbook of prescribed instructions could ever foresee every possible exigency. Water-reading must proceed in an ad hoc fashion in order to manage unanticipated circumstances.

Twain attacks the various "theories" of improvement designed to perfect nature, a theory being an a priori method for entirely and permanently mastering the river: damming, creating lakes, building levees, reinforcing the channel, or any fixed combination of these. Twain considers any theory a "contagious" disease (176), incapacitating by its very prescription men's capacity to determine appro-priate action. Yet though he scorns the claims of system, theory, or comprehensive method, Twain retains his enthusiasm for improvement; "if Congress would make a sufficient appropriation, a colossal benefit would result" (176). Even the less forgiving Mumford is eager to hedge his bets on the river's prospects: "the safe way, where a man can afford it, is to *copper* the operation, and at the same time buy enough property in Vicksburg to square up in case [the engineers] win" (174–75).

The enterprise of heroic reading is transferred after the war not only to the engineer but to the travel writer, whose province is social customs rather than the river. As Mark Twain travels south along the river, he notes a diminishing grace "in the costumes of the new passengers." He decides that "*carriage* is at the bot-tom of things":

> . . . there are plenty of ladies and gentlemen in the provincial cities whose gar-ments are all made by the best tailors and dressmakers in New York; yet this had no perceptible effect upon the grand fact: the educated eye never mistakes those people for New Yorkers. (140)

The snobbish point here is that the writer's educated eye, like the pilot's, does not literalistically accept surface features as the truth; the writer seeks the substance below appearance, "at the bottom" of garments. He apprehends what "belief" accompanies the goatee (141). As with piloting, the writer's manner of knowing cannot be fully detailed even though it has been studiously acquired over many years. Twain's conception of the writer's faculty adapts the anti-immigrationist nativism of the period. Only "the native novelist," Twain writes in "What Paul Bourget Thinks of Us," "is qualified to make a valuable report" of "the souls and the life of a people." The native novelist's "whole capital . . . is the slow accumulation [over twenty-five years] of *un*conscious observation—absorption." As the pilot's instinctive memory lays up treasures, the writer's absorptive faculty is operating capital for producing valuable reports (145–46).

While the engineer reads the protean river, the writer applies his faculties to the protean civilization along its banks, witnessing "these days of infinite change," when population figures will not "be worth much" in six months (327). One reason the site of interpretive interest has moved is that the writer commands little expertise reading nature; Mark Twain more than once admits that he has forgotten the river, as if it were rebuilt during his absence. But there is a larger historical reason that the text to be read shifts from a natural phenomenon to cultural productions. The river held prewar interest because of its commerce; after the war the bulk of the production and transport of value takes place on shore.

Nature has been superseded as the subject of interest not so much, as the nineteenth century hoped, because it has been mastered or perfected but because it has been circumvented. To this day we periodically see reports of flooding and cutoffs on the evening news.[39] The river's intractability frustrated those who had invested too much faith in the technology of control. Mark Twain was one of the disappointed, and in 1907 he harangued Teddy Roosevelt for his complicity with

> that ancient and insatiable gang, the Mississippi improvement conspirators, who for thirty years have been annually sucking the blood of the Treasury and spending it in fantastic attempts to ameliorate the condition of that useless river. . . . [I]t has always torn down the petty basketwork of the engineers and poured its giant floods whithersoever it chose, and it will continue to do so.[40]

After the Civil War, railway development permitted trade to circumvent the river's vagaries. The expansion of the railroad was extremely fortunate for river culture and the nation's commerce, for in avoiding the river's obstacles to trade, it permitted what Louis Hunter has called far greater "directness of communication between different parts of the territory served" (585).

Hunter speaks of trade here as communication. This figure, where communication is an economic form and trade a linguistic form, was employed often in the era, with the river referred to as an "avenue of communication."[41] This figure seems especially appropriate after the war, when economic communication is transacted among men rather than between man and nature. Twain reports population, trade, and manufacturing statistics in the context of this perceived relocation of the communication of value. Part 2's statistics are not a moment of bald realism, of uninterpreted and acontextual facts. They are Twain's attempt to continue the tradition of heroic reading. Reporting statistics constitutes the writer's attempt to assemble facts as the engineer assembles observations of the hydraulics of the river. Twain's conclusion about this assemblage: progress. The nineteenth century, he repeats in various forms, is "the plainest and . . . worthiest of all the centuries the world has seen" (283). This "second brand of romanticism," as it has been called (Montgomery 80), resembles in structure Twain's admiration for piloting, which was always part of the cult of progress. Piloting and manufacturing are both forms of communication. They are for Twain two historical moments in the continuing progress, or romance, of civilization.

If piloting, engineering, and manufacturing are all appropriative activities, why does Mark Twain lament the death of piloting? Mark Twain mourns the romance of the free self that underlies his mythologization of piloting. It is the transition to the "intruding" railroad (109), more specifically to the corporate trend the railroad typified and accelerated, that truly destroyed Twain's romance of independence. Whereas epistemological heroics have merely changed occupations, the organization of postwar production and trade frustrated Twain's ideal of the independent self. The railroad's sheer scale necessitated incorporation, and western steamboating, too, soon became a corporate enterprise (Hunter 566–67). It is hard to believe that you are independent and without superior when you are employed not by a local boatowner but by a large, distant office, and take orders from a local manager who answers to a district manager, and so on. The chain of command, once direct, is now oblique; responsibility, once centered and personal, now dispersed. The fate of the steamboat barkeeper dramatizes this development, which undermined, Twain thought, personal "dignity" (172). "In the old times, the barkeeper owned the bar himself . . . 'and was the toniest aristocrat on the boat.'" Now "the bars are rented and owned by one firm," which furnishes the liquor and salaries the barkeep. Individual enterprise is giving way to franchised marketing. Postwar farmers along the river, who "don't know anything but cotton," are the corollary to the franchised barkeeper. They have no interest in the legendary self-sufficiency of the farmer, and the barkeep makes most of his money selling them fruits and vegetables at a steep markup, forty-five cents on a five-cent outlay (211–12). These farmers are more properly merchants and consumers than farmers.[42]

The intruder railroad, then, intrudes mainly upon the romance of autonomy that Twain located in small-scale enterprise. Steamboating itself, Twain observes early on, once "intruded" on the keelboating industry, attracting commerce until "keelboating died a permanent death" (24). Twain is content with this intrusion, however, because the economic organization of the two industries is the same and accommodates the same sense of self. But railroad economies of scale shatter the illusion of autonomy, which may be one reason that Twain, in his generally accurate account of steamboating, omits the fact that the industry's glory days already exhibited signs of decline. He mentions as an afterthought in chapter 15 that "the new railroad stretching up . . . to Northern railway centers, began to divert the passenger travel from the steamers" (109). Only a semicolon separates this statement from the announcement of the war's annihilating effects, and the two events seem contemporaneous. In fact, however, throughout the 1850s the steamboat industry weakened and was often depressed, due largely to the growth of the railroad, which by 1860 was the supreme transport service in the valley. Steamboatmen were notoriously slow to recognize the threat of the railroad, but they could hardly have ignored the tracks lengthening along the river; moreover, during the 1850s railroad companies began purchasing steamboat lines as connecting services among train lines.[43] Nevertheless, Twain forgets steamboating's difficulties and insists that its glory days perished, as it were, with the firing upon Fort Sumter. In fact, only the mythological independent self, a notion already an exercise in nostalgia in the 1850s, died with the war.

Manufacturing Aesthetics

The fact that heroic reading has not died but undergone job retraining recasts Mark Twain's notorious, invidious comparison between southern romanticism and northern manufacturing. Twain's acerbity in *Life on the Mississippi* toward the postbellum South's continued devotion to Walter Scott's "romantic juvenilities" (327) has struck many readers as excessive and facile. Further, if it is the monarchial self that perishes when piloting dies, Twain's attack on the South's feudal romanticism seems inconsistent as well. Yet the bases of the two feudalisms differ. The feudal fantasy of piloting is intended to guarantee independence. "Sir Walter Scott disease" (266), on the other hand, strikes Twain as slavish imitation of little-understood representations. Like theories, Scott's romances exercise "a debilitating influence" (237). Both theories and Sir Walter Scott disease afflict the victim with a prefabricated system of perception and comprehension.

In Mark Twain's view, southerners fetishize the surface features of romance, simply copying its flowery language or architecture. The South's "imitation castles" merely copy surface features, without any sense that the surface, however obliquely, represents reality or beliefs or social practices (237–38). The South's productions lack the conception of dimension so crucial to reading the river. Twain's southerners are unaware that romantic values arise in exploring and exploiting the disparity between surface and substance. Like Emerson's formalist preacher, Twain's South mistakes surface for substance, form for value.

Twain's southerners are guilty not just of "pretending to be what they are not" (237) but of believing that they are what they can only pretend to be. We can see the logic and morality of Mark Twain's condemnation in the contrast between the self-deception of the "House Beautiful" and the intentional deception perpetrated in the preceding chapter, "Manufactures and Miscreants." Twain encounters "two scoundrels" (236) who hawk "first-rate imitation[s]," oleomargarine and cottonseed oil posing as olive oil. They boast that "an expert can't" "detect the true from the false" product. The counterfeits are "perfect" because "undetectable" (235–36). The scoundrels' deception exploits a paradigmatic problem in an age of mechanical reproduction, a culture of synthetics. Twain's South, on the other hand, accepts all representations and reproductions as authentic. He censures the South's ignorance far more harshly than he does the scoundrels' deception, and this judgment is entirely consistent with the values (aesthetic and cultural) of Part 1, where the pilot protects valuable cargo and lays up treasure by extrapolating the true shape of the riverbed from the misleading surface of the water.

If the pilot is the exemplary producer and protector of value from recalcitrant nature, Twain's southerners epitomize bad consumers in an emerging culture of consumption.[44] The Walter Scott South indiscriminately consumes culture, like novels, and thinks that sheer consumption produces cultural value, called sophistication. The North, on the other hand, obsessively produces new and newer products. The contrast between production and consumption underlies Twain's praise of the "sumptuous glass temple," the craft *Paul Jones,* and his ridicule of "the House Beautiful," cluttered with trinkets, cheap imitations of artifacts from Europe, and chromos of sentimental scenes and sayings (dignified as "lithographs"). Some critics have been irritated by Twain's contrast, since both house and boat are gaudy (Montgomery 80–81). Others, like Paul Schmidt, have rightly justified Twain's hierarchy on the grounds that in the House Beautiful, "Objects have lost all use and been reduced to mere commodities," while "the glass temple is valued less for itself than as the proper setting for the great skill of the pilot" (98).

Twain evaluates design in terms of what soon was to be called functionalism, albeit his functionalism departs from the ethic of Catherine Beecher's domestic designs and Louis Sullivan's office and home designs. For Beecher and Sullivan (and of course later for Frank Lloyd Wright), form follows function. Twain's is a functionalism not of form but of authority. The gaudiness of the "House Beautiful" mistakes surface for value, object for process of production. Even its inhabitants are objects in the house, without function or, as it were, functions: there is "not a bathroom in the house," and visitors don't seem to notice the deficiency (231). In contrast, the boat's gaudiness is appropriate to the pilot's authority, and Twain applies this aesthetic principle to the Cotton Exchange being built in New Orleans. "When completed . . . [it] will be a stately and beautiful building; massive, substantial, full of architectural graces; no shams or false pretenses or ugliness about it anywhere" (243). Twain gives no hint of what the building will look like, but commends its beauty and substance mainly because it will house the stock and commodities trading so crucial to the circulation of goods.

The distinction between production and unanalytical consumption partly explains the superiority in "Old Times" of the pilot's aesthetics to the passenger's. Though the cub makes "a valuable acquisition" when he learns to read the river, "I had lost something, too . . . all the grace, the beauty, the poetry had gone out of the majestic river" (67); "the romance and the beauty" of the river are sacrificed to utility (68). The romance of the uninitiated, as critics like Cox, Smith, and Leo Marx have observed, is expressed in "conventional aesthetic vocabulary" (Smith 79), to which Leo Marx contrasts the pilot's "instrumental" mode of perception (1963, 51). This difference should not be overstated, I think, for the two aesthetic economies share two important features. First, both economies transform nature into an article or even extension of human interest: the passenger's sentimentalism registers nature as the fulfillment of conventional expectations of leisure; the pilot's instrumentalism appropriates nature as an emblem of human power. Second, both the pilot's and the passenger's appreciations of nature appeal to their practitioner (in one case, Twain; in the other, the middle-class would-be-aristocratic consumer) because they erase the labor of appropriation and production.

The important difference between the two aesthetics is finally that one exercises authority and the other merely obeys it. The passenger's and the South's aesthetic of consumption cannot disguise their participation in a social order, and thus conform to the laws and conventions that the pilot, in Twain's myth, transcends.

The aesthetics of both piloting and consumption would achieve authority, finally, by disguising their embeddedness in a network of obligations. But in Part 2 the complex interrelation between art and political economy cannot be avoided. The new North's art *is* its manufacturing, the rapid changes it produces in population and statistical configurations. Because in the new North "people don't dream, they work" (321), Mark Twain sees about him "suggestions of wholesome life and comfort." Twain admires the cult of productivity because he believes that "increased . . . wealth" requires assiduity in perception and the application of labor; thus increased wealth is accompanied by "the intellectual advancement and liberalizing of opinion which go naturally with [it]" (220).

Through meeting the intellectual demands of the culture of synthetics, the self sacrificed to corporate development can reemerge. In contrast, the Mardi Gras is a "relic of the French and Spanish occupation," now practiced with "the religious feature . . . pretty well knocked out of it." Form without substance again, and Mark Twain again indicts Scott as the responsible party. The men leading the procession are disguised and anonymous, and "they hid their personality . . . merely for romance's sake" (264). Walter Scott romance kills assiduity and the self. The inhabitants along the upper river, on the other hand, "compel homage" for much the same reasons as the monarchial pilots: "This is an independent race who think for themselves" (326). But if the pilot's independence results from a fantasy about his feudal superiority, those in "the practical North" (265) sacrifice no independence even though "they live under law" and amid corporate dispersal of ownership. Twain is perhaps undeliberate in his ready equation of production with deliberateness, and he would soon desist praising progressive culture, but here the northerners' independence is not merely an idealized metonymy for a form of economic organization; rather, "they read" (326).

Internal Evidence and the Power to Find the Nub

Despite his enthusiasm for the regime of the practical, Mark Twain cannot entirely ignore its tendency to inequity and vulgar materialism. Salesmen pushing counterfeit goods are the very sign of a manufacturing boom. Twain notes as well the economics of inhumation, which make undertaking "the dead surest business in Christendom" (246). Exploiting people's fealty to the apparatus of religion, undertakers advertise the "proper" burial of relatives as the only means to obtain the "compensation" promised in the afterlife. Hence, Twain's punning name for undertaking. Nor is Twain himself exempt from greed. After "Ritter's Narrative," for example, he and his companions convince themselves that Adam Kruger does not deserve the money Ritter intended for him, and moreover would be corrupted by it. In its place, they "send him a chromo" (286).

The fate of former pilots illustrates the problem of inequitable distribution of wealth. Most try their hand at farming, an appropriate subsequent career because the farming "life is private and secluded from eruptions of undesirable strangers—like the pilothouse hermitage" (275). Although farming was already a preeminently commercial rather than subsistence enterprise, pilots farm in order to recapture the mythological independence of their former occupation. But most are unsuccessful and must subsidize their farms by returning to piloting, now just another form of wage labor, for part of each year. They fail as farmers partly because of developments in financing: credit has replaced surplus monies as the dominant method of financing. The former pilots, like freed slaves and other new landowners, were paying from 18 to 39 percent interest on their debts. Thus "although owning the land," they "were without cash capital," and were forced "to hypothecate both land and crop to carry on the business" and buy food. Here is ownership without the independence ownership was supposed to secure or signify, virtually ownership without ownership.

Yet the instances of inequity Mark Twain cites do not shatter his belief in the facts of progress, and his faith that the outward show of material wealth exhibits the substance of intellectual advancement strikes us as an a priori belief about the relation between phenomenal appearance and intellectual or spiritual substance. His celebration of progress seems to work rather as Twain thinks theory or Sir Walter Scott disease works, delivering a ready-made interpretive position. Twain suffers this disease of presupposition because, though he hopes the writer's trained unconscious faculties qualify him as heroic reader, the writer is an ignorant consumer vis-à-vis finance and production.

In the course of the book, Mark Twain recognizes his incompetence to read America's new economic developments, epitomized in the development of Chicago:

> It is hopeless for the occasional visitor to keep up with Chicago—she outgrows his prophecies faster than he can make them. She is always a novelty; for she is never the Chicago you saw when you passed through the last time. (345)

Chicago, cynosure of recent corporate development, is the new Mississippi resisting old strategies of interpretation. Mark Twain's inability to read the new economy extends as well to the river. Recall his admissions that he has entirely forgotten the river. One sign of his new, consumer status is that his descriptions of nature—for example, of the sunrise in chapter 30 or of the river and its "majestic bluffs" in chapter 57—reprise, albeit with fewer emotive adjectives, the conventional language of the passenger of "Old Times." As if narrating a chromo, Twain dissolves the bluffs into a "dreamland," with "nothing this worldly about it—nothing to hang a fret or a worry upon" (329). Twain's status as consumer explains the unpersuasiveness of his inveterate movement from statistics to declarations of independence and wholesomeness. His strategy is precisely inveterate. His consumerism helps to explain why the episodic Part 2 lacks the narrative coherence of Part 1. Each anecdote and incident forms an attempt to learn to read the culture of production, consumption, and finance. But now that the mythological, interpretive regime of the pilot has passed, the process and effort of discerning the significance of phenomena remain unpersuasively visible.

Mark Twain signals his deficiency as a reader of credit culture and synthetic culture. His dream about the otherworldliness of the river bluffs is abruptly shattered by the entrance of "the unholy train . . . ripping the sacred solitude to rags and tatters with its devil war whoop" (329–30). The alliteration and mock religious language should alert us not to take Twain literally. Through the irony of this ritualized deflation, the professional observer both is exposed as and announces himself to be the conventional consumer of nature. This irony also tests his reader. Is Twain's sentimentality genuine, or purposive sham (where the sham is the substance)? As consumers of the book, our task is to distinguish when Twain is being duped by his faith in progress, when he is reading well, and when he is duping us.

Mark Twain is dramatizing the problem of evaluating evidence, absent the magisterial authority of the pilot. The pilot's authority meant that he did not really evaluate; he just instinctively knew the hidden shape of the river. Despite his trained faculty of unconscious absorption, however, the writer's evaluative procedure—extrapolating intellectual capacity from statistics and behavior—bears less authority than the pilot's because it is conspicuous. The problem of evaluation is extensively illustrated in the episode detailing the history of a fraudulent letter written by the unregenerate burglar Charles Williams in order to arouse sentiment for his pardon. The letter purports to be written by Williams's cellmate, Jack Hunt, and recounts the uneducated Hunt's moral rehabilitation by the now-converted Williams, a Harvard graduate and son of a clergyman. Mark Twain relates how tearfully the letter was first read to him, how tearfully he listened, and how it induced many a congregation to weep "as one individual." Initially, Twain readily deems the letter an instance of bald realism: "Here was true eloquence; irresistible eloquence; and without a single grace or ornament to help it out" (295–96). Here were just the plain facts. The letter turns out to be "a pure swindle" (296). What interests Twain, however, is not the scandal of the counterfeit epistle but the difficulties determining its validity.

Charles Dudley Warner is the first to doubt the letter is "genuine": Jack Hunt was supposedly unschooled, and the letter is "too neat, and compact, and fluent, and nicely put together for an ignorant person" to have written it. "The literary artist had detected the literary machinery," applauds Mark Twain (297). Preachers, however, verify the authenticity of the letter on grounds that it "show[s] the work of grace in a human heart" deprived of Christian education. Quick to subscribe to the latest authoritative opinion, Twain is pleased that "Mr. Warner's low-down suspicion" seems misplaced, since it

> was a suspicion based upon mere internal evidence, anyway; and when you come to internal evidence, it's a big field and a game that two can play at: as witness this other internal evidence, dismissed by the writer of the note [certifying the letter]. (298)

For Twain, internal evidence (what literary critics now call textual evidence) is peculiarly unhelpful. Just as the river's surface features provide no necessary clues about the shape of the bed, internal evidence is a poor guide to the intention of the author. Thus the episode of the letter exemplifies how interpretation always translates formal features; this structure of interpretation means that evidence never bodies itself forth, and thus never in itself constrains interpretation. Consequently, even skilled readers interpret the letter's linguistic features differently or count different features as pertinent.

When (or since) internal evidence alone does not resolve disputes, perhaps one should appeal to what some modern readers call (either hopefully or nervously) external evidence? Perhaps external evidence will establish authorial intention or the meaning of words? Mark Twain's category of internal evidence is so broad, however, that no scrutiny of external evidence, such as E. D. Hirsch might call for, will supply the necessary neutral ground of interpretation. Approached to resolve the dispute, the prison chaplain judges the letter a sham. His judgment, Twain judges, "is pretty well loaded with internal evidence of the most solid description": the chaplain knows Williams to be "a dissolute, cunning prodigal." The chaplain's evidence, of course, is utterly external; he evaluates the putative subject of the letter, Williams himself, and simply disputes the letter's factual claims. Twain, however, considers the chaplain's appeal to external evidence another form of gathering internal evidence. Twain propounds the now familiar point that a living subject, here Williams himself, is itself a text; that is, we apprehend character (or any other phenomenon) in the same way we read a written text, mediated spatially, temporally, and sensuously. Interpretation gains nothing necessarily by looking to the source of a description, because the source's behavior has the same status as its description; in each case, we interpret formal features. For Twain, all evidence is internal evidence—and all interpretation risks being mistaken—because all apprehension has the same structure and entails the same interpretive problems.[45]

The dispute surrounding Williams's letter precisely signifies the demise of the magisterial interpretive authority for which Mark Twain's pilot is metonymy. Interpretive disputes erupt because interpreters must rely on formal features; that is, they have an interpretive strategy. In Twain's romance, the pilot bears authority because he proceeds so naturally that he appears to employ no strategy. His liberation from labor and interpretative strategy is a metaphor for an imagined stability of economic relations—both among citizens and between citizens and the factors to be interpreted (either nature or the economy). But in "these times of infinite change," readers are visibly empirical; their appeal to phenomenal evidence is both sign and precipitant of disputes and of competition for interpretive authority. What is at issue is who has the competence or power to perceive what Mark Twain calls the "nub" of the letter (299), a mention near its conclusion that Williams was bleeding and in mortal danger. When unnoticed, this nub stimulates efforts for a pardon; when noticed, it signals sham.

Though his instincts are correct, Warner's reasons for his opinions are (to Twain) no more compelling than the preacher's. Why does evidence persuade?

Significantly, Mark Twain, duped throughout, notices the nub only after he knows of the attempted sham. This chronology of persuasion reverses the ordinary one, where evidence precipitates interpretation. Here, the relevant sentence counts as evidence of the sham—indeed, is noticed—only after the swindle has been revealed. Twain admits that "an indifferent reader would never suspect that [the nub] was the heart and core of the epistle, if he even took note of it at all" (299). Though he could never have identified it as such, Twain's account of interpretation is fundamentally pragmatist. For him, any reader is like De Soto discovering the Mississippi: a great river, or a nub in a letter, is apprehensible only in relation to one's categories of comprehension. Twain is doubly disposed to notice a nub, because concealing the nub, he writes in "How to Tell a Story," is his habitual technique in storytelling (8). Once aware of Williams's imposture, his instinct tells him there must be a nub, and he finds one. When the nub, the sign of artifice, is identified, Williams's literary attempt fails.

We are probably not convinced that his nub is the telling stylistic feature. But the unpersuasiveness of Mark Twain's appeal to the nub is also evidence for his unwittingly pragmatist point: since no metaphysically neutral system for selecting and evaluating evidence exists, no evidence will persuade a reader who lacks the terms and values in which it can be convincing. Interpretive technique is always habit or trained instinct.[46] Notice that Twain does not term his *ad hoc* and *ex post facto* rationalization a method. He is merely a reader trying to figure out why he believes what he believes. But because his technique, effort, and motivation are conspicuous, Twain's conclusion is disputable. For the same reasons, his conviction of the spiritual progress of his era readily appears a faith too hastily purchased.

Mark Twain's faith in the romance of progress marks his anxious attempt to keep interpretive pace with the measurable, or perhaps immeasurable, changes occurring around him.[47] His faith in the cult of production is of the same kind as his worship of the pilot's office, but the throne of natural interpretation has been overthrown by changes in commercial organization and in the shift in the production of value from transaction with nature to manufacture and finance. Twain's pilot was a mythological figure for the autonomous self conducting business with nature. Lacking piloting's mystical authority, new professional readers debate not only the meaning of the changes they confront but even what features merit scrutiny. This mediation that attenuates power is figured in Part 2 of *Life on the Mississippi* by Mark Twain himself, the consumerist writer struggling (at times arrogantly) to understand the spirit producing the scenes he consumes.

The travel writer of 1882 is a novice in the new economic order. Later in his life, however, while experiencing severe financial difficulties, Twain will encounter a man competent to read and exploit the complexities of a corporate economy, Henry Huttleston Rogers, a vice-president of Standard Oil. In an 1895 letter to Rogers, Twain discusses the contracts he has signed as exercises in interpretation.

Well, I *am* a pretty versatile fool, when it comes to contracts, and business and such things. I've signed a lot of contracts in my time; and at signing-time I probably knew what the contracts meant—but 6 months later everything had grown dim and I could be *certain* of only two things, to-wit: 1. I didn't *sign* any contract; 2. The contract means the opposite of what it *says.* (*Correspondence with Rogers* 149)

The contract's meaning eludes Twain much as the river's shape eludes the cub's, and later the writer's, memory. The contract has consistent formal features, but the novice businessman is uncertain what intention they represent. The inexperienced reader's relation to water and contracts is the same: he must expend effort trying to apprehend the underlying shape or meaning of surface features or legal and economic jargon.

Mark Twain describes his education by Rogers in the same terms he commemorates Bixby's lesson in water-reading: watching Rogers is "a valuable lesson" (149). Rogers's ability to negotiate deals resembles the pilot's facility negotiating the river. Like the pilot's negotiation, Rogers's mediation in 1896 of the contractual dispute among Twain, *Harper's,* and the American Publishing Company regarding an edition of Twain's complete writing, is "a heroic achievement" (234). The two men form a negotiating "team," and for a time Rogers seems "the most useful man I know, and I . . . the most ornamental" (310, 21 December 1897). Slowly, Twain sheds the role of "commercial somnambulist" by "watching you [Rogers] with all that money in your hands," and begins to feel comfortable with the language of contracts (389, 19 February 1899). In a eulogy of Rogers, Twain compared his financier friend to a pilot who reads the waters of commerce and furnishes instruction in the mysteries of finance: Rogers's "wise guidance" and "cool steering . . . enabled me to work out my salvation and pay a hundred cents on the dollar—the most valuable service any man ever did me" (710).

In the corporate financier, Twain discovers the successor to the heroic, monarchial pilot who, miraculously interpreting and appropriating the infinite changes about him, at once resolves and underscores the conflicts informing the liberal vision of self and value. Remedying the too conspicuous and too individual economics and ethics of reading in Part 2 of *Life on the Mississippi,* Twain's admiration for Rogers updates the protectionist romance of "Old Times." Twain's monarchial ideal of effortless reading and natural value rather crudely intuits the notion of combinatory or transcendent agency that, in the age of incorporation, was coming to inform both critiques of and apologies for developments in American finance and culture. But I will postpone examining this conception of action and value in order to probe the difficulties of incorporating persons into corporate structures encountered by prototypical realists, William Dean Howells and contemporaneous analysts of value.

The Sordid and the Organic: Character and Economy in Howells and Economic Theory

Fiction and the Economic Chance World

The linkage between cognition, ontology, and authority in Mark Twain's romance of piloting illustrates both the progressive and regressive aspects of the liberal paradigm. By *progressive* I mean not contractarianism per se but the movement from cognitive authority to independence to social and economic power that underwrites contractarianism. But this lineage includes a limiting, corollary condition: the individual enjoys independence only within sanctioning matrixes—habitual, axiological, and institutional. Contractarian independence, that is, as its eighteenth-century formulators explained, implies as well obligations and limits. Twain's romance of independence compensates for and protects against the self's subjectification in networks of authority. But Twain's protectionist romance also has regressive implications: it is most ideally realized in a feudal hierarchy that obscures the very processes of production and power that liberal thought developed to justify. The inveterate appeal to nature is both the index of this contradictory state and the rhetorical instrument for resolving it. In Twain's romance, appeal to the law of nature secures autonomy and power through sovereign imperative and the elision of labor. Twain's romance is one way of dramatizing the oft-noted antinomies in liberal thought, specifically the potential tension between individual action and the social authority that legitimates and legislates it. The antipodal terms are at once complementary and conflicting. This relation constitutes the central problem addressed in liberal thought; and this problem is underscored by, as it is intended to be resolved by, social contractarianism: individual reason and power underlie and compose the state, yet the state's existence limits the scope of individual authority.

Contractarianism seeks to resolve this paradox through the notion of consent, by which some natural rights are relinquished in order to protect others. But as Hobbes anticipated and Hume saw, consent is always an *ex post facto* attribution, rarely given historically, and ascertained only retrospectively and putatively. That is, the social contract is constituted and legitimated by the putative consent of hypothetical free agents not yet enmeshed in a conventional and social order. The moral quality of the natural state may be variously characterized, as Hobbes, Locke, and Rousseau did. But whether the natural state is vicious or harmonious, and whatever the motivations for entering (or preserving) the social contract, its legitimacy—and the legitimacy it establishes for subsequent contracts—depends on imagining a precontractarian possibility. Contracts, however, are always merely legal, entered into by social rather than natural beings. Therefore, contracts fail to mark the prior *dis*entanglement and individuation of free agents from convention; rather, individuation and independence are functions of and have meaning only within a particular social order, and thus can neither absolutely nor historically serve as its genesis.

This chapter will examine the ramifications of this problem in a period and genre of high liberalism: realism. The specific aspect of the problem I will look at is the relation between character and economy, between the free agent and the network of relations that condition its agency. As Warren Susman and Amy Kaplan have demonstrated, the late nineteenth century saw a transformation in what Susman nicely calls "modal types of persons" (273). Since the late seventeenth century, persons had been conceived to have "character," a principle of "integrity" coordinating conduct and consciousness, which valued work, thrift, and responsibility (Kaplan 24). But industry's escalating capacities to produce goods impelled a transformation to a consumption-based, rather than production-based, economic ethos (see Rodgers), and with it a transformation of the type of person bearing this ethos. "Personality" began to supplant character as a principle of personhood. This notion emphasized idiosyncrasy, viewing the self more as a series of discrete rather than coordinated acts. Such a self was better able to consume a wider variety of goods.

Like any transformation, this one proceeded unevenly, often outstripping attempts to understand it, and it made visible certain conflicts in notions of value. Even as this transformation was taking place, custodians of culture and policy exhorted readers to secure or retain their "character." For example, character was the principal term and entity pursued in literary and economic inquiries. Ideally, character was a natural ground of values, but attempts to discover adequate and just forms of value ultimately discovered character to be at once cause and result of the circulation of values. For a number of reasons, I choose as my fulcrum William Dean Howells's *A Hazard of New Fortunes.* Howells, of course, was the most powerful manager of and spokesman for the realist movement in America. Moreover, both Howells and Henry James cited *A Hazard* as Howells's most ambitious and (to quote James) "prodigious" work.[1] Finally, written during the years of Howells's personal politicization and his politicization of his writing, *A Hazard* explicitly concerns, in narrative and theme, the relation between individual will and social authority. It thus engages the problem that Howells rightly took to be central to economic discourse of the day: What is the generative relation between character and value?[2]

This issue is central in *A Hazard*'s most famous passage, where Basil March delivers his complaint against "this economic chance world in which we live." "No doubt" a universal, divine "law" governed the accidental death of Conrad

Dryfoos at a streetcar strike riot, but it seems absent from the economic chance world "we men seem to have created." To make human economy resemble God's economy, Basil would naturalize "human affairs": the "inflexible" patterns of cause and effect that organize "the physical world" should be adapted to human economy as a natural law of merit, "that if a man will work he shall both rest and eat. . . . Nothing less ideal than this satisfies the reason." But this ideal, labor-based, natural law of value eludes us. Value is constituted haphazardly, "by the caprice, the mood, the indigestion" of the capitalist. With security contingent upon the spices in the capitalist's diet, the outcome of human effort is not teleological but desultory, oscillating between "pushing and pulling." Moreover, it is harshly competitive: "Someone always has you by the throat, unless you have someone else in *your* grip." Life in the chance world is a life "lived sordidly to no purpose" (436–37).

Basil's discovery of the sordid, chance economy is especially disheartening because its effect on character impedes reform. In order to find a way to alleviate the "real suffering"—both impoverishment and nervous distress[3]—that competitive capitalism causes, he tries to locate responsibility for sordid competition. If men and women are the source of imperfect behavior, then we might know where to begin a program of social improvement. But one can't blame people, because "conditions *make* character" (437). In a sordid economy, people don't consciously make ethical choices; instead they imitate the conventional patterns of behavior that "are held up to them by civilization as the chief good of life." Ontology, here, suffers the same fate as value. In the chance economy, "the dollar is the measure of every value" (173); more happily, value should confer worth upon the dollar. Analogously, what we might call the sordid selves of this sordid world, its personalities, are merely products and reflections of the gaudiness and chance that mobilize the economy. If persons merely reflect external values, how can reform be pursued?

The autobiographical basis of Basil's disillusion with competitive economics is well documented. The Haymarket riots and the public outcry and trial that followed catalyzed Howells's social education and activism. The only gentile intellectual figure to censure the proceedings against the accused Haymarket conspirators, Howells in subsequent years read and reviewed numerous reformist or radical tracts, displayed interest in Edward Bellamy's Nationalist Movement, spoke before W.D.P. Bliss's Society of Christian Socialists, served as a vice-president of Bliss's later (more moderate) Social Reform Club, wrote articles with titles like "Equality as the Basis of a Good Society," and published his utopian romances and what have been called his "economic novels."[4] The question Howells's activism addressed, and still raises, is how individual action, whether aesthetic or more explicitly political, affects events and the polis. That is, his activism engages the problem of the relation between subject and structure that remains central to postmodern critical analysis, most explicitly in the work of Foucault and neo-Marxists.

This issue has directly informed critical assessment of *A Hazard.* The term *economic novel,* coined in the 1930s, was an index of the period's interest in the social agency of art. Very much in response to Howells's denigration as bourgeois

by H. L. Mencken and others, this criticism abounded with the language of collectivism: Howells was working out a critical aesthetic practice modeled on the
reform, socialist, and communitarian programs of Bellamy, Henry George, William Morris, Lawrence Gronlund, Tolstoi, and others, in order to examine how
the individual could inform, without being thwarted by, the social organism.[5]
Though intended as vaguely Marxist, the premises of this work were rather sim-
plistically liberal, taking for granted that individual agency composes and is fulfilled in the social organism. Later, new critical premises displaced this reflection-
ist homology between the individual and the social, and rejected the term
economic novel precisely because it narrowed the distance between the individual
and the social order. The term sounded determinist. Howells himself wrote in
1912 that "fiction can deal with the facts of finance and invention only as the
expression of character," and he demanded that the "wonders of the outer world
be related to the miracles of the inner world."[6] These remarks led George Bennett
to argue that "realism is most valuable [when] it reveals character" (1973, 5); foregrounding economic issues, in contrast, poses a "real threat to Howells' art"
(1959, 37) because it subordinates inner character to outer events. Subsequent
criticism has generally evaluated the novel primarily in terms of its success in
exploring individual psychology and the "sovereignty" of free will (Walsh 164).[7]
To the extent that individual agency is seen to structure its narrative, *A Hazard*
is called, in the familiar term, organic; to the extent that the novel relies on accident and on formal conventions—on the accidental killing at the streetcar strike
and on its various failed resolutions, marriage or travel abroad—it is charged with
failing to meet the essentialist demands of organicism.

One virtue of poststructuralist critical approaches is their recognition that the
relation between subject and structure is not tangential to or a compromise of
aesthetic concerns but crucial to any aesthetic economy. In this view, the subject
does not ontologically precede the social but emerges within a social order. Howells echoes just this view. His dictum that fiction treat the facts of finance "only
as the expression of character" is not intended to preclude economics and social
questions from fiction; rather, Howells is warning that elaborations of detail alone
do not make good fiction: it is precisely the productive relations between (or the
specific economy linking) inner and outer miracles that novels must investigate.
By Howells's definition, then, all novels are economic novels, just as they are
novels of character, because they investigate the relation between subject and
structure.

Recent critical emphasis on the sociological basis of representation has had
the effect, especially among Foucauldians, of replacing the language of organism
with the language of mediation, technology, and discipline. Some work grants the
historical fact of the language of the natural in order to translate it into the language of disciplinary technology and expose its constraining (or carceral) function.
This enthusiasm risks two problems, one formal and the other historical. The formal danger is that the carceral vision (ideological and imperialist are more Marxist
terms for related configurations) implies an impossible alternative: independence
prior to social organization, the very idealization of the liberal vision fueling
Twain's romance of piloting. Without this premise as an ideal alternative, it

makes no sense to think of the subjectification of the self or the production of subjectivity as "carceral" in any strong sense. Subjectivity acquires positivity only within social structures. What could the alternative be?

Historically, entirely transposing the language of the natural into the techne of control risks overlooking the specific force that naturalist language could have. As a result, heterogeneous discourses may be too quickly identified. My case in point here will be American economic discourse of the late nineteenth century, the work Howells was very much addressing in *A Hazard* while he was reviewing it in *Harper's*. Only just coming under the influence of utilitarian precepts from the Continent, American economic discourse explicitly resisted reducing human interests to mechanisms of marginal return. Albeit positivist and normative (though not always orthodox), American economic discourse sought (much like Emerson in the opening of *Nature*) to reform the flawed relation between subject and structure, and it deployed naturalist rhetoric to resist utilitarian reductionism. *A Hazard* marks Howells's greatest fictional intervention in this debate, whose terms he found consonant with his aesthetic concerns.

The Still, Small Voice

It is because economic values are psychological facts that Howells found Thorstein Veblen's *The Theory of the Leisure Class* "an opportunity for American fiction," as he titled his 1899, two-part review of Veblen's book. Howells's was the book's first favorable review, and it was largely responsible for intellectuals' sudden enthusiasm for Veblen. The economy of leisure that Veblen identified—of pecuniary emulation and conspicuous consumption—"is the most dramatic moment, the most psychological moment" upon which fiction might capitalize. "If he is psychologist enough [the literary author] will be fascinated by the operation of the silent forces which are, almost unconsciously, working out the permanency of a leisure class."

A Hazard anticipates the project Howells finds in Veblen's treatise. But the novel's inquiry into the interaction between character and economy disturbs Basil because the sordid, chance economy appears determinist: conditions *make* character. For both character and economy, as Howells would later write, "hazard apparently overrul[es] law" (EEC, March 1912, 636): accident determines the substance of value and character. Just as the economy oscillates haphazardly between "pushing and pulling" (or growth and recession) without resolving into patterns of necessary law, so experience is composed of discrete moments or actions without necessary sequence or purposive motivation: "we live from one little space to another," Basil remarks, "and only one interest at a time fills these" (369).

The psychological economy Basil describes is discontinuous because it is not governed by intention. The unscrupulous painter Angus Beaton epitomizes this mode of action. Beaton's intentions are never more than "inchoate, floating"; they are "the stuff of . . . intention, rather than intention" (488). As chance constitutes values in the sordid economy, "chance determined his events." Beaton denies responsibility for the pain caused by his frequent visits to Christine Dryfoos, which lack the intentions (of courtship) they seem to represent, by denying intentionality in human action: "a good deal goes on in life without much thinking of consequences" (461).

Behavior is not intentional because, as Hobbes counseled, the will is subject to desires. Beaton acts only according to "wayward impulses" (486). Metaphorically, his desires thrust his psyche into debt: "He only knew that his will was sick; that it spent itself in caprices" (394). (More literally, his wayward impulses keep his father in debt, since Beaton supports his art by borrowing from his father money his father has borrowed.) Basil, too, recognizes the regime of desire, though for him desires undermine necessity rather than the will. (Of course, these are complementary substitutions, because the economics of necessity are actualized by volition.) Despite Basil's attempt at "the simplification of his desires," he realizes that obvious luxuries begin through habit to appear "necessary." In a chance economy, "the need which is" gets replaced by "the need that isn't" (438–39), in an infinite chain of substitution of new desires for old needs.

The substitution of desire for needs and intention undermines the possibility of heroic action, stock feature of serious art and culture. This effect is illustrated during the banquet heralding the success of the magazine *Every Other Week*. At one point, Beaton suggests that the journal publish "The Career of a Deputy Hero" or "The Diary of a Substitute," a series of sketches on the paid substitutes who served the Northern cause in the Civil War: "You might follow him up to the moment he was killed in the other man's place, and inquire whether he had any right to the feelings of a hero when he was only hired in the place of one" (336). Purported to be written "by substitutes" themselves, these fabricated, tabloid sketches exemplify a sordid economy's radical usurpation of character. True character would be heroic. But now only deputies or substitutes parade for heroes and authors. The principle of substitution, which Basil feels as a "solvent in New York," "brings to the surface the deeply underlying nobody" (243).

Near the novel's conclusion, Basil offers a contrasting economy of character. Jacob Dryfoos, apparently out of grief at Conrad's death, has offered Basil and Fulkerson control of *Every Other Week* at the generous terms of 4 percent interest. Since Jacob has formerly been a dispassionate speculator in gas and stocks, Isabel March judges that the tragedy has "changed" and "softened" him. Basil disagrees. No "great" or "cataclysmal" events, he argues, nothing "from without" can change us. Instead,

> it is the still, small voice that the soul heeds, not the deafening blasts of doom. I suppose I should have to say that we didn't change at all. We develop. There's the making of several characters in each of us; we *are* several characters, and sometimes this character has the lead in us, and sometimes that. . . . The growth in one direction has stopped; it's begun in another. . . . (485–86)

Basil proposes here an organic self that doesn't change but develops independently of contingency. Indeed, any "event" is "an effect" of this character's "being born" and "coming into the world." In the organic economy of the still, small voice, value and event radiate from an Aristotelian essence of character.

In contrast to sordid personality's service to wayward desire, organic character subordinates desire to need: "Now if we were truly humane," Basil remarks during house-hunting, "we would modify our desires to meet [others'] needs" (61). Serving others' needs is the behavioral principle of an economy of Christian sacrifice. In an "Editor's Study" column published in *Harper's* around the time he began work on *A Hazard*, Howells proposed such an economy as a communitarian ideal by which aesthetic practice (realism) could reform both social practice and economic discourse. Tolstoi is the exemplary realist practitioner of what Howells calls "Christmas" (read Christian) literature. "The dismal science of political economy . . . has left something out of [its] account, and that . . . forgotten factor is Christ himself." Christ "entreats each and every of us first to love his neighbor as himself." Applied to economic behavior, this ethical doctrine means that to seek "personal happiness" is to seek but a "phantom," for "there is and can be no happiness but in the sacrifice of self to others." Some form of socialist ownership of the means of production would seem the corollary to "Christmas" sacrifice. Such Christian economics "is at the moment inspiring the literature of the world as never before, and raising it up a witness against waste and war." "All good literature," Howells concludes, "is now Christmas literature" (ES, December 1888, 158–59).

The Christmas aesthetic resurrects the hero. "Risking all they have in the world for the sake of justice," "true heroes," or true Christians, "are willing to suffer more now that other poor men may suffer less hereafter" (419). A common businessman risks money for profit; the Christian hero, or Christian businessman, risks himself for the benefit of others. Conrad's death epitomizes Christian economics. Like Christ, Basil observes, "it was [Conrad's] business to suffer . . . for the sake of others" (452).

A Hazard elucidates the effects of organic character and Christian sacrifice. The sordid, desirous self engages in cutthroat competition, in which one person's profit arises at the expense of another's loss. Dryfoos unflinchingly admits that "whatever is won is lost" (222). Organic character seems to encourage what was called "cooperativism," the form of organization adopted by, say, the Grange movement (McCabe 471–504), and proposed by many to improve the lot of labor. *Every Other Week* is undertaken by Fulkerson as an experiment in cooperativism, with contributors "paid according to sales" (10). Symbolically, Dryfoos's offer allows the journal to continue its profit-sharing arrangement, which Fulkerson hopes will have a twofold effect: making "the relations between the contributors and the management . . . much more intimate than usual" (36), and making individuals in the venture "exempt from the losses" (10). Labor relations are harmonized and individual risk is minimized. Cooperativism alleviates the conflicts most central to capitalism. To Dryfoos's assertion that one man's profit means another's loss, Conrad counterposes an account of exchange without loss: "when we make . . . money here, no one loses it" (222). When the "the still, small voice" of the soul is heeded, profit occurs at no one's expense.

Organic Economies and the Hedonistic Calculator

As a corrective to the dismal science of political economy, Howells's Christmas literature is typical (both in its objectives and difficulties) of American responses to the tradition of Adam Smith and David Ricardo, christened by Marx the classical school. Carlyle had famously dubbed this school the dismal science mainly in response to the work of Ricardo and Malthus. It is obvious what is dismal about Malthus's projections that population growth will outstrip growth in the food supply, but Ricardo's thesis of diminishing marginal returns was, if subtler, at least equally dismal, because more central to economic debate. Ricardo assumed (1) that the more fertile land is farmed first, (2) that consequently there is an increasing scarcity of good land, and (3) that landlords become more powerful as good land becomes scarcer. From these premises Ricardo deduced that capital investment receives incrementally diminishing returns, more and more of which go to landlords. The result is increasing conflict between the classes of labor, capital, and landlord, and a steady progress toward the "stationary state" Smith worried over, an inertia in capital investment precipitating massive starvation and violent conflict. Inspired by the work of Henry Carey, American economic thought sought to transform the dismal science into the "potato gospel," as Arthur Lapham Perry put it in a textbook that saw eighteen editions in twenty years and sold fewer copies than only Smith's *Wealth of Nations* and Mill's *Principles.*[8]

The grounds for this revisionism was America's abundance of land, which Carey had argued would avert the inevitability of diminishing returns on investment and thus defer class conflict.[9] But the availability of land did not mean its appropriation and distribution needed no administration; hence, the numerous debates about land improvement, protectionism and free trade, and public land policy, like homesteading and the disposition of Indian lands. Whatever positions partisans occupied in specific debates, all held that the potato gospel would follow from an economy of organic character, in which citizens possess and express the right character, and self generates material conditions.

Interest and Organism

If modern economic science appears statistical rather than humanist in its concerns, in the latter third of the nineteenth century economists still conceived their discipline as the study and cultivation of human nature and motivation. Though a conservative spokesman for textile and insurance interests, Edward Atkinson is typical when he writes in 1889 that "the mind of man is the prime factor in material production," and therefore economic study seeks "the development of the character and capacity of the individual" (1890, 27). Economic arguments were very much arguments about what the nature of man is or should be. John Bates Clark, a progressive with whom Veblen studied at Johns Hopkins University, makes the point concisely:

> The value of the results of economic reasoning depends on the correctness of its assumptions with regard to the nature of man. If man is not what he is assumed to be there is no certainty that the conclusions will be even approximately correct.[10]

Extending this premise, Washington Gladden, most prominent of the reform Christian Gospel economists, and for some years Howells's neighbor, concludes that what economists do most of all is teach "temper," which is at once index and function of the health of the system of exchange (296).

American economists inherited the classical tradition of economic analysis, inaugurated by Adam Smith's challenge to mercantilism in *The Wealth of Nations.* Smith's analytical innovation—redefining wealth as the circulation of products for consumption rather than as the hoarding of riches—was made possible by his redefinition of human nature. Before Smith, self-interest was deemed unchristian and unethical (in the Aristotelian sense of the word). Economic activity ought to be altruistic, aimed at the benefit of the community, and hence of God. Private wants and interests were considered vice. The most preeminent oracle of this view was Mandeville's *Fable of the Bees* (published frequently from 1705 to 1724). Mandeville is disturbed that "the very wheel that turn'd the trade" is "that strange ridiculous vice," self-interest, "the grand principle that makes us sociable creatures, the solid basis, the life and support of all trades and employments without exception" (see Cannan lii–liv). Mandeville's attack on self-interest finally undermines his Pauline program, for by his own logic, purging desire would effectively end trade and thus end society.

Smith removes the viciousness from "self-love" by arguing that self-interest is already Christian, precisely because it does supply, through trade and the invisible hand, the inevitable wants of the community as well as of the individual. Typically, Smith justifies his social vision by comparison with the natural state, about which he was more sanguine than Hobbes, certainly, but also than Locke and even Rousseau. Individuals in a "natural state" provide for their needs independently, and therefore have no "occasion for the assistance of" others, or to assist them. But in society, "man has almost constant occasion for the help of his brethren, and it is in vain for him to expect it from their benevolence only." It is only by appealing to self-interest

> that we obtain from one another the far greater part of those good offices which we stand in need of. It is not from the benevolence of the butcher, the brewer, or the baker, that we expect our dinner, but from their own interest. We address ourselves, not to their humanity but to their self-love, and never talk to them of our own necessities but of their advantages.

Trade and, consequently, the division of labor are natural, since economic activity appeals to our "trucking disposition [for barter and exchange] which originally gives occasion to the division of labor." The greater productivity enabled by the division of labor provides not only for subsistence needs but also for needs developed through "habit, custom, and education," "those desires which cannot be satisfied but seem to be altogether endless" (14–15). (These are the same habitual desires that Basil discovers replace old needs with new ones.)

By redefining desire as productive, Smith could redefine wealth not as mercantilist accumulation of riches but as per capita consumption necessitating further production. He thereby shifted the focus of economic analysis from accumulation to exchange. But the regular panics and recessions of the late nineteenth century, including the thirty-five-year deflation following the Civil War, compelled American economists to reevaluate the merits of self-interest. Three major groups debated. At one extreme, conservatives wished to preserve unregulated competitive individualism. At the other, radicals advocated equalitarianism of a socialist or communitarian nature. Between these poles, a range of progressives generally advocated reform through cooperativism and profit sharing. Many reformists were clergy or aligned with the Christian Gospel economists like Gladden, the Reverend John Bascom, Henry Wood, and Richard Ely, the first organizer of the American Economic Association. Though often nonsectarian, all American economists, whatever their partisan commitments, were avowedly Christian, and felt that for God's design to secure the general welfare, an economy must develop as an organism.

Collectivist and progressive criticism of capitalism focused on excessive self-interest that engendered conflicts among individuals. To reconcile competing interests, reformists sought to institute cooperation in industrial organization. Adhering to a narrow interpretation of Smith, conservatives argued that competition is already cooperative and harmonious, since any individual exchange belongs to a greater organic whole, imagined as a cell or as a solar system, where discrete bodies move in different directions but in perfect harmony. Atkinson writes:

> There is a law of harmony in the universe ultimately controlling the relations of men to each other under the conditions of mutual service, as truly as the planets in their courses are bound by the supreme law. (1890, 389)

If for conservatives heavenly harmony is immanent,[11] critics of capitalism employ the organic analogy to illustrate present disharmony. Society is "a physical organism," writes Gladden, "formed . . . of organized cells" (187–88), but unrestricted competition impedes the interdependence necessary for each part of the system, and thus the whole system (since each member of an organism is identically important), to be healthy.[12] All discussants held that the ideal cellular unit on which to cultivate social relations is the family, the most intimate extension of the individual.[13] But critics characterized present relations (marked by poverty, conspicuous consumption, trusts and other combinations, and severe labor disputes, with thousands of strikes each year) not as familial but as "cutthroat competition" (Clark 120) and "warfare," charges made by Fulkerson and Basil in *A Hazard* (85, 455). As a result, industrial progress leads not to the satisfaction of wants but to waste, "waste in some form or other," Howells writes (following Veblen), "being the corollary to wealth" ("Opportunity" 362).

What specifically is wasted is human exertion, "mental energy," as Henry George terms it in his famous *Progress and Poverty* (1879), the inalienable capacity for reason that defines the divine self of the organic order (508). Radicals, as well, championed his unique, rationalist, essentialist self. The Christian socialist, W.D.P. Bliss, whose Church of the Carpenter in Boston Howells periodically attended and addressed, writes that "the most highly developed social organism" realizes "the highest individuality for every member" (350). This celebration of individuality entangles the discussants in the contradictions we have already encountered in the liberal vision, which Roberto Unger has characterized as an antinomy between reason and desire, and which here is also an antinomy between form and substance. In order for individual capacity to be the genesis of value and of the social organism, it had to be defined paradoxically: at once realized in its alienable, material productions, and inalienable from itself because independent of and prior to the material environment and the products of labor.

Like Horace Bixby and the 1802 Mississippi settlers, this vision of the self somewhat revises the metaphysics of Locke's axiom about the origin of property. For Locke, we have seen, labor applied to nature (resources) is the cause of property and value because it is the expression of divinely granted reason; the expression of reason engenders property because reason is the "property" that "every man has ... in his own person" (*Second Treatise* 134). The self owns itself because the prototype and source of property inheres naturally in it; the Lockean self is at once a form of property and a principle of proprietorship. The assumption that one has an inalienable property in one's own person was, of course, widely current. For Henry George, the inalienable "right of man to himself" is the reason that the fruit of a man's exertions should be his own. "As a man belongs to himself, so his labor when put in concrete form belongs to him" (334). Exemplifying organic logic, George believes that "form," here wealth and value, derives from the individual essence. This principle both sanctions and is sanctioned by the words of the Apostle Paul: "the husbandman that laboreth must be the first to partake of its fruits," a sentiment cited by Howells as well as by many economists.

But late-nineteenth-century economists recognized that once the self is conceived in terms of property it becomes a commodity like any other, with no special entitlement. They therefore radicalized Locke's reasoning. John Bates Clark, of Columbia, an influential proponent of profit sharing, distinguishes the definitive human capacity to labor from the wealth it produces. A commodity "is capable of appropriation," but "knowledge" and skills are only potential labor, "subjective qualities," and these are inalienable.

> In every exchange two commodities are alienated and transferred to new ownership. Nothing can be subjected to this process which is an inseparable part of one man's being. ... Man produces wealth and consumes it; but man himself is always distinct from it. (3–7)

"Wants," the sign of incomplete self-possession, are extraneous to the inalienable self. Exchange alienates the laborer from his products, which satisfy these inessential desires. But Clark's self remains distinct from its external and alienable products. Conservatives believe no less in the independence and inalienability of character. Perry proposes that "character, though it has points of contact with things to be sold stands on ground very distinct from them" (114).

Self-possessed and autonomous character is the source of value in an organic economy. For the reformist Henry Wood, the positive element of political economy is the "subjective," "the human heart and consciousness"; "mind is the active worker," and "the outward must correspond to the inner." Material forms, that is, must be the "outward articulations" of the mind.[14] The Christian Socialist George Herron deems the self the basic unit of value, the standard that by unconstrained self-expression confers value on nature as well as goods:

> ... the individual is all that has any worth. The individual man, his wholeness and liberty, is the unrivaled concern of the universe and all that gives it any worth or meaning. Nature and economic things have a value just to the extent that they are the materials by which the human soul may freely express itself. (372)

In this androcentric and proto-imperialist vision, individual worth precedes and generates the value of the universe, and its wholeness and liberty are protected by free, unconstrained self-expression. The question of individual or personal worth is very important to Howells. In *A Hazard,* the disposition of Margaret Vance, Conrad's fellow charity worker, to alleviate suffering is said to result from her encounters with literature, presumably Christmas literature. "Her reading" makes "common people . . . seem so much more worthwhile than the people one met" (177). Christmas literature increases the personal worth of its subjects and readers. In *Altruria* personal worth is described as the failed ideal of America: "if America means anything at all it means the honor of work and the recognition of personal worth everywhere" (11).[15]

Personal worth is achieved when the self is independent of the material arena and enjoys unencumbered volition and free choice. It might seem contradictory that collectivist and cooperative writers, who advocate interdependence rather than individual competition, should urge an individualist basis of value. Yet they, too, view the organic self of self-generated volition as indispensable to cooperation. This inclination belongs to an American myth of self-reliance, and Wood begins nearly all of his twenty-five chapters with an epigraph from Emerson. Man's inalienable ability to choose both his desires and the means of satisfying them must be maintained. "Economic laws depend on the voluntary action of men," Clark writes (32).

Manufacturing Desire: Converting the Zulu

Anxiety about self-determination accompanied (and was used to justify) suspicion of European socialism. Journalism and popular writing of the period regularly excoriated the red flag and its foreign, violent, anarchic influence. Economists' rhetoric was seldom so vehement, except in the work of Bellamy and Richard Ely;[16] still, economists shunned the violence predicted by Marxian socialism and the repression they believed necessary to maintain the socialist state. Henry George admires socialism's ideals and even thinks they are realizable, "but such a state of society cannot be manufactured—it must grow. Society is an organism, not a machine" (321). The danger, that is, is historical materialism, which celebrates the artifactual nature of history and the social order. Bliss likewise censures "manufacture." The "divine state" in which man's free will shall flourish "cannot be manufactured" (350). George and Bliss's distinction resembles the one Basil March draws between divine economy and the economy "we men seem to have created." God's state grows, man's is manufactured.

Distrust of manufacture has specific historical motivation: the unsanitary and unsafe conditions of most factories, and the unfamiliar customs and immigrant origins of many industrial workers. But the manufacture of the socialist state—the production of men and women—seemed dangerous in principle because it violates organic character by prescribing opinions and beliefs. And "liberty," Howells writes, "is only another name for choice" ("Nature of Liberty" 404). In his review of *The Cooperative Commonwealth,* by Lawrence Gronlund (one of the first importers of Marx to America, via his native Denmark, and whose work was influential to the Socialist Party), Howells cites Ely to question Gronlund's need for a "unity of belief" as the precondition and foundation of his commonwealth (Gronlund 163).

> "With the best will, we cannot avoid the fear that in the socialistic state public opinion would exercise a tyranny now unknown . . . which would repress as with an iron hand any divergence of belief or action from a low prescribed level." (ES, April 1888, 803)

When differences are repressed by communal need, Ely and Howells fear, character reflects the organization rather than vice versa. To obviate this predicament, Bellamy's utopia "depends in no particular upon legislation, but is entirely voluntary, the logical outcome of the operation of human nature under rational conditions"; each man is allowed to discover "what his natural aptitude really is" (89, 59).

Socialism's compromise of volition was also a compromise of desire. Gladden feared a "prescribed . . . uniformity" of "individual tastes and preferences." Prescription of desire is at least as important as the prescription of volition, because, following Hobbes, economists believed that the will operates to satisfy desires. "Choices and responsibility," Gladden writes, are "narrowed only subsequent to the limiting of desire" (90, 96). For this reason, Bellamy's utopia, modeled on Gronlund's commonwealth (both inspired Socialist Party planks), allows "free play to every instinct" not intended to exploit others. It might appear contradictory to license desire in a utopian commonwealth, but in Bellamy's utopia "true self-interest . . . appeal[s] to the social and generous instincts of man" (121, 184).

We have seen that defining self-interest as already communal interest is also apologist logic, and was the innovation of Smith's invisible hand. Unregulated self-interest, like utopian self-interest, is, Perry writes, "an enlightened self-interest that sees another's loss cannot be permanently one's gain." Because "character, if genuine, . . . will never compromise for the sake of gain," "the spirit of free trade is the spirit of peace and good-will" (203, 114). Because a creature of enlightened self-love and unencumbered volition, "every man makes his own rate of wages by the amount of intelligence that he puts into the work that he is called upon to do" (Atkinson 1890, 249). In the organic state, compensation is always fair because it is an epiphenomenon of individual natures.

Conservatives view the individual as a self-willed calculator, able to determine needs and the most efficient means to gratify them. Because uncompromised volition guarantees the capacity to estimate and satisfy interests, a system that works by exchange among self-interests can achieve a natural harmony. At the end of the nineteenth century, with the importing of Jevons's marginal utility theory (modeled after Bentham's maxim that men attempt to acquire the most by the least exertion), economists tried to discover a "hedonistic calculus." The search for the hedonistic calculus went so far that an English economist, Francis Y. Edgeworth, devised what he called a "hedonimeter." Because the will governs the hedonistic calculus, society is not warfare, the capitalist account goes, but a naturally self-regulating system. This natural harmony is best achieved without the "wretched legislation of men" (Perry 110), which conservatives view as no less manufactured than stock manipulation. In order to establish political economy as an inductive and just science, Atkinson writes, we must study unregulated exchange, "with free commerce of the exchange of services established under organic laws" (1890, 6).

Lacking the utopian idealism that inspires capitalists as well as collectivists, the reformists doubted that self-interest can be the *same* as "generous instincts." If liberty is the freedom to choose desires, nevertheless, Wood writes, "an excess of liberty to some individuals may prove a tyranny to others" (1901, 130). Howells concurs: "liberty . . . is always in danger of becoming Tyranny" ("Nature of Liberty" 409). Yet if self-love and communal beneficence are often "conflicting" (Gladden 32), reformists nevertheless considered them in principle to be "entirely compatible" (Wood 1901, 13). Gladden argues that "society results from a combination of egotism and altruism." (Hence, Howells's "Altruria.") The "self-regarding virtues" supply the economy with "vigor; without the benevolent virtues it would not cohere" (32). Ely explains the Christian (Protestant, distinctly non-Pauline) position:

> Self-love is not an evil. It is a good and its exercise is commanded. He who is recognized as first among moral teachers, placed it on a par with love for one's neighbor. The two loves are not contradictory; they find their union in the highest love—love to God. ("Past" 34)

Most reformists espoused profit sharing to strike the delicate balance between desire and rational altruism (keeping them distinct but cooperative) that founds society's organic functioning. To illustrate this balance, they adduce the analogy between the economic system and the solar system. Rather than a simple harmony, the heavens embrace by counterpoint a harmony of conflicting forces. Self-love and altruism, Gladden writes,

> are like the centripetal and centrifugal forces of the solar system. These forces are, so far as our measurements can ascertain, perfectly balanced; therefore we have the rhythm and harmony of the heavenly bodies. When self-love and benevolence are perfectly balanced in human conduct, we shall have on earth the beginning of the thousand years peace. (180)

In the moment of "perfect equilibrium," Henry Wood writes, "labor and capital
. . . melt into each other" (1901, 26–27), and labor is reunited with both its fruits
and its inalienable, organic self.

But if counterbalanced desire promises economic and ontological salvation,
nevertheless a fundamental, twofold problem remains. First, although profit sharing circumscribes the exercise of self-interest, the institution of this program must
nevertheless be voluntary; otherwise, reform will commit the sin of "manu-fac-
ture" or "paternalism," a charge progressives hurl at collectivists, and which Bellamy and Gronlund in turn hurl at European socialists. During the banquet in *A
Hazard,* Colonel Woodburn accuses the German socialist Lindau of paternalism
(344). How does one measure at what point limiting desire becomes paternalism?
Second, if "rivalry in giving is the essence of legitimate competition" (Clark 155),
how can one successfully distinguish legitimate rivalry, serving harmony, from the
excessive rivalry of warfare? Critics impugned capitalism precisely because its
fealty to desire destabilizes the harmonious equilibrium. But once desire is admitted into the model, how do we guarantee that it will (voluntarily) keep its Christian place? Since desire is by definition the impulse to exceed or alter present
boundaries, how can desire regulate itself?

Imbricating the will with desire, and then making desire the instrument of its
own limitation, is the fundamental (though finally productive) contradiction of
the American liberal vision. The goal of the American system, constituted in the
year *Wealth of Nations* was published, is the maximum satisfaction of personal
desires, which every economist, even the communitarian Gronlund, considers in
principle unquenchable, "absolutely unlimited" (88). Indeed, desire's insatiability
is heralded as the very mechanism of salvation, since technological progress is
spurred by desire's constant outstripping of current forms of satisfaction. Henry
George proclaims that "restless desire" instigates "higher forms and wider powers" (136–37). Divinely inspired, such desire in turn inspires invention to increase
production faster than Malthus and Ricardo promised population and land developments would outstrip subsistence levels. This is why, in the American imagination, the motor of Carlyle's "dismal science" really generates the "potato
gospel."

Not only does desire produce progress, the progress of civilization produces
desire. Because society's salvation lies in desire, desire may be society's key, even
its founding, product. Gladden illustrates this point through an imperialist anecdote, in which the effect of Christian civilization on a primitive is the production
and then proliferation of desire:

> . . . its uniform effect upon human nature is to create in man many of those wants
> which it is the office of wealth to supply. The savage has few wants; the fully
> developed Christian has many; the progress of the savage from barbarism up to
> Christian civilization consists largely in the multiplication of his wants. A mis-
> sionary lately returned from Africa testified that the great difficulty with the
> natives . . . was that they had no wants; "their greatest want was a want." How to
> develop in them the sense of want—that was the problem for the missionaries. It
> was a great encouragement when one day a Zulu found out that he wanted a
> wash-basin. Pretty soon he wanted a shirt and a pair of trousers, and, after a little,

a house with a chimney, and a hoe, and a plow, and by and by he wanted a book to read; and when he had got all this property he was a wealthy man compared to his neighbors. So Christianity always has the effect to develop faculties that require for their exercise the possession of property, and to waken desires that can be gratified only by the use of those material goods whose aggregate we call wealth. (7–8)

Inverting the logic of ancient Greek disparagement of barbarians, Gladden ascribes the natives' moral deficiency to their lack of wants; converting them to Christianity means developing and then multiplying wants that require a Christian (consumerist) economy for gratification. When the Zulu wants a wash-basin, he's got the right idea; when he wants a book he's wealthy and righteous.

In this anecdote suggesting a reciprocity between postbellum racism and anxieties about fluid markets,[17] Gladden anticipates by a decade the contribution to the conception of imperialism that, Martin Sklar argues, Americans made well before Continental critics of imperialism like John Hobson, Lenin, and Rosa Luxemburg: the surplus capital theory of modern capitalist imperialism. Jeremiah Jenks, Charles Conant, and Paul Reinsch were the first to argue that capitalist economies could not remedy their tendency to cycles and panics by opening new markets for products (the model of traditional colonialism); instead, they needed to develop avenues for investing surplus capital. Simply imposing products and regulations on markets was insufficient. Investment expansion must, Sklar stresses, utterly transform the social and political relations and institutions of assailed cultures. That is, investment expansion must catalyze a full-scale cultural transformation of nonindustrial cultures into capitalist societies.[18]

Gladden's remarkable anecdote of conversion intuits how annexation involves as well the (re)formation of self. Christian civilization, for Gladden, is the paradigm of the need to create markets by advertising. Christian society not only creates particular demands, however; it produces the principle of demand that makes a market possible.[19] Gladden understands that new markets are not the real issue. He is fascinated mainly with Christianity's/capitalism's production of identity, specifically an identity that wants and can adopt and perpetuate the cultural (literary) as well as financial institutions indispensable to the consumerist self Christianity cultivates.

Gladden's anecdote of the Zulu concisely exhibits the problem that structures the liberal self, at once reasonable and desirous. The multiplication of restless desire is supposed to harmonize and redeem society, but it is precisely desire's proliferation that manufactures sordid disequilibrium. Desire is what supplants the "intention" of labor with that of capital, in Perry's term; it is what threatens volition and autonomy, what distends liberty into tyranny. The contradictory understanding of desire by Gladden and others has affinities with, but is finally different from, Gilles Deleuze and Felix Guattari's argument that capitalism has a schizophrenic relation to desire, which it produces but then must contain. For Gladden and others, desire clearly is disruptive of habitual and institutional structures. (So was capitalism as it emerged historically.) Desire displays, that is, what Deleuze and Guattari call a "disjunctive" capacity (76). But Deleuze and Guattari propose a generic ontology of desire that needs, I think, historical modification.

They argue that desire, like some primal force, "wells up" and spills out of "breaks" in the capitalist regime; it threatens to "short-circuit social production," then is anxiously decoded and regulated, and thus contained, by capitalism (37, 31, 33).[20] For the economists and Christian moralists who celebrated the economy of desire, however, desire does not have such a two-step relation to capitalist culture. They think desire needs managing, not containing. Desire and its disjunctive effects are produced within Christian capitalism, and desire accomplishes important cultural and psychological work. The discrepancy between desire and satisfaction stimulates profit and progress. Now, there is a specific, historical antinomy here: Christian economists avow the goal of reducing this disparity, but the logic of Christian progress in fact requires perpetuating it. The dual assumptions of, on one hand, the volitional, satisfied, organic self and, on the other, the multiplication of desire are discordant. Christian economists inconsistently hailed the self sometimes as self-contained, sometimes as hedonistic.

The Emulative Self: Veblen

Thorstein Veblen criticized the classical and utilitarian schools of political economy precisely for this contradiction in their "faulty conception of human nature." The "hedonistic terms" of economic inquiry posit a "substantially inert and immutably given human nature." Veblen continues, with his distinctive irony:

> The hedonistic conception of man is that of a lightning calculator of pleasures and pains, who oscillates like a homogeneous globule of desire of happiness under the impulse of stimuli that shift him about the area, but leave him intact. He has neither antecedent nor consequent. He is an isolated, definitive human datum, in stable equilibrium. . . . Self-imposed in elemental space, he spins symmetrically about his own spiritual axis until the parallelogram of forces bears down upon him, whereupon he follows the line of the resultant. When the force of the impact is spent, he comes to rest, a self-contained globule of desire.

Because by definition desirous, hedonistic man cannot be an "intact," "self-contained globule of desire," willing his destiny "in stable equilibrium," "self-imposed . . . about his own spiritual axis." Rather, he inhabits disequilibrium, ever wanting something different from the present self. Hedonistic man cannot fulfill the classical demands for the organic self: autonomy, permanence, organic articulation. Such autonomy could be achieved only by a being like Gladden's preconverted savage—without desire, without civilization, but also, in essence, without life; for as imagined, hedonistic man "is not the seat of a process of living."[21] The indispensability of desire spoils "the perfection of character" necessary to conventional organic harmony.[22]

We have seen that for classical school writers, the fact of desire, which subjectifies the self in a nexus of relations and contingencies, makes impossible a systematic science of character. Veblen is not so thwarted. He believed that orthodox economics cannot realize its systematic ideals because of its idealized and teleological premises.[23] Instead, for Veblen, forgoing the teleological notion of character is to make possible inquiry adequate to the institutional quality of all personhood, value, and experience. Veblen writes that character is "a coherent structure of propensities and habits which seeks realization and expression in an unfolding activity." This practical coherence consists in, and is expressed in, "habits of thought" that are "cumulatively wrought out under a given body of tradition, conventionalities, and material circumstances" ("Why" 74–75). These habits that constitute the self are coextensive with the "fabric of institutions";[24] institutions are, Veblen writes in *The Theory of the Leisure Class,* "prevalent habits of thought"; "they are of the nature of an habitual method of responding to . . . stimuli."[25]

Like character, "a standard of living is of the nature of habit" (*LC* 82). We experience our habits, and thereby ourselves, only in relation to extant standards of conduct. This phenomenology proceeds "by an invidious comparison with other men" (40). Because the conspicuousness of conspicuous leisure is the yardstick of value, we discover value "not so much in substance as in form" (77), and character necessarily fails the liberal test of organicism, for substance is inextricable from form and self-worth is pursued by "pecuniary emulation." It is important to specify that emulation is not simply imitation of others; it is imitation to supersede others, "a straining to excel others," "to outdo those with whom we are in the habit of classing ourselves" (81). Emulation never ceases, because an equilibrious perfection of character is never reached. Veblen's economy of emulation is a hyperkinetic version of the Kantian sublime.

Where Twain fancied a self embodying primary narcissism, Veblen sees the self occurring in an activity that combines features of repetition-compulsion and Lacan's mirror stage. The emulative self is perpetually sought in the other; one self is ever striving to be born and to exceed both the present standard and that of the other. Veblen's self is a kinesis of desire, always moving to appropriate and supersede another. Because "each class envies and emulates the class next above," "the standard . . . which commonly guides our efforts is not the [one] . . . already achieved; it is an ideal of consumption that lies just beyond our reach" (*LC* 81). Emulative desire "can scarcely be satiated in any individual instance; satiation . . . is out of the question" (39). Emulative desire is not abated but, rather, stoked by being fed. As the old French saying has it, Fulkerson remarks in *A Hazard, "l'appetit vient en mangeant"* (284).

Veblen's emulative self consists of precisely those insatiable desires of habit or custom that Adam Smith observes arise when subsistence wants have been met, and that Basil identifies as new needs replacing old. Kinetic and emulative, Veblen's notion of self is intended as a corrective to the absolutist premises of classical organicism. Specifically, it offers an American pragmatist alternative, defining essence, value, and nature in terms of practical and institutional effects. If the emulative self is Lacanian in structure (always known as incomplete and different from itself), it is nevertheless "a coherent structure of propensities which seeks realization and expression in an unfolding activity" ("Why" 74). Like William James and C. S. Peirce, with whom he studied at Johns Hopkins University, Veblen understands coherence to be practical and unfolding in effects rather than metaphysical. Nor as a pragmatist does he abolish the organic as a criterion of value. But the organic is known only in the context of habit and institutionality:

"individual habits of thought make an organic complex" (*LC* 173), and furnish the "scheme of knowledge" for studying the "sequence of phenomena" from which we deduce "natural laws" ("Evolution" 61). Veblen promulgated this pragmatist account of the status of nature in his first published article, "Kant's Critique of Judgment" (1884): because "objects are conceived to stand in such relations of dependence and interaction as correspond to the logical relations of the concepts we have of them," concepts make "a part of our knowledge of nature" (*Changing Order* 183–84).

Deductive economists complained that Veblen's institutional notion of self and of the organic vitiates economic analysis. Veblen, however, thinks it is the condition of economic science (indeed, of any science). Like some contemporary poststructuralist (especially Foucauldian and neopragmatist) analysis, Veblen probes the institutional formation of value and character. As institutions and material conditions change (though *not* teleologically), character adapts to new exigencies. To become an evolutionary science, economics must become a science of institutions and their evolution, and Veblen's work inaugurated America's institutional school of economics.[26]

Impartial and Hazardous Taste

Veblen's critique of liberalism's self-contained globule of desire is anticipated in *A Hazard*. Howells, too, we have seen, recognized that a self constituted by desire cannot possess the autonomy demanded by the hedonistic model. Howells is disturbed that the moral claims of the hedonistic self can never be absolute. He attempts an express correction to the sordid economy of character through the sacrifice of the selfless Conrad. By inducing Dryfoos's offer to Basil and Fulkerson, Conrad's death secures *Every Other Day*'s cooperative ideals, thus inspiring Basil's organic account of character. Conrad's sacrifice causes "greater kindness" rather than "any business obligation" (492) to compose the relations among the various characters. Such effects redeem the loss of Conrad. A Pauline purification of Smith's already Christian vision of the invisible hand, his death is meant to realize the liberal ideal: the harmony and justification of interests by their transcendence and abnegation.

This definitive maneuver of classical liberalism structures an aesthetic and critical ideal Howells delineated in "Editor's Study" columns published just before he undertook *A Hazard*. "Our own ideal," he wrote, speaking as both author and critic, is to provide a "mirror of impartiality and balance of justice" (ES, March 1888, 643–44). In quest of this impartiality, Howells had addressed the December 1887 "Editor's Study" to "the question of a final criterion for the appreciation of art, or a 'unity of taste.'" Sounding very like Thoreau, he seeks to locate a "solid ground" of "simplicity and naturalness and honesty" in taste, so that just as economic fluctuation should not constitute value, fluctuations of mood or fancy, the emotional manifestations of a chance economy, should not define the beautiful. People who follow shifts of fashions "represent the petrifaction of taste": they are not "real" but "artificial," "card-board" characters. True characters generate their own value and standards and thus have taste (153–55).

A Hazard tests this account of taste: Basil is said to have "taste . . . and conscience" (9). In this model, the propriety of taste seems a function of the interests of true characters, but in fact, taste arises from the evacuation of interest. The March family's taste is "very sympathetic," "altruistic"; this taste without self-interest expresses their "democratic instincts" and seems to Basil the basis by which "he did full justice to the good qualities of . . . other people" (26–27). This vision of taste as altruism and justice posits epistemological and ontological neutrality. Like the positivist objectivity sought by American economists, Basil seeks to observe phenomena disinterestedly, so that he may conceive of things with the "no doubt" that characterizes divine economy. In the uncertainty of the economic chance world, one becomes one of the "populace" (412), formed by and heeding the flux of representations. The danger of being "populace" leads Basil periodically to express his reluctance to commit himself, so that he might remain "anonymous" (124), that is, not perceiving phenomena in terms of interest. He tries to remain a "philosophical observer" in order to "do justice" to the people and events he observes. To Basil this "philosophical position" is the condition of both the production and appreciation of art. For example, he and Isabel, during their inveterate strolls through New York, attempt to maintain a "purely aesthetic view of the facts" (65), exercising a refined "literary curiosity" rather than a sordid, exploitive interest in their subject.[27]

Conceiving their taste in the rhetoric of divine economy, democratic and sympathetic, the Marches ascribe to it an organic ontology. Basil feels his cultivation is made possible by "the simplification of his desires." Because purged of excess desire, the Marches think while in Boston that their taste, hanging from the walls of their home, "expresses their character." This organic structure, where articulation unproblematically represents character, contrasts with the Dryfooses' situation in New York: the gimcrackeries that decorate, or rather clutter, the Dryfoos home *form* the family's character and attitudes. The vulgar populace "can't give character to their habitations. They have to take what they can get" (67). The tasteful Marches, on the other hand, produce their own character.

If Conrad's sacrifice is meant to protect and redeem this economy of character and its attendant, "final" aesthetic standard, the novel's resolution nevertheless displays discord. For one thing, the redemptive effects of Conrad's sacrifice are incomplete. Margaret Vance remains unconsoled. She believes herself responsible for his death, assuming that in an earlier conversation she inspired him to help the strikers. Jacob Dryfoos is equally distraught, believing his son visited the riot in a mood of filial defiance following a quarrel. Thus the novel's incomplete recuperation is managed via inaccurate interpretations of Conrad's intention, since he, like Basil, arrived at the riot purely by accident (421). The event, that is, was not caused by his intention. Indeed in general, we will recall, Howells reduces chance, risk, and caprice in personal and economic relations through the introduction of accident, the coincidental appearance of Basil, Lindau, and Conrad at the riot. Basil realizes that he "was there purely by accident" (429), and, he concedes, "in the cause . . . of conjugal disobedience" (452). His appearance transgresses both the etiology of event and the family unity essential to organic harmony. Add to this the fact that the cooperative conclusion includes Beaton while sacrificing Lindau, and it seems that Howells achieves organic necessity through the strategies of the economic chance world.

Howells's dependence on coincidence has struck most critics as unrealistic, the stuff of sentimentalism, and as a compromise of his express political objectives.[28] Amy Kaplan has more fruitfully argued that the unsatisfactory quality of the novel's final hundred pages reveals the ideological function of Howells's realist project—his declared communal objectives finally seek to harmonize social diversity and a plethora of details in unified narrative form. But the novel's failed resolution releases the conflicts realism is intended to repress, and therefore *A Hazard* "both fulfills and exhausts the project of realism" (59–64). In this view, Howells's criticism and practice are inadvertently middle-class, more liberal than he knows, and he pursues justice by means that contravene it. I think Howells's strategy is more subtle. He comes to recognize during the composition of *A Hazard* that his organic ethic of realism, indeed his very liberalism, very much inhabits a chance (capitalist) economy of character and event. The accidents and failed or usurped intentions that make up the novel's conclusion and his response represent more than a poorly conceived egress from the ontological, economic, and formal problems Howells has set himself. They constitute a pragmatist rethinking of the nature and status of organic organization and liberal commitment.

Howells concluded his essay on the search for a final criterion of taste with the "hazardous . . . conjecture" that "possibly there is no absolutely beautiful, no absolutely ugly" (ES, December 1887, 153). While the Marches initially cherish their Boston aesthetic sense as absolute taste, *A Hazard of New Fortunes* is an appropriately titled novel, in which Howells, himself just arrived in New York, the newer American metropolis, confronts the hazardous conjecture that taste, character, and economy are always contingent. He comes to comprehend that the opposition between the absolutely and hazardously beautiful is not so firm as he desired it to be in earlier reviews; final criteria always emerge in the interstices of hazardous, fluctuating, socially produced taste. In *A Hazard* Howells anticipates Veblen's idea that "requirements of pecuniary decency have, to a very appreciable extent, influenced the sense of beauty" (*LC* 94).

Most American economic writing posited the liberal dualism between the organic and the conditioned or sordid, between free will and what Foucault calls subjectification. Logically, the typical demand for organic articulation through selflessness (Conrad) and disinterest (Basil) is materially and socially unrealizable. Self without desire and perception without perspective make social relations supererogatory; there could be no point in reforming them, and no way to do so. Without desire there is no need to interact; without perspective there is no perception. But Basil's notion of the organic does not, finally, require that ontological and epistemological autonomy wholly and necessarily condition outward forms. Rather, his conception is Veblenian, and includes accident, chance, and conventional inculcation in the makeup of the organic. That is, his conception of the organic already includes the sordid.

Even from the outset, and despite the Marches' attempt to imagine their taste as absolute, *A Hazard,* a novel about a family's adaptation of absolute standards to new commercial exigencies, presents taste as interested and socially implicated. An incident during one of the Marches' treks through New York ironically indicates the embeddedness of the Marches' prized tasteful sympathy, pure aestheticism, and literary curiosity in an invidious class structure. Walking through an immigrant neighborhood, they respond like aristocrats slumming for a day: "They met the old familiar picturesque raggedness of southern Europe with the old kindly illusion that somehow it existed for their appreciation and that it found adequate compensation in this." Seeing a tenement street, they censure "artists for their failure to appreciate it" and "content themselves with . . . wondering why nobody came to paint it" (65). The Marches' aestheticized sympathy would compensate poverty by making it an artifact to be appreciated by the comfortable. If "kindly," such compensation is still an "illusion," not eradicating but merely eliding class distinctions.[29] The immediate cause of Basil's "tacit sympathy" is his "letting a little Neapolitan put a superfluous shine on his boots" (55). Basil's sympathy—the money he pays the shoeshine man—is superfluous, a bit of Marxian surplus value, the profit Basil lives on while the Neapolitan labors, in this instance, for him. The Marches' aesthetic values now appear as exploitive as those of the Dryfoos daughters, who "unconsciously imputed merit to themselves from the number and violence of the [gas] wells on their father's property" (257).

Basil is realizing that the literary curiosity by which he would divorce himself from market and class contingencies is defined and sanctioned by social exigency. Specifically, his taste works by the inexplicitness of its interestedness. Veblen adapted from Kant the observation that "any [subjective] interest spoils the judgment of taste, and takes from it its impartiality" (Kant 138).[30] While Kant would purge private interest from impartiality, he nonetheless recognizes taste and judgment to be social constructs, impartial only insofar as the shared categories of perception generating taste and judgment remain essentially inexplicit. To translate Kant's terms into Veblen's adaptation of them, taste emerges in a network (a necessarily invidious network) of institutions that condition habits of thought, so habitual that even while they generate habits of "expression" they remain unexpressed (*LC* 83).

Its inexplicitness of interest, or disguise of interest as impartiality, marks this concept of taste as capitalist. Gronlund, for example, following Marx, argues that the slavery of feudalism and the exploitation of the "wage system" differ morally in the degree to which exploitation is disguised. Feudal lords appropriated "the lion's share from [labor] without disguise." Under the guise of free exchange, expropriation no longer takes place "frankly." Laborers seem to sell their labor to capital of their own free will, but in fact this exchange is just an alternate, "softened" expression for slavery.[31] In this view, the mark of an exchange economy is its disguise of self-interest as the communal.

In *A Hazard,* taste clearly belongs to a disguised economy of imbalance. The Marches deem the "bric-a-brac" objects that fill the Dryfoos home vulgar because "their costliness was too evident; everything meant money too plainly, and too much of it" (151). Their decor would be tasteful were the cost of its design invisible. Objects are vulgar when their expense is explicit, whereas taste effects a deflationary disguise of expense. Conversely, Beaton (of course it is Beaton) devises a method of illustration to impress the journal's patrons. This method is inflationary: "the illustrations look as expensive as first-class wood-cuts, and they're cheaper than chromos." This inflationary effect is called "style" (201). Whether inflationary or deflationary, taste and style inhere in an object when it appears to cost other than it actually does.

A Hazard knows taste to inhabit a capitalist harmony. It is part of economic surplus, and complements imbalance and inequity by disguising them. Concomitantly, since aesthetic distinction and value result from an object's social function, instead of extending from an a priori perfection of character, taste violates the absolute organic articulation of social event by character necessary to a Christian economy. On all counts, taste is sordid, a hazardous phenomenon.

Preferring the Freed Will

Basil comes to recognize the fallacy of the simple organic. "No matter what whimsical, or alien, or critical attitude" he takes, he nevertheless feels "a sense of complicity with the forces at work" (306).[32] Basil perceives that his very ability and tendency to appreciate the "spectacle" of the "forces of pity, of destruction, of perdition, of salvation" (306) is both sign and source of his complicity. Basil's confession of complicity signifies his Veblenian recognition that his tastes and judgment emerge and make sense only within the current network of institutions and habits. (Today we associate this insight with the work of Foucault or Stanley Fish.) The fact that Basil labels the "speculations" of his literary curiosity "comfortable reveries" suggests their membership in a leisure class. But Basil feels the complicity of his taste as a moral problem. If he, like everyone, is complicit with the forces at work, he must accept some responsibility for the suffering that partly composes the conditions of his taste. Yet this is a painful responsibility; hence, at the nadir following Conrad's death, the absolute dualism of the economic-chance-world speech offers itself as a moral pardon, and Basil, like Beaton, employs the logic of conditions-make-character in order to displace (even his own) responsibility.

When Dryfoos's offer facilitates the cooperative spirit of the magazine, Basil's spirits improve: he can again acknowledge complicity, and he proposes his organic model, in which events obey the still, small voice of the soul. But the organic logic of this proposal diverges from the comfortable and absolute dualism of most economic models and of Basil's own earlier model. Here, the organic condition of the self is a function of, rather than wholly constitutive of, belief and faith. The clue that Basil is surrendering absolute dualism is found in his rhetoric of faith: "*I* don't know what it all means . . . though I believe it means good" (486). Belief now supersedes knowledge as the ground of the organic; the organic's ethical structure is that of taste. Basil might now (as Howells will later) appreciate Veblen's sentiment that "the canons of pecuniary reputability do, directly or indirectly, materially affect our notions of the attributes of divinity, as well as our notions of what are fit and adequate manner and circumstances of divine communion" (*LC* 93). Basil's very next remark—that Christ himself argued "that if one rose from the dead it would not avail. And yet we are always looking for the miraculous!"—indicates his sense of the contingent rather than final efficacy of Conrad's death. Conrad's death feels redemptive only under scrutiny of a particular kind of faith and trust.[33]

Viewing divine law as an aspect of institutional habit—that is, understanding that there is, in Veblen's words, a "habitual and accredited spiritual attitude for the members of the group" (*LC* 32)—modifies the opposition between free will and determinism that fuels both nineteenth-century economic and also many modern critical debates. This challenge is consistent with the pragmatist critique of rationalism and dualistic philosophy, which rejects the free-will/determinism question and views the will not as an essence but as a power whose specific agency is known only in its exercise. In his 1891 review of William James's *The Principles of Psychology,* Howells addressed the question of the relation between the will and habit. The soul, for Howells, evolves and dwells in this relation. Rationalist and idealist psychology and philosophy, Howells follows James in recounting, considered the will an absolute faculty. As economics leaves Jesus out of its account, so in psychology "the soul is left out of the account" (ES, July 1891, 314). The soul is to psychological economy as Jesus is to social economy, and the question of the soul can be approached only by exploring the will as a habitual activity.

James, of course, based psychological study on the study of habit, which is not just "the enormous fly-wheel of society" (1:121) but the very principle of nature. He tempered a proverbial exclamation by the Duke of Wellington— "Habit a second nature! Habit is ten times nature" (1:120): "The laws of Nature are nothing but the immutable habits which the different elementary sorts of matter follow in their actions and reactions upon each other" (1:104). As the law of nature consists of habits of Newtonian physical relations, individual natures consist of habits of thought and social interaction, and the will arises out of the intention of habits toward a goal by an "effort of attention." As Emerson (we saw in chapter 2) defined nature and value as phenomena, not substances, James, whose Emersonian heritage is well documented, views the will within the transcendentalist tradition, as not essence but agency, not a state of being but an exercise of power. Like Emerson, James sees no intrinsic conflict between the discourses of power and the soul. "The strong-willed man . . . is the man who hears the still small voice" of his soul and can obey it, as can Basil's organic character. The weak-willed lack this power, and therefore "never get to holding their limp characters erect" (2:563, 547).

Imagining will as a principle of masculine potency, James (as had Locke) rejects as incoherent the notion of a free will.[34] The will might be considered free if it could be said to operate unobstructed by resistant forces and contrary impulses, and "devoid of any sense of coercion." But this concept of free will, or "free effort," as an "independent variable," exists only in an ideal world and acts only upon an "ideal object." Free effort cannot exist in any absolute sense; it is an oxymoron. Freedom, in the pragmatist view, is a relation of work, its particular quality experienced only in differential relation to forces of resistance, and defined as freedom only in relation to an experience that doesn't feel free.[35] Therefore James dismisses the logical and metaphysical question of free will, for the idealist demands the free-will/determinism debate makes on the nature of the will are unsatisfiable and incoherent (2:562–63).

While agreeing that the debate over free will stalls at a logical bottleneck, Howells insists on addressing the issue, because it does circulate in ethical traffic:

> In fact the will is *not* free; but the will of the strong man, the man who has *got the habit* of preferring sense to nonsense and "virtue" to "vice," is a *freed* will, which one might very well spend all one's energies in achieving. It is this preference which at last becomes the man, and remains permanent throughout those astounding changes which everyone finds in himself from time to time. (ES, July 1891, 315)

Will and preference are not antinomies. The will that composes character is precisely preferential, a form of moral preference that "at last becomes" the essence of "the man." Thus freedom, as Howells comes to understand it, is an attribute of action, not an ontological condition. As preference, the will cannot be free in an ideal sense but can be freed in a habitual or pragmatist sense, its permanence experienced amid "astounding" fluctuation.[36] But if clearly contingent, the freed will is nonetheless a real freedom, the only freedom of will available. This is why we can be "creatures of our own habit," and have a pragmatist rather than idealist form of self-possession.

Howells believes that the contextuality conditioning the freed will animates psychological science as well. As Veblen contends about economics, Howells maintains that psychology can never claim the ideal objectivity of science "but must always remain a philosophy," concerned with the logic of beliefs; and though it certainly holds its present beliefs to be eternally true, "it can always change its mind." Veblen (who because of his focus on habits of thought has been called the William James of political economy) thinks no inductive enterprise can "reach a final term": "because its prime postulate is consecutive change," "every term" of "modern scientific inquiry . . . is transitional" ("Evolution" 33). Howells extends this point, adducing Kant's notion of organic unity to suggest that knowledge of the organic, and thus knowledge in general, are, though sufficient, never absolute. By an act of faith, "Kant felt the moral law within him one in meaning with the starry heavens above him"; that unity is one "we can perpetually know better, but never wholly know." Here, Howells posits nature (typically figured, as in *Nature,* as the stars) as an absolute exceeding human comprehension, but he also accepts the disparity between eternal truth and human comprehension as the possibility of power and progress.

For Howells, James, and Veblen, the will, the organic, scientific inquiry, and knowledge lack final, objective and independent status. Yet these are not then useless; indeed absolutist demands for the organic, because unrealistic and false, imply a truly useless, because unrealizable, ideal. Though not final, these categories are contextually "coherent," in Veblen's term, substantial and even stable within the context of their enabling conditions, the only kind of stability available. Responsibility and freedom in behavior possess a similar stability.

Recognizing the inferred and socially embedded nature of the organic, character, and the will, Basil can express a certain irony toward that faith that grounds the organic. Lacking both Basil's comprehension and his irony, Isabel cries "fatalism!" when Basil formulates his organic model. Basil's irony surfaces a bit earlier, when he proposes his Christian economy. Isabel worries that such an economy—heroic doing *for* others rather than *to* them—could be practiced only in hospitals and charity homes, not in ordinary families. Christ's heroic example should teach us not only how to live out of the world (in hospitals or fatal sacrifice) but "how to live in it, too." "But perhaps you don't think the homes are worth minding?" she challenges Basil. He rejoins, "I think the gimcrackeries are" (452). A Christian economy is desirable, but it, and the self inhabiting it, cannot be grounded in anything more stable than gimcrackeries. The Howellsian organic self discovers the very conditions of its organicity in the needs of a sordid, gimcrack economy.

Basil's conversion to the sense that Christian economy and organic character are already, as it were, "sordid"—that is, contingent—organizations concludes with the novel's final sentence. The Marches pass Margaret Vance on the street. Despairing the futility of Conrad's death, she has entered a convent, and in her nun's habit she has effectively left this world: she is "at rest" in "the peace that passeth understanding." Her response seems to confirm Isabel's fear that a Christian economy is realizable only by abandoning the mundane. Isabel correctly suspects that even now Margaret's grief is not assuaged, that she remains unpersuaded that Conrad died "for God's sake, for man's sake." Perhaps with a grin, Basil advises, "Well, we must trust that look of hers." The modal "must" carries great weight here, for it signals that trust is not a necessary response to present events; the modal designates the subjunctive or imperative nature of the desirable condition. We must, but we might not. Basil's "must" concludes *A Hazard*'s pragmatic reconception of the liberal self and the organic. Though practically distinct from the sordid, the organic grows out of the contingency to which classic liberalism would like to oppose it.

CHAPTER 5

"To Find the Value of X": Speculation and Romance in The Pit

An Awkward Marriage

The Pit is not a great novel. In making this evaluation, I am concurring with most critics of Frank Norris's last work, who deem it interesting but flawed. It is interesting in that it represents if not the first, yet hitherto the most direct fictional confrontation with speculation, which by the turn of the century was unmistakably, and to many disturbingly, coming to determine economic values, to affect many other cultural activities, and to inform citizens' consciousness of daily experience. *The Pit,* that is, frankly explores the nation's transformation to a credit culture, an ongoing transformation that is a central feature of contemporary life, and in which we have all, broadly, been forged.

The novel's flaws, critics have argued, are twofold. First, Norris's "epic theme" of the wheat, the foundation of his projected trilogy, "lacks force," Donald Pizer writes, because we have no "direct contact with the wheat."[1] Although conceptually "the chief actor" in Norris's drama, the wheat is lost in speculation, writes Franklin Walker, since "the manipulation of it upon the board of trade [is] the manipulation of paper representing paper." Thus, the essential "force" that the wheat represents—that is, the natural law of supply and demand that the wheat obeys—"must remain an abstraction."[2] There is also a structural flaw: the plot regarding Curtis and Laura's marriage "seems unrelated to" the story of wheat speculation; "there seems to be no common theme in the two," yet Norris not only yoked the two together but "allowed the love story to gain the upper hand" (Pizer 1966, 165; Marchand 86).

There seems no connection between these thematic and structural flaws: Norris's attention to wheat speculation has seemed to enervate the epic subject; and the wheat narrative is combined for no apparent reason with the love plot. I want to argue the obverse of the same coin. Norris integrates the narrative of speculation with the marriage plot too well, perhaps too predictably. Speculation or gambling has long figured in novels as the complement, if a hostile complement, to domestic careers. This is certainly evident in some works of Dickens and Eliot, and was a common theme in French realism. In America, the integral narrative and rhetorical tension is central, for example, to Cooper's *Home as Found.* The typical thematization of speculation as a threat to marriage and domestic life is evident in Francis Edmund's popular 1852 painting *The Speculator* (figure 19), in which a speculator tempts a couple to turn from their hearth and to neglect the family table. As the largest element of the composition, the hearth invites our attention. Moreover, it is tangible and nurturing, like the corn the husband scrapes. Yet our eye is distracted from the hearth's formal and thematic centrality, just as the couple is distracted from its domestic warmth, by the goateed speculator's offer.

Norris fancied himself a formal innovator. One of the main "responsibilities of the novelist," as Norris titled one of his essays on writing (the title under which these essays were posthumously collected), was to reform fictional convention and popular expectations. Not only did he specifically discuss speculation in these essays, it was his central trope for both the emerging cultural order and the office of the new novelist. By Norris's own conception of the speculative responsibility of the novelist, speculation *ought* to violate the epic subject of wheat, and the plots of marriage and speculation should be finally incompatible. But *The Pit* goes out of its way to exorcise what turn-of-the-century Americans considered scandalous about speculative activity. Absorbed and vanquished by natural law, speculation readily suits the novel's wheat theme, and the marriage and speculation plots are harmonized and become versions of each other. Ultimately, the various aspects of the novel, which critics have so well identified but whose formal and rhetorical principles they have not fully understood, are not disjunctive flaws but are all too harmonious. The novel's true failing is its obsession with harmony, which marks Norris's (and many Americans') fear of the cultural transformation to a credit economy he insisted it was the artist's job to narrate.

The Poet as Financier

Henry Adams, in his 1869 essay "The New York Gold Conspiracy," vividly depicts the "speculative mania" that captured the nation after the Civil War:

> The Civil War in America, with its enormous issues of paper currency, and its reckless waste of money and credit by the government, created a speculative mania such as the United States, with all its experience in this respect, had never known before. Not only in Broad Street, the centre of New York speculation, but far and wide throughout the Northern States, almost every man who had money at all employed a part of his capital in the purchase of stocks or of gold, of copper, of petroleum, or of domestic produce, in the hope of a rise in prices, or staked money on the expectation of a fall. To use the jargon of the street, every farmer and every shopkeeper in the country seemed to be engaged in "carrying" some favorite security "on a margin."

Before the war, speculative activity—investing money in anticipation of the direction of fluctuation of exchange values—was the province of "regular brokers." Now, "every one speculated," "until the 'outsiders,' as they were called . . . rep-

FIGURE 19. Francis William Edmonds, *The Speculator*, 1832. Oil on canvas. 25⅛ × 30⅛ inches (63.7 × 76.4 cm). National Museum of American Art, Smithsonian Institution, Washington, D.C. Gift of Ruth C. & Kevin McCann in affectionate memory of Dwight David Eisenhower, thirty-fourth President of the United States (1976.114).

resented nothing less than the entire population of the American Revolution" (Adams, *Erie* 101–2). With this last allusion Adams reproves his compatriots, his metonymy intimating that America's revolutionary ideals have found their fulfillment merely in speculation.

One might remark that Adams is naive in his irony, that speculation has always incited America's soul. After all, as Charles Beard argued in 1913, the Constitution was written with the interests of creditors and land speculators in mind. Michael Paul Rogin has amply demonstrated, with Andrew Jackson as his model, the broad influence of land speculation in the early decades of the republic. And he follows Douglass C. North in arguing that between 1815 and the early 1840s, a market economy, replacing an economy of household subsistence, became firmly established (Rogin 251–52; see North v–vii, 1–14, 61–74). Thus the drastic changes that seemed the immediate offspring of the Civil War were in reality the long-nurtured descendants of antebellum developments.

Such an analysis rightly amends the Beard-Hacker thesis that the Civil War was the unique precondition or even cause of subsequent and radically novel economic developments;[3] and of course speculation has been important throughout American history. Economic development, in America as elsewhere, is best understood as more often a continuous transformation than a succession of discrete stages. Yet perhaps we can reaffirm a fundamental difference between prewar and postwar economic activity by replacing the common distinction between a subsistence and market economy with a threefold distinction, first expounded in America by Thorstein Veblen. Adapting the terminology and conceptual framework of the German historical school of economic analysis, Veblen distinguished among a "natural economy" (premarket, subsistence agriculture), a "money economy" of a "goods market," and a "credit economy" of a "capital market."

Veblen's money economy is the market economy that Rogin and North identify. Its "characteristic feature," Veblen writes in *The Theory of Business Enterprise,* "is the ubiquitous resort to the market as a vent for products and a source of supply of goods." The businessman of this epoch exchanges goods, owns "industrial equipment," and keeps "an immediate oversight of the mechanical processes"; he sees to "productive efficiency." In a credit economy, the market no longer channels goods but is "a vent for accumulated money values and a source of supply of capital." "Traffic," once "in goods," now takes place "in capital," whose definition and substance have profoundly changed. Formerly "a stock of material means by which industry is carried on," capital in a credit economy "means a fund of money values ' that "bears but a remote and fluctuating relation to the industrial equipment," a "shifting" or "shifty" relation, Veblen often writes. Concomitantly, the businessman no longer directs the production of real goods but manipulates "putative" value, or "earning capacity," in "an interminable process of valuation and revaluation." This credit stage of the economic process, Veblen argues, has taken hold since the Civil War, and especially since 1880.[4] Previously, traffic in goods enabled the flow of credit; now credit underlies the exchange of commodities.

With value putative and immaterial, an interminable and shifting process of valuation, the distinction between credit and capital vanishes, and all business activity becomes in some respect "speculative," inferring distant and future trends and needs (Veblen 1904, 165). Thus speculation now becomes "one of the chief directive forces in trade and industry," Henry Crosby Emery of Yale writes in 1896. Indeed, "the stock and produce exchanges are the nerve centers of the industrial body . . . in themselves as necessary institutions as the factory and the bank." "By making prices [speculation] directs industry and trade, for men produce and exchange according to comparative prices. Speculation then is vitally connected with the theory of value," Emery concludes (8, 9, 12). But with "real values" a function of, and often indistinguishable from, "speculative values," as Senator Algernon S. Paddock from Nebraska put it in 1892,[5] America faced a crisis of value and valuation. In 1898, Alexander Noyes, financial editor of the *New York Evening Post,* calls speculation (specifically in silver and gold) a "contest over the standard of value." But the violent fluctuations of such contests render, in Adams's words, "all values unsettled" (Noyes 68; Adams, *Erie* 123).

This sense of speculation's destabilization of value is not entirely new, of course. James Fenimore Cooper was typical when he complained in 1838, in *The American Democrat,* that "abuse [of] a system of credits" and "the unrestrained issue of paper-money, with its attendant contractions," keep "the value of property unsettled" and "produce that instability that so peculiarly marks the condition of American trade." But Cooper's diatribe against credit insists on its accidental and supplementary relation to commerce. He complains of speculation as an *abuse* of a credit system, whose function ought to be merely "to facilitate the operations of trade," and whose derangement of values can be remedied by firmly basing all currency in specie and by limiting the amount of paper circulating and the magnitude and scope of credit (170–71). In his view, the disorder of credit is a temporary, confined, and curable affliction.

By late in the century, however, credit was regarded as a constitutive, and perhaps chronic condition, a view evident in literary production. For example, Henry Blake Fuller's 1895 novel of Chicago, *With the Procession,* opens with the death of "poor old David Marshall," a wealthy grocery merchant "whose sole function was to direct the transmutation of values . . . into the creature comforts demanded" by his family (3–4). It is as if a mode of valuation dies with Marshall, and the novel narrates the subsequent unleashing of unpredictable events and vicissitudes of personal fortune.

The result of the passage Veblen describes from tangible to putative and shifting value is not just unpredictability but a pronounced opacity of values. In his preface to *The Spoils of Poynton,* Henry James describes this effect concisely: values and literary subjects are "not directly articulate" (xlvi). Thus art must practice a "sublime economy" of excavation. "Life being all inclusion and confusion, and art being all discrimination and selection, the latter, in search of the hard latent *value* with which alone it is concerned, sniffs round the mass as instinctively and unerringly as a dog suspicious of some buried bone" (xxxix–xl). Norris also views art as a quest for the value obscured in visible forms. The "purpose" of the novel, he writes in *The Responsibilities of the Novelist,* is "to find the value of x." This is the achievement of great writers. "Shakespeare and Marlowe," for example, "found the value of x for the life and times in which they lived."[6]

Note the historical limitation of Norris's proposal: the composition of value is not eternal, and can be sought only for a given era. In addition, imagining with James that art is an enterprise of disclosing value, Norris shares as well James's conception of value as hidden or "buried." Norris feels that one should distrust stated prices, value's external trappings, for these can deceive. Ideally, one should penetrate "the clothes of an epoch and [get] the heart of it." Norris thus desires, in the search for value, "not the Realism of mere externals (the copyists have that), but the realism of motives and emotions" (*R* 199). In a comparison that recalls Twain's distinction between the heroic, authoritative romance of piloting and the surface imitation of southern romance, Norris favorably contrasts excavatory, epic romance with "copyist" realism.

"Copyist" realism follows principles of objective representation that Norris believes Howells sets forth. It "confines itself to the type of normal life" and "notes only the surface of things." In this way, "Realism stultifies itself," for the normal and external do not impel one after deeper truths. The realism of romance, on the other hand, "takes cognizance of variations from the type of normal life," because here, in what is often "the sordid, the unlovely," one can discover the primary motives of behavior, "a complete revelation of my neighbor's secretest life" (*R* 280–81). This penetration of persons is achieved by the selection from the confusion of life that James espouses, and one seeks truth, not accuracy. A "merely accurate description" supplies only the accidental details of a particular circumstance, whereas the "ludicrously inaccurate" may convey the real truth of a circumstance and its effect upon a perceiver (author). "Accuracy is the attainment of small minds"; "To be true is the all-important business" of romance (*R* 284–85), and one succeeds in this enterprise not by settling for surface values but by seeking the flows of the "elemental forces" of business and pecuniary traffic which determine value, the "motives that stir whole nations" (*R* 204).

Applying these principles to business, Norris compares the financier to the great writer: "the genius of the American financier," here "Mr. Carnegie," does not differ "in kind from the genius of" the writer (*R* 244). Norris ramifies this comparison: the financier is to the "mere businessman," who simply markets goods, as the poet (or romancer) is to the mere writer. "You must be . . . something more than just a writer. There must be that nameless sixth sense or sensibility . . . back of" your work. This is "the thing that differentiates the mere businessman from the financier (for it is possessed of the financier and poet alike—so only they be big enough)." The mere businessman and mere writer practice copyist realism, handling goods and external details, exchanging only stated values. The financier is the extraordinary type of the credit economy of 1900 America, who, like the true poet, "deals with elemental forces" of valuation that "stir whole nations" (*R* 201).

Real and Fictitious Values

Norris's financier poet in *The Pit* is Curtis Jadwin, who attempts to corner Chicago's wheat market. Chicago is the center of produce speculation because, as Twain saw, it is the cynosure of the transportation lines conveying "Trade—the life blood of the nation" (*Pit* 61).[7] As conduit of the forces of nature and commerce, Chicago seems "civilization in the making, the thing that isn't meant to be seen, as though it were too elemental" (63). The city embodies those unseen elemental forces that lie behind events, the forces that it is the office of romance to reveal. And the wheat pit, the bench-enclosed area on the floor of the board of trade, is where, so to speak, civilization is transacted, and the nation's blood circulates, "roar[ing]" "in the heart's heart of the affairs of men" (80). The financier's job is "to watch, govern, and control the tremendous forces latent" in his investments, forces that are "reshaping" the nation (280–81); Jadwin is a "successful speculator" because he possesses "that blessed sixth sense" (191) that distinguishes great authors and financiers from copyist writers and mere businessmen.

Although speculation and its financier embody the epic ideals of romance, nevertheless the novel inexorably leads "to the conclusion of Curtis Jadwin's career as a speculator" (396). Why? And by what logic is the end of speculation combined with the salvation of the Jadwins' marriage? One might propose that the novel must terminate speculation because it creates hardship in some sectors of society. But as Ernest Marchand observes, "protest against economic injustice" occupies "but a minor" place in *The Pit* (168). The novel does acknowledge the at times unhappy consequences of speculation: Jadwin's manipulation causes various speculative failures; and sometimes farmers receive a low price for their product, or European consumers must pay high prices for bread. But Norris only notes these effects in passing, essentially accepting the argument that profits generally balance costs; and of course failure is precisely what a speculator risks in order to make money.

The real reason *The Pit* must, in Emery's phrase, "suppress speculation" (194) has to do with the crisis of value I spoke of earlier. Speculative activity violates what Jadwin's friend Charlie Cressler calls "legitimate value." It does so because "those fellows in the Pit don't own the wheat; never even see it. Wouldn't know what to do with it if they had it. They don't care in the least about the grain" (129). Cressler's accusation typifies the concerns of many of Norris's contemporaries. An "act of taking advantage of fluctuations in the prices of property," speculation was disturbing because it "deals in invested capital instead of consumable commodities."[8] That is, speculators make money by, in Emery's words, trading on the "fluctuation in price"—called "trading for differences." They "are not concerned" with actual commodities or their production; rather, "the purpose of [speculative] transactions is to secure the difference in price" (57). Because speculation so disregards the thing that a security represents, discussions of speculation stressed the distinction between "intrinsic," "real," or "actual" value and "fictitious" or "counterfeit" value.[9] "Fictitious value" applies to sales "representing nothing," to cite debate on the Senate floor (*CR* 1892, 6642), sales that do not "contemplate an actual exchange of stocks" or goods, or of "real values" (Hamlin 413–14; Emery 101). Fictitious sales are often transacted by "selling short," selling what you do not own in anticipation of a fall in prices, at which point traders buy actual securities to cover their shorts. Selling short is the essence of "artificial" or "unnatural" manipulation of prices.[10]

When speculative transactions are entirely fictitious—that is, when traders have absolutely no interest in exchanging actual securities and seek only to profit from a chance fluctuation—they are bluntly called "gambling." A transaction is considered gambling, Emery writes, when it "has no reference to actual trade" and "consists [only] in betting on the course of prices" (98). Some observers noted (rather nervously) that all business transactions involve some degree of speculation. Risk, after all, is part of any business undertaking, but then is not all business gambling? It is not just the presence of risk, one critic responds, but the presence of nothing but risk, that defines gambling. A trade in price fluctuation is gambling because, "contemplat[ing] no exchange of real values," it "is all risk and no work."[11] Another critic is even more blunt: because trading for differences is indifferent to the production and distribution of goods, it "is not a 'transaction' at all" (Eliot 18).

Speculation for price difference on legitimate exchanges seemed precariously like the trading that went on in "bucket shops," which were local (often rural) establishments offering very low margins (3 percent as opposed to the normal 10 percent) to small speculators, who often did not know that the bucket shops had no official connection with the legitimate exchanges or even, sometimes, that licensed exchanges existed elsewhere. Cedric Cowing, in his excellent history of speculation from 1890 through the New Deal, describes how bucket shops worked:

> The shops only pretended to buy and sell in the market; actually they merely booked the transactions and carried the risks of fluctuation themselves. Acting upon the axiom that "the public is always wrong," they assumed they could make a steady profit from the hordes of amateur speculators. (28)

Clearly, patronage of a bucket shop was sheer gambling.

As much as the legitimate exchanges wanted to distinguish themselves from bucket shops and from the general charge of gambling, even their defenders felt obliged to grant that "there seems to be no clear-cut line between gambling and legitimate business,"[12] for fictitious "dealings are in no way different in form from any other dealings," Emery observed. "What men are after may be the 'differences,' what they *do* is to buy and sell property"; all trading contracts stipulated the sale of property (58). Without an indefeasible substantive or formal distinction between real and fictitious trading, some were tempted to differentiate legitimate from illegitimate trading according to the intention of traders. They would ask, Do traders intend to exchange goods or merely to trade on differences? Yet finally, critics and regulators found it difficult "to adopt the somewhat shadowy distinction of 'intent and purpose,' in other words the spirit pervading [contracts]" (Emery 99, 217). For one thing, appeal to these immaterial conditions is likely to induce interpretive squabbling requiring adjudication and might easily lead to interminable litigation. For another, "the vast majority" of contracts are resold before any securities are exchanged.[13] Were intention inferred from this pattern, most transactions would be judged to be gambling. Regulators were forced to concede that stated property obligations in contracts constitute intention and must remain the basis for distinguishing legitimate business from gambling.[14] This strategy, however, leaves intact rather than solves the problem of what distinguishes legitimate from illegitimate trading.

Commodity speculation, the subject of *The Pit,* posed a special version of the ethical dilemma engendered by investments with oblique or no reference to actual goods or values, by investments, that is, "representing nothing." Stock speculation involves, in a phrase I cited earlier, "invested capital instead of consumable commodities"; therefore it always deals with immaterial earning capacity and makes no claim actually to handle objects. Commodity or produce contracts, however, do claim to trade not paper securities but bushels, bales, and pounds. Yet, as Cowing points out, "In only 3 per cent of the futures trades was there actual delivery; in fact, to demand delivery was to brand oneself a miscreant and led to ostracism by the brokers" (14). Since speculative commodity contracts did claim to exchange real, tangible values, this statistic suggests that that accusation of fictitiousness seems especially appropriate to commodity speculation.

Cowing alludes here to "futures," which differ from "options," another kind of contract: a futures contract obligates future delivery, whereas an "option" signifies no exchange of property but only the "privilege" to buy wheat at a future date, depending on the relation between the future market price and the present option price. Not actual transactions, then, options aroused, needless to say, even more scandal, and precipitated various legislative attempts to curb such practices. The first and most prominent of these, the 1892 Hatch Anti-Option Bill, sought to end options because they were an "unnatural mode of determining prices," as William Hatch, the bill's sponsor, writes in his report on congressional hearings held in regard to the bill.

Hatch objected that, utterly divorced from labor and production, "the limitless offers of fiat products by the 'short seller,' regardless of the value of or the volume of actual product in existence," do not constitute the trading of an "honest market" but assume "the form of swapping contracts" (Hatch 1, 3, 6, 7). By stipulating "that unless the party selling . . . obligates himself to deliver in the future it is a gambling transaction," Hatch's bill was intended to "restore . . . the unfettered operation of the law of supply and demand." Thus, listed value would be determined naturally by the actual value, by the product itself, with no troubling gap between listed and actual value. Yet Hatch, too, realizes that even most (legitimate) futures deals, because they so rarely issue in the transfer of property, are by his own definition no more than contract swapping, and not, in the term employed in Senate debate on the bill, "legitimate commerce" (Hatch 6–7; *CR* 1892, 6439, 6442).

Clearly, a well-established tradition lies behind Franklin Walker's discomfort that wheat speculation in *The Pit* is merely "the manipulation of paper representing paper" (292).[15] Since ancient Athens, the objective and substantive value of tangible property has often been rhetorically opposed to the cunning enabled by representation. This distinction structures the composition and theme of Edmunds's *The Speculator.* The painting's family is averted not just from hearth and food preparation to speculation but from tangible, consumable concerns to a piece of writing. Speculation is a practical metaphor for abuses potential in representation. Norris shares this discomfort, despite his declarations that manipulating the immaterial forces that generate the value of *x* constitutes both the subject and method of epic romance. Jadwin at first is "not opposed to speculation" (10), but mainly because he doesn't think he's speculating. Even as he enters modest speculative schemes with his broker, Gretry, Jadwin denies that he is speculating: "this wasn't speculating. . . . It was certainty . . . sure." He compares his first deal to knowing "a certain piece of real estate was going to appreciate in value" (110–11). "I never bet," he will later insist (198).

The distinction between real estate investment and speculation reveals the invidious comparison between tangible value and representation to be, finally, moral. Jadwin has earned his initial fortune in real estate: "He was one of the largest real estate owners in Chicago. But he no longer bought and sold. His property had grown so large that just the management of it alone took up most of his time" (75). The distinction between property management and trading in differences derives from Aristotle's distinction between chrematistics (wealth-getting or usury), and economy, which in Greek means the management of a household.[16] Aristotle's distinction hinges on one's responsibility to a physical object of value. Economy maintains its property; chrematistics is indifferent to it. *The Pit,* like debates about commodity contracts, reproduces this distinction about obligations: "legitimate business" (131) manages property, establishes values according to the "Visible Supply," and does not just bet on "rumour" (100). The frenzy of speculation, on the other hand, is "no time to think much about 'obligations'" (328). Indeed, speculation in the novel is said not only to rupture obligations to the grain but finally to render Jadwin indifferent to the money gained, which would seem the goal of speculation. Rather, he becomes obsessed with the manipulation itself; as he explains to Laura, "Oh, it's not the money. . . . It's the fun of the thing; the excitement" (231). Within this framework, it is appropriate that when he fails, Jadwin loses both his real estate holdings and the house he bought to commemorate his marriage.

Herein lies the essence of what I am proposing is the novel's real failure, and it is a failure of commitment. Norris propounds an aesthetics of scope and magnitude—finding the value of *x* in the forces of a credit economy. He glorifies the financier for possessing the genius of the poet, able to discover (which is to say, manipulate) culture's elemental forces. But in *The Pit* Norris balks from pursuing the value of *x* according to the mode of valuation his notion of romance exalts. As Carl Smith has written, the novel "evades its most interesting complexities" (69). It repudiates the economics of romance for what Emery calls an "objective idea of value, that is, the idea that value may be determined by certain physical facts" (113).

In a credit economy, we will recall, value is constituted not by physical objects but by shifting market forces and attitudes toward these objects and forces. Hence, Emery observes, "the idea of value, which was at the basis of much of the early struggle to control [speculative] prices by law, is entirely inconsistent with the conception of value which gives speculation its sole justification" (Emery 193). It is at best contradictory to wish to suppress "gambling" in the same breath that one declares speculative trading legitimate because it facilitates trade in a complex and widespread society. Both gambling and legitimate speculation operate according to the same notion of value; the form that seems excessive (speculation) merely exposes the true operation of legitimate exchanges. In neither is value "intrinsic," inherent in objects; in both it is "putative," in Veblen's term, a function of conceived earning power, and it is known precisely by the kind and extent of activity surrounding an object (see Emery 150–52). Putative value is no less real (or less intrinsic) than so-called intrinsic value, however, only differently composed; as Emery writes, "there is a real increase or decrease in the value of property due to outside causes" like speculative exchange (101). To insist upon what Emery calls a "fetich of delivery" (69), as if actual goods were the exclusive vessels of value, is to wish to inhabit an obsolete (and probably fantastic) era or mode of economic behavior.

With all Norris's emphasis on "forces," not only in this novel but in all his work, he is anxious in *The Pit* (though not, for example, in *McTeague,* which associates gold fetishism with wife beating and worse) to locate the forces of value in one identifiable, stable object: the wheat. He naturalizes and simplifies the historical forces of supply and demand. Thus Jadwin believes that his corner is beaten not by market exigencies but by the wheat itself: "The wheat cornered me, not I the wheat" (419); he had been "fighting against the earth itself" (347). "Why the Wheat had grown itself; demand and supply, these were the two great laws the Wheat obeyed" (374). Supply and demand, here, are asserted to be natural laws, immutable forces of the earth, finally autonomous, independent of any market or human action; the wheat is a pure embodiment of these laws, which is why it can grow itself.

Many, like Congressman Hatch, joined Norris in according the forces of supply and demand natural status (see Hatch 1–3; *CR* 1892, 6881, 6885). But it is not, perhaps, self-evident, nor did everyone believe, that the elemental forces circulating through a society are natural. "Prices on the exchanges," Emery writes, are certainly "determined by the existing demand and supply. But the existing demand and supply are both speculative, and depend for their strength on the conditions in other markets" (114). Emery's meaning is twofold. First, a substantial portion of supply and demand results precisely from speculators testing the market. However, both consumption and even "the production of the raw material" are always influenced by speculative price. Farmers generally decide what quantities of a crop or even what crop to plant according to market price levels (Emery 148; see also White 530). This proposition is well demonstrated in *The Pit,* despite Jadwin's misinterpretation of the situation. His corner is broken by a huge influx of wheat, an influx created precisely by the extraordinarily high price that his corner has induced. To a great extent, Jadwin's corner has grown the wheat, but he insists on thinking it has grown itself.

Since I am accusing Norris of fetishizing objects and naturalizing market forces, it is interesting to note that he chose as a model a corner on wheat that was as atypical in its operation as it was archetypal in its magnitude—Joseph Leiter's 1898 corner. Charles H. Taylor writes, in his 1917 *History of the Board of Trade of the City of Chicago,* that Leiter's "was one of the greatest manipulations ever attempted," one that "demanded the attention of the entire civilized world." At the same time, "a newcomer in the field of grain speculation, Mr. Leiter's tactics baffled the old-time traders." He refused to resolve transactions by accepting the "forced settlements" on the difference in price "to which [the 'shorts'] were accustomed." Rather, "Mr. Leiter actually bought real grain" and demanded actual delivery of wheat owed him. (We should note Taylor's surprise in the redundant "*actually* bought *real* grain.") Like Jadwin, Leiter claimed that he "for the most part allowed prices to take their own course"; wheat "was very cheap," and the "conditions of supply and demand throughout the world" would force up the price.[17]

Norris, Walker tells us, was greatly puzzled by the principles of speculative trading. We may therefore assume that Norris was unaware that Leiter's techniques were anomalous, that other traders regarded Leiter a "miscreant" (to use Cowing's term) with a fetish of delivery who downplayed his manipulation as but an outgrowth of the natural flow of supply and demand. Whether consciously or not, Norris was attracted to a manipulation that justified itself as natural because he "had difficulty comprehending how a man could sell a thousand bushels of wheat without owning them" (Walker 275). Short selling confused Norris because it so obviously violated the intrinsic, substantial, objective determination that he sought for the value of x. Because of this desire for objective determination, the financial plot rejects the values of romance and exhibits the sentimentality that critics have sensed in the love plot.

Economies of Marriage and Flirtation

Before her marriage, Laura—as her sister Page observes, and as she herself admits—likes to flirt (*Pit* 68, 132), and she quite naturally contrasts flirting with being in love and, especially, with being married. If this opposition strikes us as customary, it is at the same time part of a purposive rhetorical strategy. The love and marriage plot incorporates the same terms as the financial plot, and in this conjunction we can see how the novel of speculation can also be the story of what Norris once called in a letter Laura's "career" (cited in Pizer 1966, 165). This conjunction may also lead us to question the naturalization by which marriage and speculation are opposed.

The novel presents marriage as a specific kind of transaction. For example, before their marriage, Jadwin makes a "bargain" with Laura, which he hopes will consummate their relationship. He will add a conservatory to their house in exchange for a sudden, surprise kiss that will signify the depth and finality of her love for him. Marriage is an equivalent exchange of goods: a kiss for a conservatory, or, as expressed in this remark by Laura, marriage itself for the pretty gifts that are an index of love—"I would marry a ragamuffin if he gave me all these things . . . because he loved me" (170). If thinking of marriage as a balanced exchange seems somehow vulgar, consider this comment by Sheldon Corthell, the novel's aesthete, who more than anyone is spared vulgarity: "When I offer myself to you, I am only bringing back to you the gift you gave me for a little while. I have tried to keep it for you, to keep it bright and sacred and unspotted" (25). This offer of marriage as repayment for the gift of self is the novel's prime example of sincere love.

Because love and marriage are imagined to be direct and equivalent exchanges, it is important, once love has been consecrated in marrige, to eschew credit and avoid debt: "Dear me, Laura, I hope you pay for everything on the nail, and don't run up any bills" (217). In short, marriage does not "buy on the margin" and always entails an actual exchange of goods. Marriage is conceived here as an Aristotelian economy, where value is produced by actual and proper handling of objects. To be married means, as with acceptable futures contracts, to "be willing to put [one]self under obligations to" one's spouse (140). Obligations are fulfilled, meaning that behavior signifying love is exchanged and, in ethical terms, that one realizes and accepts "certain responsibilities." In this economy of marriage, the inner substance of the partner surfaces for cognition: "She began to get acquainted with the real man-within-the-man that she knew now revealed himself only after marriage" (206).

Flirtation, on the other hand, is emotion in the chrematistic mode. Late in the novel Laura explains that her relationship with Landry Court "was all the silliest kind of flirtation," for "he never really cared for me" (376). Like speculators, who "don't care in the least about the grain" (129), flirts do not care about the objects of their desire, or about those who desire them, only about the excitement of flirting. Similarly, Jadwin admits that in his speculative dealings he does not really care for the money earned but only for the "fun" and "excitement" of speculating (231). Nonetheless, like commodity speculation, flirtation pretends commitment to natural objects rather than just temporary interest in them. Flirtatious Laura "let[s] every man she meets think that he's the one particular one of the whole earth. It's not good form" (68).

Page's concluding remark here recalls, in the realm of human relationship, the problem economists encountered when trying to determine formal distinctions between legitimate business contracts and gambling contracts, both of which stipulate the same intentions to handle objects. Flirts let their beaus think themselves the sole object of love, the same thing real lovers want their partners to believe. The difference becomes a matter of intention, as exemplified in this exchange, when Laura coyly rejects one of Landry's proposals:

> "As if you really meant that," she said, willing to prolong the little situation. . . .
> "Mean it! Mean it!!" he vociferated. "You don't know how much I do mean it. Why, Laura, why—why, I can't think of anything else." (53)

Of course, if intention seems the factor differentiating love from flirtation, the course of Laura and Jadwin's marriage makes it clear that good intentions alone do not suffice; they must be inferred from behavior or symbolic exchanges, and inference is risky. Jadwin fully intends to fulfill a late "bargain" that Laura demands, in which as a sign of love he will celebrate her birthday alone with her; but although lovers are supposed to think of nothing but their love, Jadwin, immersed in his corner, forgets the birthday arrangement. Laura decides Jadwin does not love her and nearly consummates an affair with Corthell; Norris keeps pledging that Jadwin "loves" Laura, but these assurances only call further attention to how easily his love neglects its beloved.

Thus, in dealing with people and with obligations in general, flirtation follows a speculative economy. At the moment Laura becomes disgusted with her flirting, she imagines this activity precisely in the terms economists applied to gambling, with herself as the object of speculation: "No doubt they all compared notes about her." "Now it was time to end the whole business," for "in equivocating, in coquetting with them," Laura "had made herself too cheap," which is not only a colloquial expression for improprietous women but the actual effect of much speculation, for to be successful, short selling must drive prices down. Laura wearies of the "spirit of inconsistency" prompting her flirtations (127). When she is so "changeable" (165), her emotions fluctuating according to the same "indefinite . . . 'sentiment'" that makes stock prices unstable (290), she cannot accept the obligations and responsibilities of marriage, indeed can "love—no one" (127). Marriage promises a "final" stability to an emotional state that otherwise wavers "capriciously" (205).

Even more than just emotional stability, marriage promises connection to the world rather than the self-centeredness, the virtually masturbatory self-referentiality, of flirtation, where acts are produced by caprices for the sake of excitement, with scarce concern for the people with whom one interacts. Jadwin appeals to Laura because he represents the real to her, the "real, actual" drama "in the very heart of the very life in which she moved" (34). Although she seems to understand that the actual consists of the movement or interaction of the forces inspiriting objects, she nevertheless thinks these objects are the actual itself rather than, more properly, the phenomenal convergence of these forces. She mistakes the nature of Jadwin's business dealings; even when he explains that he speculates for neither the wheat nor the money but for the excitement, she continues to think that he deals not in price differences and speculative trends but in the "wheat—wheat—wheat, wheat—wheat—wheat" (231). She falls in love according to the theory of objective value that Jadwin brings to his speculative ventures: "I only want to be loved for my own sake" (173). If she is loved in this antispeculative manner, she feels, she can be "understood to her heart's heart" (187).

As in most fiction of the period, like Emerson Hough's popular historical novel, *The Mississippi Bubble,* about the notorious John Law (also 1902), marriage and speculation cannot coexist in *The Pit,*[18] and for more fundamental reasons than the mere contingency that Jadwin's speculation keeps him away from home. These authors require the economy of marriage to be radically opposed to that of speculation. In contrast, in *A Hazard of New Fortunes,* Basil March "consider[s] the enormous risks people take in linking their lives together after not half so much thought as goes to an ordinary horse trade." He wants to entertain the risks of marriage and expects that "by-and-by some fellow will wake up and see that a first-class story can be written from the anti-marriage point of view; and he'll begin with an engaged couple, and devote his novel to *dis*engaging them . . ." (479). Basil is registering here his sense—based on his own experience—that marital harmony is an achievement rather than a natural condition, and elsewhere Howells presents Silas and Isabel Lapham's marriage similarly, as a "tie [that] bears a daily strain of wrong and insult." The "divinity of the institution" abides in its parties' capacity to adjust to stresses (43). Basil March attributes the common, "popular demand for the matrimony of others [to] our novel reading" (479). Like Howells, *The Pit* disparages "trashy" sentimental novels, and Jadwin admires Bartley Hubbard (*Pit* 216), whose actions ruin a marriage in Howells's *A Modern Instance*. Nevertheless, Norris expends considerable energy eliminating the risks of marriage, metonymy for the nonnaturalness of all valuation and human relations. *The Pit* attempts to suppress risk in its world in a way the real world neither could nor would, and in so doing, the novel adopts the very sentimentality that Norris intended romance to combat.

Taking Control

Yet I think we must still ask, Why? Why does Norris so fiercely eliminate risks and seek a stable and objective determination for immaterial forces that are precisely forces of change? And why this policy when it clearly contradicts Norris's theoretical program? Moreover, as I mentioned earlier, there is nothing natural or necessary about the alignment of marriage and legitimate business against flirtation and speculation, even though the former allies rhetorically claim nature's sanction by meeting obligations to people and actual objects. Similarly, both Jadwin's success and his failure in speculation are justified by a naturalization of economic forces that are clearly market-variable. If these naturalizations and the opposition of marriage to speculation are not necessary but polemical, why the polemic? If the injustices of speculation matter only tangentially to Norris, why does he care so much that flirtation and speculation violate objective value? Why, that is, does Norris practice a protectionist aesthetic economy like those we have encountered in so many forms throughout this study?

The answer, I think, lies in the implications of Norris's aesthetic principles. I have been arguing that Norris, while explicitly casting his lot with the financier, actually practiced the "mere" businessman's economy of balanced exchange. In *Looking Backward,* Edward Bellamy articulates the logic that helps us to understand Norris's attraction to the operation of a market or money economy. "Money was a sign of real commodities," Bellamy writes, "but credit was but a sign of a sign." As a "conventional representative of goods," money facilitated their exchange, but once people "accepted promises for money," they "ceased to look at all behind the representative for the thing represented" (161). Bellamy accuses a credit economy of fostering laziness and gullibility. With tangible values available only in the form of promises, people compliantly accept those representations of values. If Bellamy's complaint is moral, Emery suggests its epistemological basis: "The causes influencing prices are too many to permit of tracing the effect of a single cause easily" (119). The apprehension of value in a credit economy is difficult not because people lazily cathect on representations but because when representations are signs of signs, one looks for the signified only to find another sign; one cannot trace the determinants of value to a single cause but finds only matrices of forces that may always bear further examination and that are always subject to interpretation.

Discussed in language about signs of signs and interminably tracing causes, the operation of credit sounds like an economic correlative to Derridean propositions about language and interpretation. As cultural phenomena, "trading for differences" and the credit capitalism it exemplified incarnate the Derridean trace and *différance*. At least two things may be said about this coincidence, the first having to do with the political status of a Derridean account of representation, the second with its usefulness in understanding *The Pit* and other responses to speculation.

First, many proponents of deconstruction—indeed, poststructuralist critics in general, even if they reject the most extreme claims of deconstruction—have argued for the subversive political implications of deconstructing the classical account of representation (as correspondence between signified and signifier, and as full presence—presence of meaning, signified, and self to themselves). In opening to scrutiny closed systems of signification and authority, deconstruction also opens them to challenge and change. But few who defended trading for differences and advocated disengaging value yet further from its tangible vehicles were progressive; most opposed populism and unionism. Moreover, through the 1880s agrarian and labor interests were suspicious of money not tied to specie, precisely because of greenbacks' capacity to diverge from tangible value and to be manipulated by bankers; hence, activists like William "Coin" Harvey opposed greenbacks because they were not "real value" or God's good coin, and, ironically, were for a time allied with conservative goldbugs like William Graham Sumner, A. L. Perry, Edward Atkinson, and Charles Francis Adams, Jr.[19] In the historical instance I am discussing here, policies premised on principles of representation consonant with Derridean principles were hardly subversive of conservative policies. This historical coincidence may seem to confirm charges from the academic left as well as the right that deconstruction (especially de Manian deconstruction) is nihilist and intrinsically conservative (see Lentricchia 1980). But such an indictment explains neither the character nor causes of the conceptual overlap between the Derridean trace and late-nineteenth-century speculation. Trading for differences was a live political issue, and proponents and opponents alike of speculation and its regulation employed parallel rhetorical formulations. Rhetoric displaying features of what we now call postmodern notions of representation was deployed to achieve very real persuasive and political effects, and as we will see in the next two chapters, could be used by unionists attacking the trusts as well as by defenders of developments in credit capitalism.

If observing a historical instance of poststructuralist notions of representation discloses no intrinsic political implications, noting this coincidence can nevertheless help us to understand Norris, in *The Pit,* as a positivist attempting to contain the interpretive and practical problems potential in his vision of romance. For romance to fulfill its aims, it must transgress the bonds of objective value and mimetic representation in search of the elemental but veiled forces of value. "Accuracy is the attainment of small minds," Norris wrote in one essay, and indeed, "in fiction, [accuracy] can under certain circumstances be dispensed with altogether" (*R* 285). But to scorn surface externals as the province of the small-minded is to inhabit willfully the world of Veblen's "shifting" or "shifty" relations, where the rules of perception are never mere givens.

If value exists in shifting relations, verifiability, tracing causes, is problematic, requiring effort. *The Pit* depicts an excellent example of this problem. Page attends the board of trade on the day of Jadwin's failure. Unfamiliar with the operation of the pit, she profoundly misinterprets the events she witnesses, and believes that Jadwin achieves a great victory. Given these difficulties, copyist Realism—even though Norris explicitly disdains it—is far more compatible with Norris's economic ideals than the method of the romancer. The copyist mode satisfies because words signifying external objects unproblematically signify the things themselves as autonomous entities, and because such mimetic accuracy furnishes the essence of a thing's truth.

Describing the problem, however, does not fully explain the anxiety it provoked in Norris and many of his contemporaries. For further explanation we must look at two practical consequences of a credit economy. First, speculation and flirtation threaten certain concepts of the self. Ideally, one should be "master of himself" (179), but speculation "seems to absorb some men so" (245); "this trading in wheat gets a hold of you" (232). Self-control feels diminished and becomes an aspect or instrument of the social forces speculation traffics in, until it seems as if the forces of wheat execute the trading without the aid or agency of men. Although he too conflates the wheat with the forces of exchange, Calvin Crookes, the allegorically named leader of the bear clique that breaks Jadwin's corner, perceives this redistribution of human agency: "They can cheer now, all they want. *They* didn't do it. It was the wheat itself that beat him; no combination of men could have done it" (396).

In her flirtatious mode, Laura lacks the self-control and certainty of self said to anchor love and marriage. She cannot marry, she argues, because she is "not sure" of any man's love. "Even if I were sure of [a man], I could not say I was sure of myself." She is sure only that she is not sure enough of herself to love anyone. This uncertainty of the "spirit of consistency" (127) is clearly metonymy for the uncertainty of market fluctuation, and if one is changeable by the hour (291), or even is "one girl one minute and another another" (163), the self becomes as putative and insubstantial as the value of commodities. Thus when Laura says, "I don't know myself these days" (163), she means that her self is unknown. And we can also interpret her remarks, "I love—no one" (127), or "I love nobody," quite literally. She feels she loves no one, not even herself, an entity of which she is uncertain. Or if she is as selfish and self-centered as she keeps accusing herself of being, she still loves no one: she loves herself, who is no one, unknown, not an identity in the objective way she requires.

This loss of self-control can develop into insanity. As Jadwin's speculation intensifies, he is overcome by headaches that feel like "an iron clamp on his head" (348). He begins to wonder, "Were his wits leaving him?" (349), and finally during the "violence" of the collapse of his corner, "something snapped in his brain" (392). Similarly, when Laura is her theatrical, flirtatious self, as opposed to her steady, "sincere" self, she is "moved by an unreasoning caprice" (171). In her loneliness produced by Jadwin's concentration on his deals, she inhabits this self of unreasoning caprice more and more, and her identity begins to seem like "a pit—a pit black and without bottom," which lies at "the end" of the "current" of whimsy that has "seized her" (360). Ultimately, because she is wholly uncertain of her identity and tempted to infidelity with Corthell, "a kind of hysteria animated and directed her impulses, her words, and actions" (401).

If the notion that speculation leads to insanity seems melodramatic, it was nevertheless a commonplace, especially with reference to the national economy, as registered in the word *panic.* Writers often allude to the "speculative manias" or "frenzies" that periodically overcame, as it were, the mental health of the national economy. Adams writes that during the 1869 manipulation of the gold market, "all business was deranged" (123). In "The Ethics of Gambling," T. L. Eliot speaks of speculation as the activity of an "unbalanced public mind" (29). Speculation is seen to affect the mental health of individuals as well. Senator Paddock argued that, when speculating, "understanding," "judgment," and "intelligence" are "worse than valueless" (*CR* 1892, 6881–82); rather, speculators, in Cowing's paraphrase of Paddock, live "like victims of schizophrenia" in a world of "rumor and secrecy" (20). Eliot complains that the "fascination" and "fever" of speculation cause an "inflamed state of mind" and threaten the "dethronement of reason." The gambler is "intoxicated" to "madness" (18–22). In *The Pit,* Charlie Cressler rehearses this scenario: "The Chicago speculator . . . raises or lowers prices out of all reason"; at the individual level, the "fascination" of speculation is "worse than liquor, worse than morphine," and the least taste of it addicts a man until "finally he is so far in that he can't pull out"; he loses "the very capacity for legitimate business" and "is ruined, body and mind" (*Pit* 129–31).

This addictive and schizophrenic effect of speculation is paradigmatic of the second effect of credit capitalism that disturbs Norris: the attenuation of property relations. As Laura's hysteria builds, she finds that even as she gratifies the many caprices that strike her, "she felt none of the joy of possession; the little personal relation between her and her belongings vanished away" (353). Ideally, one should exercise final control over the things one owns. Jadwin's corner represents to him this ultimate goal of financiering: to "control the tremendous forces" of valuation. Jadwin imagines a corner as "a master hand, all powerful, all doing," directing the flow of value (259). At the climactic moment before his failure, Jadwin appears on the trading floor for the first time, in hopes of daunting "his enemies" by "direct assumption of control." But it is in the nature of the market to "run clear away from everybody." When his corner breaks, Jadwin thinks that "the Wheat had broken from his control," though he "once had held the whole Pit in [his] grip" (389–93).

An Eden of Mimesis

This goal of direct control was also the aim of Joseph Leiter's corner: "to control the price of wheat single-handed against the world" (C. H. Taylor 948). Yet if the manifold currents of value are not finally traceable—and "current" and "flow" are the dominant images in the novel—we may wonder if "the speculative corner" can be any more than "incorporeal" and at best "temporary," as Emery puts it, "Many of the most active securities," he explains, "represent a capital of such

enormous proportions, and so widely distributed, as to make individual control
. . . impossible. No corner . . . could occur in such securities" (Emery 174, 182).
In a credit economy, where value is clearly not intrinsic or objective, the "property
relation" is, Veblen writes, "attenuated" and individual control is abridged.[20] Norris from the outset misconceives the role of the financier. The financier does not
"control" anything in the absolute, objective, or individual way that Norris, a
novitiate following Leiter's failed corner, imagined; rather, he attempts to glimpse
and anticipate currents. We might say that Jadwin's true madness lies not in his
speculation but in his desire to gain direct possession and control by it. As Gretry
cautions Jadwin, when he announces that he is going *"Into the Pit"* to "play my
hand alone," "J., you're mad, old fellow" (390).

Like Twain, Norris fixes on a fallacy of individual control potential in the
liberal vision of self and property, but which Emerson had rejected and which
Howells had slowly if painfully relinquished. Given his notion of the novelist's
financier-like responsibilities, it would seem incumbent upon Norris not to
renounce speculation's dispersal and attenuation of self and proprietorship but to
examine how they are reforged in a credit economy. The next chapters of this
book will investigate formal enterprises, literary and organizational, whose visions
of value, representation, agency, and self more frankly exemplify the cultural
transformations of evolving capitalism. These developments, I will argue, are profoundly transcendentalist and transmute the terms and premises of liberalism.

The innovative Standard Oil Trust will illustrate the emerging reconception
of property relations and agency. A fictional financier who deploys credit capitalism's transcendentalist conception of self and property relations is Dreiser's Frank
Cowperwood, whose exploitation, by hypothecating collateral, of the possibilities
of attenuated ownership exemplifies what we might call a hypothecated self. Control, here, is actualized through hypothecation, and self always looks elsewhere for
its substance and is not controlled or collected in any direct way, as Jadwin wishes.
Property and self are not the self-sufficient objects or locations in which they are
perceived but rather are dispersed in the matrices of forces, debts, and influx that
compose them. Eugene Debs's industrial unionism displays a correlative understanding of hypothecated self and representation.

But *The Pit* retreats from the effects and risks of hypothecation, a retreat typified in its account of marriage. Marriage is imagined in *The Pit* as a condition of
absolute ownership. The bargain for a kiss that Jadwin proposes is intended to
certify Laura's love for him so that he will know that she is his "own girl." When
Laura is caught in the "insidious drift" of her moods, she cries out for the property
relation that is most immediately attenuated by Jadwin's speculation: "I want my
husband. I will have him; he is mine, he is mine. There shall nothing take me
from him; there shall nothing take him from me" (360). As Twain's pilot achieved
authority through primary narcissism, so in this novel marriage is secured from
the vicissitudes and uncertainties of speculation through narcissism. Love succeeds here when it and one's partner are reflections of the ego, thus belonging to
it absolutely. Laura exhorts Jadwin not to "think of anything else but just me,
me" (362); all intrusions of the world must be excluded, with the result that the
lover is wholly possessed and self-possession is consummated. In the ideal of
marriage,

all the noisy, clamourous world should be excluded; no faintest rumble of the Pit would intrude. She would have him all to herself. He would, so she determined, forget everything else in his love for her. . . . She would have him at her feet, her own again, as much her own as her very own hands. (402)

The novel's conclusion restores the marriage, and thus symbolically restores the ontological and proprietary dislocations of speculation. Spouses will once again belong to each other, and identity will be stabilized in self-possession: there will "be only Laura Jadwin—just herself, unaided by theatricals, unadorned by tinsel" (403). These relations are naturally secured, no longer drifting "off the stable earth" (387). Jadwin explains what the restoration has effected:

"I understand now, old girl, understand as I never did before. I fancy we both have been living according to a wrong notion of things. We started right when we were first married, but I worked away from it somehow and pulled you along with me. But we've both been through a great big change, and we're starting all over again." (417)

In this account, the narrative of their marriage follows an innocent stage, a fall during which they behave according to a wrong notion of things, and a postlapsarian redemption. If we need any more reminders of *Paradise Lost,* Laura paraphrases its conclusion: "The world is all before us where to choose" (414); and when they approach the carriage that will take them to their new life, they walk hand in hand (419), as do Adam and Even exiting Eden.

To appreciate the fantastic quality of this ending, we might compare it to the at once similar yet very different conclusion to Harold Frederic's 1896 *The Damnation of Theron Ware.* The young minister Theron Ware is not a commodity speculator, but under the seductive influence of Celia Madden and the authoritarian, typological (which is to say, in this novel, manipulable) interpretive procedures of Father Forbes, he becomes a profligate, squandering his salary and parish funds, and buying goods on credit. After his trip to New York, Theron and his wife, described as "pure and good as gold," reconcile and prepare to journey to Seattle to begin anew. The details of the Wares' departure anticipate the Jadwins'; in both novels, for example, a carriage arrives to transport the couple to their train west. Considering Theron's former occupation, analogies to the fall from Eden would seem tempting, but Frederic alludes to the story only as a "joke God plays."

Instead, Frederic focuses on the struggle to overcome temptation. Theron, his face "older and graver," hopes to rejuvenate his marriage, his psyche, and his fortunes, but before the carriage arrives he daydreams of addressing a political audience. He imagines receiving the authority and adulation that eluded him as a minister but that he believes Catholic priests receive from congregants. The scene dramatizes the resilience of his speculative imagination and reminds us how laborious reconciliation is, how easily tempted is Theron, and how readily his old profligacy may repeat itself (281, 345, 353). The Jadwins, on the other hand, as in Frederic's later *The Market-Place,* seem to think that the mere vow of fidelity and the disavowal of speculation recreates their marriage as an accomplished fact.

I think most of us would agree that Laura and Curtis live, or want to live, in a dream world, if only by virtue of their misreading of Milton; Adam and Eve are not starting and entering upon a time of understanding with a "right" notion of things, but are continuing a process of learning to understand. The Jadwins seem to feel they are not leaving Eden to learn to manage in the world but are leaving the pit of hell for an Eden where all has already been settled. Jadwin "abjure[s] the Battle of the Street" (402) just as Norris abjures the speculative economics of romance for the balanced exchange and objective valuation of the businessman and the copyist realist; hence, we might say that Laura and Jadwin now seek, or believe they enjoy, what Eric Sundquist in a different context has called an "Eden of mimesis," where there is "an emblematic relationship between image and thing" (1979, 131). In the terms of my discussion, Edenic contracts represent an exchange of real and equivalent goods and are not just trading for differences; obligations to objects and people are necessarily fulfilled; and behavior signifies its face value and is subject to none of the interpretive problems structuring futures contracts and options.

The Jadwins seem to dream of a return to an Edenic money economy, a world without loopholes. Although with this fantasy Norris renounces the romance he advocates in his essays on writing, *The Pit* still represents an important dramatic grappling with the nature of business, marriage, and self in a credit economy; it coherently establishes and weaves the terms of the struggle it would like to resolve or escape. Moreover, it typifies a common attitude toward speculation and its implications, evident in the fact that *The Pit* was for decades Norris's best-selling work.[21]

Finally, however, it is not clear that a money economy of trading effectively obviates the obliquity of credit. Emery notes that once contracts for delivery and not solely actual commodities can be purchased, "full fledged speculation is at length made possible" (39). That is, money, as we saw in chapter 2, already signifies the indirectness and nontransparency of relations and representation. As A. L. Perry observed of value in his popular textbook, value is "not an attribute of matter" or "a quality of any one thing." It is "a relation subsisting between two things," and a "consequent of exchanges"; like a marriage, Perry proposes, value results from the "mutual *action* of two persons" (42, 38–39). But this fact about value—that it is an effect and affect rather than a quality—means that every economic agent is an "instrumentality," as Edward Atkinson wrote, always conductive; therefore, Atkinson concludes, all economic behavior is "metaphysical," by which he means imaginative in forecasting the future (1885, 188–89).[22] Because it is an effect of social relations, value is in principle speculative. It is this potentiality that leads Bellamy to call money the "original mistake," for it "necessitated needless exchanges." "The confusion of mind which [money] favored, between goods and their representatives, led the way to the credit system and its prodigious illusions" (211, 160–61). Once money is acceptable, one can easily accept credit promises for money, because the enabling principles of interminable exchange and oblique representation are already operative. This, indeed, is the force of Rogin's and North's arguments that the antebellum market economy set the stage for postwar credit development. Once trade constitutes economic organization, credit cannot be far behind. And if one, like Norris, considers credit and speculation, in their effects on property and identity, a pit of hell, then a market economy is no Eden but is precisely the original mistake, the halfway house to hell.

Norris attempts in two ways to defuse the troubling fact that trade turns out to be, however safe-looking, a prelude to speculation. First, Laura and Jadwin hold hands while walking to their carriage, a physical unity that obviates exchange or mediation among different elements; thus nothing stands between them as Jadwin "stood between two sets of circumstances" (419), mediating between producer and consumer. That the Jadwins drive to their train "in silence" reinforces this interpretation; their unity requires (or rather depends upon there being) no exchange of words, no verbal contracts. Second, because radical selfishness, without obligations to others, is the horror of speculation, we may surmise that once there are distinct selves who exchange behavioral and verbal intentions, speculative selfishness is always a possibility, and not effaced if one becomes "only" oneself. Norris addresses this problem when he resolves Laura's crisis of self: she recognizes not any true self but "identity ignoring self" (405). Here, identity is neither discrete nor known by normal forms of social intercourse; it is an intuitive and immediate identification with the beloved.

The world the Jadwins are entering, then, seems safer even than an Eden of mimesis, the idealization of market exchange to which so many Americans were attracted. Their carriage conveys them to a premarket Eden that is prior to exchange and representation, where there is an immediate identity of lovers undistracted by endless interests and exchanges. Norris must evacuate the world for love to flourish. "Everything in life, even death itself, must stand aside" (491), and "all the noisy, clamourous world should be excluded" (402), if lovers are to "forget everything else" in their love for each other.

But if this vision of marriage is intended to end the career of speculation, or to forestall its inception, it nevertheless inhabits the structure of speculation it is intended to avert. In the idyll of love the novel's conclusion rehearses, spouses consider and care for each other so intently and exclusively that other obligations are willfully ignored. Is this not the masturbatory danger of speculation? There is a limit to this comparison, because of course speculation resonates more widely in the world than does any one marriage. At the same time, however, a person's social, economic, and legal obligations are not severed in marriage but altered. A marriage that forgets the world that enables the marriage contract reprises speculation's scandalous insouciance. How redemptive is it, after all, to forget the clamorous world when, by virtue of money won in speculation, one boards a train for a home further west? If the Jadwins' hermetic love is offered as the best way to escape the difficulties of a credit economy that is but obliquely and shiftily related to phenomena, it is no escape at all. In the following chapters we will visit other formal attempts to end or elude the vicissitudes and volatility of the market: the Standard Oil Trust, Cowperwood, Debs's industrial unionism. These enterprises are less romanticized, however, for in the tradition of Emerson's transcendentalism, they forge protective economies by inhabiting the vicissitude they seek to manage.

PART THREE

TRANSCENDENT AGENCY

Transcendent Agency: Emerson, the Standard Trust, and the Virtues of Decorporation

Are Trusts Transcendental?

At a crucial moment in *The History of the Standard Oil Company,* Ida Tarbell feels compelled to assure her readers that John D. Rockefeller was not a transcendentalist. The "uncanny" ability of Standard Oil to "know every detail" of the oil trade exalted his image in the oil regions.[1] Rarely seen in public, unavailable for interviews, notoriously taciturn in business and personal relations, Rockefeller was regarded "with superstitious awe" (125):

> this man, whom nobody saw and who never talked, knew everything—even unexpected and trivial things—and those who saw the effect of this knowledge and did not see how he could obtain it regarded him as little short of an omniscient being. There was really nothing in the least occult about Mr. Rockefeller's omniscience. He obtained part of his knowledge of other people's affairs by a most extensive and thoroughly organized system of news-gathering, such as any bright businessman of wide sweep might properly employ. But he combined with this perfectly legitimate work . . . sordid methods of securing confidential information. . . . Certainly there is nothing of the transcendental in this kind of omniscience. (127)

Tarbell demystifies Rockefeller's power by declaring what we probably take for granted: the immensity and sordidness of his business affairs constitute a departure from American transcendentalist concerns with truth and spiritual law. At the same time, however, Tarbell's anxiety to draw an invidious distinction between trusts and the transcendental suggests the very connection she would deny, one she herself makes by quoting "Self-Reliance" in the book's epigraph: "An institution is the lengthened shadow of one man."

Invoking self-reliance as the moral measure of the businessman, Tarbell's attack is prototypical of the tradition in which the age of the robber barons is the era of the powerful individual, the vulgar appropriator of the tradition of self-reliance. With his Nietzschean contempt for the conventional, Frank Cowper-

171

wood, hero of the *Trilogy of Desire,* whom Dreiser often calls self-reliant, has been seen as the fictional model of this radical individualism (see Henry Nash Smith 1964, 99–102). Yet this was also the age of incorporation, as Alan Trachtenberg has recently emphasized, and as such it exhibits another posture toward individualism. The controversy provoked by the trust, the era's real innovation in business organization, focused on its "tendency," Grover Cleveland charged, to "crush out the individual" (quoted in Ely 1897, 268). Thus the era seems to undermine even as it celebrates individual power, and it is regularly censured for this apparent contradiction: unbridled exercises of individual power obstructed and expropriated the agency of others, and were immoral for doing so.

This critical tradition has as its premise a "metaphysics of natural liberty" that Thorstein Veblen contended was already anachronistic by 1880 (1904, 273). Liberal metaphysics developed within a handicraft economy of small business and private contracts, Veblen argued, but it was unsuitable to a machine and credit economy, and from 1880 to 1915 the culture was experiencing what Martin Sklar calls the passage from proprietary to corporate capitalism (4).[2] This often painful metamorphosis in all sectors of the economy and in all institutions of the culture was labeled The Trust Problem, the phrase entitling the influential book by Jeremiah Jenks, professor at Yale and Cornell, a principal advisor to Theodore Roosevelt, and the leading expert on many investigative commissions, including the Industrial Commission established by Congress in 1898 to probe irregularities in interstate commerce.[3] That the culture understood the intensifying concentration of capital as the trust problem is illustrated in the March 1897 installment of Theodore Dreiser's "Reflections" column in the magazine *Ev'ry Month.* Although his essay only generally describes "officered combinations of huge aggregations of capital"—which were not always trusts—Dreiser terms popular discontent over this trend "worry over trusts" (1977, 112).[4] The cultural transition was termed the trust problem because its peculiar characteristics were not sheer concentration and the existence of corporations (a centuries-old feature of commercial life), but rather, the controversial emergence and innovative organization of the trusts.

The enduring liberal tradition that scorns the trusts' transgression of individuality understands individuality as discrete, autonomous agency, and clear individual agency as the original and only sufficient ground of moral judgment. As Veblen believed, however, to evaluate the age of The Trust Problem by such a principle of individuality is historically inappropriate. We can begin to see why if we consider a remark by Cowperwood, who, Dreiser tells us, aspired to the power of the trusts and anticipated their methods (*Financier* 104). Early in *The Financier,* Cowperwood ponders what it means to be a man: "A man, a real man, must never be an agent . . . —acting for himself or for others—he must employ such" (44).[5] Cowperwood is reminding himself that safety lies in not acting visibly on the floor of the commodity exchange. But his formulation can be taken more strikingly to mean that to be a person one should never be an agent, even for oneself. Cowperwood's curious logic of the self, in which one achieves selfhood by lacking the quality—agency—so central to strong individuality, is not a contradiction but a double movement in which the self is substantiated by the disappearance of its agency. This concept of self informed debate over the trust problem, and, I will urge, was central to the moral understanding of the era. The reforging of self and agency was a critical achievement of the cultural transformation that the trust form both signaled and advanced. The dream of the trust, shared even by its radical opponents, is to become a powerful person by not being an agent, or rather by being merely the agent or instrument of transcendent forces.

This logic of selfhood constitutes the true transcendentalist foundations of the Standard Trust and it is what finally makes Tarbell wrong to gainsay the affinity she senses between trusts and transcendentalism. Indeed, we should extend Tarbell's insight and consider corporate practice and literary or philosophical practice as affiliated cultural formations demanding contiguous analysis. For all their obvious differences, Emerson's thought and the organization of the Standard Trust addressed related (albeit not identical) historical problems and anxieties through a common logic: the morality of action is justified by the transcendence of personal agency. Emersonian self-reliance, epitomized in the notorious transparent eye-ball figure, seeks virtue and self-perfection in self-eradication. Similarly, the scandal of the trust for Tarbell is not just that it was immoral, but that the agent (and hence evidence) of its immorality was obscured in the shadow of the institution he spawned.

If Emerson's account of selfhood is recapitulated in the form and logic of the trust, we should not conclude that his transcendentalism was the servant of a developing capitalism. It is more significant that the trust, both the culmination and the scandal of corporate development, evolved within archetypal terms of American thought, specifically within what Richard Poirier has called the American enthusiasm for "quotidian self-erasure" (*Renewal* 192) whereby "human will and power" are affirmed in the very dissolution of individual agency ("Writing" 132).

The Conventionality of Intuition

The most important sentence in the passage with which I began is the one containing Tarbell's moral condemnation: "But he combined with this perfectly legitimate work . . . sordid methods . . . of securing confidential information." It is the dubious moral quality of Rockefeller's business practice that, in Tarbell's mind, disqualifies it from being transcendental. The fact that Rockefeller uses ordinary news-gathering techniques indicates to Tarbell that his omniscience is not "occult," which seems here a purely epistemological category. She concludes that "there is nothing of the transcendental in this kind of omniscience" only after adding the charge of ruthlessness to the observation that his methods are conventional. Her emphasis falls on the "But" that qualifies his use of legitimate methods. Rockefeller is an unfit candidate for apotheosis not so much because his sources of knowledge are conventional but because the morality of his organizational practice is questionable.

We might want to remind Tarbell that the superior morality of transcendental action should derive from superior knowledge of the world and not simply from the moral disposition of the actor. The antebellum ministers who were called transcendentalists famously insisted, against liberal Unitarian empiricism, that virtue and faith derive from unconditioned, intuitive principles of knowledge. Rather than "But," Tarbell's conjunction ought to be "furthermore" or "more-over," adding information, not shifting the terms of critique. But in the logic of her paragraph, conventional methods of gaining knowledge can qualify as transcendental if they are "properly employ[ed]." I argued at length in chapter 2 that for Emerson spiritual truth and any understanding of it are conditioned, conventional, and contextual. Hence, Tarbell's apparently inadequate grasp of transcendentalist principles—her inability to ground moral preference in epistemology—is in fact consonant with Emerson's thought. If spiritual truth is always contextual, however, rules for virtuous action cannot be derived from epistemological principles. Nor, consequently, can the relation of virtue to market behavior be specified.

Emerson is concerned, I argued, not with an a priori freedom from convention but with what we might call the economy of conventionality. The question of self-reliance centers on the determinative relation between self and convention. To what extent are judgment and action determined by conventional and phenomenal form? Emerson's formulation, let me submit, is somewhat more materialist than the Kantian transcendental sublime (discussed in the second section of chapter 1). Kant's objective in probing sublimity is to establish the independence of experience from objective determination. The inadequacy of faculties of comprehension to grasp sublime phenomena (waterfalls, mountains, storms—generally, excessive magnitude and might) is paradigmatic of our inability to intuit the thing-in-itself and the universal principles motivating sensible phenomena. But this sensible alienation from universal principles also makes possible the moral basis of judgment. Since sensation cannot comprehend natural purposiveness, but since we inevitably intuit purposiveness, he says, intuition transcends sensible determination; which means, more importantly, that intuition of universal principles is independent of material determination. Judgment, then, is moral for Kant, because it exceeds sensible determination and signals our intimation, if not possession, of divinity.

Emerson's idealism seeks, to paraphrase Stanley Cavell, a greater intimacy with existence—that is, with materiality and convention—than does Kant's (1981, 126–28; 145–47). As Emerson explains in "The Transcendentalist," wherein he addressed charges that he and his circle were transcendentalists, the idealist "concedes all that the [materialist] affirms." He "does not deny the sensuous fact . . . ; but he will not see that alone." Rather, he looks to "the *other end*" of phenomena, to the spiritual fact they complete.[6] Antinomianism is not exterior to convention; rather, like Rockefeller's omniscience, it uses conventional means to see further and thus to surmount them. It is Emerson's insistence on the inevitable imbrication of intuition with materiality that leads him, I think, to specify that what he describes is "Idealism in 1842," a particular practice in a particular place, not simply the wholesale importation of continental romanticism.

A question remains, however: if the idealist "admits the impressions of sense, admits their coherency, their use and beauty," on what basis can he "affirm facts not affected by the illusions of sense, facts which are of the same nature as the faculty [the spirit] which reports them"? Beyond faith, how has he any greater "grounds of assurance" than the materialist "that things are as his senses represent them"? In his account of sublimity Kant begins from the idealist base of liberal and Enlightenment thought: the self is finally an a priori self-possession by virtue of its possession of the divine faculties of reason and intuition. But Emerson cannot forget liberalism's materialist premise, that the self is known and realized through its transactions with material forces.

The materialist foundation of his idealism creates the drama of the "Idealism" chapter of *Nature,* wherein he poses to himself the "noble doubt . . . whether nature outwardly exists." In a manner later christened pragmatist, Emerson tries to (and finally does) answer his skeptical query: because nature is "useful," his "impotence to test the authenticity of the report of [his] senses" finally makes no difference. Since nature is a "phenomenon, not a substance," the efficacy of beliefs provides all the proof possible of their adequacy. Nevertheless, at times, especially in an age of science, "the presence of Reason mars this faith" (26–27). The senses possess an "instinctive belief in the absolute existence of nature. In their view man and nature are indissolubly joined. . . . The first effort of thought," however, "relax[es] this despotism of the senses which binds us to nature as if we were a part of it" (27). Thought, then, as the essay "Experience" stresses, signals the preeminent nonidentity of experience with world, and recalls Emerson's noble doubt about the existence of the world, or more precisely, his skepticism about the validity of experience and value.[7] Recognizing the undeniable materiality of phenomena suggests to Emerson at once his alienation from nature and the consequent danger of being utterly mistaken about it. He figures mistake as a violence against nature, mother of ideas because the source of sensation: "I do not wish to fling stones at my beautiful mother." If estranged from nature's nurturing materiality, intuition is invalid and degraded. To recuperate the basis and faith of idealism, Emerson must conceive his relation to phenomena as purely physical, and thus momentarily imagines himself as a child, with an undeveloped reflective capacity to "expand and live in the warm day like corn and melons" (33).

Perfecting and Annihilating the Self

The materialist foundation of intuition is, for Emerson, then, the possibility of both solipsism and salvation. To adapt Kenneth Burke's terminology, self-reliance or transcendence is a project in bridging the gap between the two, or between inner experience and materiality. As I argued in chapter 2, this is a powerfully Lockean conception, but it is also a reconception of Locke (whom Emerson claimed not to find a powerful mind). Emerson goes to nature to learn a very Lockean "lesson of power" (22), the "art" of "the mixture of [man's] will with [Nature]" (4). Moreover, this appropriative art is an interminable metabolic and metaphorizing power. The "Power" Emerson seeks to experience and command emerges not in repose but only in "the moment of transition from a past to a new state" (142). But if "virtue subordinates [nature] to the mind" (33), the twofold problem of Emersonian transcendence remains. Virtue—the act of self-reliant power—has no limiting content; nor will Emerson specify universally the conditions of the claim *not* to have violated nature. Therefore, any claim to be legitimately powerful may simply be a disguise for despotism.

The charge of despotism has recently been leveled against Emerson by Michael Gilmore, Sacvan Bercovitch *(American Jeremiad),* and Myra Jehlen *(American Incarnation).* These scholars argue that Emerson's notion of power, in which all is concentrated in and subordinated to mind, is a monism that preempts dissent. Donald Pease, on the other hand, has defended Emerson's vision as capturing the dynamic, metaphorizing quality of a visionary compact between mind and nature, between self and other, which by implication is not simply an epistemological transaction but a social compact as well (1987). Although all these arguments explicitly challenge liberal conceptions of the autonomous self, they nevertheless respect a version of the same, notorious liberal premise. Whether finding Emerson villain or hero, these interpretations hold that for action not to be despotic there must be proportionate representation among different parties to an action, either between self and nature or among citizens. These critical analyses would, finally, promote a balance between private and public interests, even though—or, rather, precisely *because*—residual conflicts are recognized. The goal, as it was for Locke and Adam Smith, is an operative harmony of interests, although contemporary criticism disowns Smith's urge to absorb conflicts within an invisible hand. Critics continue to uphold the integrity (albeit not the autonomy) of self and other, private and public precisely by emphasizing the simultaneous interdependence and potential discord among coordinated spheres.

It seems to me that Emerson's vision is superliberal (to import Roberto Unger's term for his social theory in *Politics*), in that it attenuates the discrete individuality of entities. The harmony, what he calls the "original relation," that he projects between self and other, the Me and the Not Me, is never a balanced negotiation but, rather, an endless exchange and appropriation. This is why he says at the beginning of *Nature* that "there is a property in the horizon which no man has but he whose eye can integrate all the parts, that is, the poet" (5–6). Only through the poet's integrative faculty can discrete selves and phenomena, "private purposes" and ends, ineluctably serve a "public and universal function" (23). The problem, as those critical of his purportedly monistic politics intuit, is that the appropriative impulse of the poet's combinatory vision is never exhausted. The self must act, and to act is to appropriate the Not Me; gratification is out of the question, and every new situation entails another impulse to appropriate. "The perpetual presence of the sublime" that Emerson goes to nature to experience recapitulates but never resolves the appropriative tension of the Kantian transcendental economy.

Emerson's notorious figure for transcendence, the transparent eye-ball, exemplifies this problem, and begins to suggest its bearing upon the trust problem.

> In the woods, we return to reason and faith. There I feel that nothing can befall me in life—no disgrace, no calamity (leaving me my eyes), which nature cannot repair. Standing on the bare ground—my head bathed by the blithe air and uplifted into infinite space—all mean egotism vanishes. I become a transparent

eye-ball; I am nothing; I see all; the currents of the Universal Being circulate through me; I am part or parcel of God. (6)

Carolyn Porter has aptly remarked that this passage performs a decapitation of the observer, his head floating in space; in this decapitation, the movement from "I am nothing" to "I see all" seems contradictory—that which is nothing can neither see nor be a subject. For Porter, the contradiction between being nothing and yet seeing obstructs Emerson's goal of eradicating the self and marks his fundamental confusion. He purports to be a neutral observer of nature even as he senses that he inevitably sees only within the "clutter of culture," from the alienated perspective of private ends; he thus fails to glimpse the universal design and reconstruct it on earth (105–7).

Like most accounts of American transcendentalism, Porter's charge requires that transcendence be an absolute break with the past and all institutional forms.[8] In fact, however, the decapitation Emerson figures enables him to regard the private ego as already universal. It figures his assault on liberal Unitarianism, and on what Roberto Unger calls "the antinomy between reason and desire" that structures liberalism (1976, 51). The decapitation kills the self's "*mean* egotism" (emphasis mine), egotism commensurate with the "mean and squalid" use of the commodity, which accepts visible form as essence. Emerson decapitates himself to inspire his essential egotism: the self as site of the circulation of volatile universal forces. In this circulation, particular, material values are purified by being universalized.[9]

Being nothing while seeing all is not a contradiction but a sublimation of agency. In the figure of the eye-ball, Emerson kills the "particular man" who has only "private" ends (160). The sublimation here is generally, then, catachrestic, as Donald Pease has characterized the eye-ball figure (1980, 59). Let us recall that catachresis is the rhetorical figure that Willard Phillips used to defend antebellum protectionism: by perfecting the self as no longer merely individual but as vehicle of universal forces, he argued, protectionist policy finally makes customhouses supererogatory. Emerson's eye-ball figures a peculiar catachresis, with different ends from the protectionists, yet the coincidence helps elucidate the economy of Emerson's thought. In protectionist policy and Emerson's eye-ball figure, the tenor—in both cases the self—disappears as a discrete entity; but in both as well, the vehicle—body or tariff—also disappears, as it prefects the self whose mean ego disappears. In the profoundly protectionist economy of the eye-ball figure, universal forces are apprehended in private ends because the contours of private action and desire become invisible. Meaning is transparent to the eye-ball because the eye-ball itself is transparent: it becomes the site where universal meanings are viewed as the physiological limits of the self and perception disappear, because annihilated. The individual isn't entirely evacuated, but the self becomes God "by transferring . . . my body, my fortunes, my private will" (*Journal* 5:336). No longer perceived as mere body or as agent, the transcendent self incorporates the circulation of universal forces by transcending particular representations of them and of itself.

In the outrageous physiology of the eye-ball figure, Emerson wants to lose his

self and have it too. This logic, only an apparent equivocation,[10] is the mechanism
for reforming the insulating cult of "the single person" (62), or what Hegel called
in his early essay, *Natural Law,* the popular "fixation of individuality" (102). In
Emerson's transcendent exaltation, the cause of self-reliance is God, the beneficiary the powerful man, his power the concentration of divine intention. The
result is not exactly what Harold Bloom has called a "dialectic of power," mediating between the material and the divine, the Me and the Not Me (20). Dialectic
implies dualism, and as B. L. Packer argues, "when the soul is really present . . .
all *sense* of dualism ceases" (145).[11] The Emersonian self is "conductor of the
whole river of electricity," possessed of and by an inexhaustible "power transcending all limit and privacy" (McQuade 323); it is proprietor because its particular
agency is transcended. When "our action" and will are "overmastered" (163) by
the sublime's "power to charm and command the soul" (70), the "body [is] over-
flowed by life" (310) and we "escape the . . . jail-yard of individual relations" (317)
to "own the world" (73). The word "conductor," less mechanical than, say,
"channel," suggests the advent of the agent in the dissolution of its agency: "The
man who renounces himself, comes to himself" (69).[12]

Sublime subjection is also universal entitlement, and transcendental self-
eradication is simultaneously a self-perfection. Following Packer, who develops
an observation by Cavell, we might call this state "the art of abandonment" (147),
while keeping in mind the Lockean, appropriative basis of Emerson's notion of
art. Though its ineluctable corporeality is acknowledged, the self experiences its
properties not as limits or constraints but as resources subject to the will. Thus the
materiality of the self is protected, while the self is protected from its own
materiality.

Transcendentalist reform requires a self not wholly fungible with its properties; otherwise, as Thoreau admonished, we could not imagine an alternative
foundation of values, for "essential laws" would be indistinguishable from the
"uncertainties" of "market fluctuations" (*Walden* 7, 43). Accordingly, the character of Hawthorne's Holgrave, clearly modeled on the young-man-from-Ver-
mont passage in "Self-Reliance," for which Thoreau was the model, has "true
value" because, with his "consciousness of inward strength," "amid all [his] personal vicissitudes, he had never lost his identity" (38, 180, 177). Holgrave's consciousness earns him, through marriage, aristocratic fortune, which is not a compromise but a sign of his inner integrity.

We should note that this ontology of wealth is not antithetical to but consonant with nineteenth-century defenses of capitalism, a social Darwinism, popularly articulated by Andrew Carnegie, in which a "a special talent" intrinsic to the
self and independent of material circumstances causes an individual's economic
power (1904, 15). From its earliest formulation in the work of Locke, William
Petty, Quesnay, Say, and Adam Smith, the liberal capitalist project, like the transcendentalist, seeks to identify a self of intrinsic value distinct from a market
destabilized by legislation and accidents that obstruct the self's genesis of market
value. Smith's conceit of "the invisible hand," positing that a divine will governs
market behavior and that therefore the self-interested acts of individuals are in
fact altruistic manifestations of God's will, nicely anticipates the ideal Emerson

formulates in "The American Scholar": when "the single man plant[s] himself indomitably on his instincts" he will discover that he is "inspired by the Divine Soul which also inspires all men" (McQuade 63).

At the same time, Emerson crucially reformulates liberal ontology and its corollary notion of freedom. The transparent eye-ball figures a transcendent self commanding material and social authority by utterly subjecting itself to it. The passage thus remedies anxieties like those de Tocqueville remarked in a society of individualism, anxieties about the dependence of identity upon alienable property, fears that fluctuations in the marketplace, and thus in one's property, effect what William James later called "vicissitudes in the me" (*Psychology* 1:371). Our "instinctive impulse" to "identify ourselves" through our bodies and material sensations, James observes, "drives us to collect property" in a kind of material accrual of the self. As a result, a loss of property produces an anxious "sense of the shrinkage of our personality" (1:292–93). James here combats anxiety about the shrinkage of identity by appeal to the idealist premise of the liberal vision; he invokes a Kantian "I," "pure consciousness" and "self-identity" that cognizes the material and alterable "me" (1:373). This "Pure Ego," vehicle for "the passage of thought," secures one's sense of identity even while—or better yet, because—it "is simply nothing" (1:365), and therefore exempt from shrinkages of the me.[13]

Emerson's peculiar figure would convert such anxiety to faith. The self, here, need not be exempt from shrinkage, nor need Emerson resort to the idealist notion of a pure ego; the materiality and volatility of identity are the very grounds of its transcendence. The eye-ball figure, then, at once perfects and revises the logic of its kindred cultural developments, liberalism and capitalism; this departure, however, retains the ontological and consequently moral difficulty the ethos of liberalism sought to resolve. It will be instructive to pursue the differences between Emerson's view and Carnegie's sense of the "special talent" that is expressed and realized in capital expansion. For Carnegie, special talent governs an entire system and justifies its hierarchies, which if they do not actually reflect individual talent, are absolutely predicated upon it. Emerson's self is never so complete nor authoritative.

Carnegie's vision supremely manifests the fixation of individuality that Hegel explicitly opposed in his early treatise on natural law. Translated into Hegelian terms (albeit utterly foreign to Carnegie's thinking), Carnegie's "special talent" epitomizes the "absolute ethical life," which Hegel says takes "a negative attitude" toward any external system of fixed determinacy, whether physical or moral (98). Anxious to overcome the contingency and particularity of experience, the ethical urge "must give its anxious mind the impression that its possessions are secure" (107). Standard principles of contract, however, retain the individual differentiation (and hence conflict) that provokes persons to enter into contracts to secure property. Only "the will to power" (107), Hegel asserts, "real absolute consciousness," an absolutely antinomian urge that comprehends ultimate relations rather than expresses particular properties, can transcend the finite authority of contract. Absolute consciousness can transcend particularity because it is a model of synecdoche: like the "most perfect mineral," it "displays the nature of the whole in each part." That is, through consciousness, separateness "is absolutely and perfectly taken up into the universal" (109). Here is conclusive synthesis, if not the de facto end of dialectic, albeit experienced in time. Ethical nature, in its "absolute indifference" to formal and material particularity, can "resume the complete equality of all its parts and the absolute oneness of the individual and the absolute" (110).

However critical of conventional individualism, Hegel's vision of transcendence reprises, and even would consummate, the ideals that he argued the liberal vision was inadequate to fulfill. And we can better understand Emerson's vision of self by examining Hegel's more individualistic vision a little more. There are, of course, striking differences between Hegel and Carnegie's terminology, specific motives (philosophy, or an autobiographical celebration of great wealth), and larger purpose (criticism, or an apology for property distribution). Nonetheless, Hegel's transcendental and Carnegie's individualist visions share two important, liberal conclusions. First, Hegel's consciousness and Carnegie's special talent both transcend and thereby govern systems and principles of action. Second, what amounts to equality—Carnegie titled one of his books on great wealth *Triumphant Democracy*—results from the "absolute identity" of "the universal and the particular," as Hegel puts it (115, 122). As Hegel projects harmony between individual and communal, in *The Empire of Business,* Carnegie forecasts harmony between labor and capital.

Hegel's critique of liberalism is finally Carnegian individualism writ large, Carnegie's aspirations secured by far more than mere wealth, secured, indeed, by consciousness's absolute unity with, and thereby governance of, world. Emerson's vision, in contrast, is closer to, though not identical with, Kant's sublime economy, in which the possession and expansion of self that is the outcome of the sublime retains the feeling of finitude, sacrifice, and sublation in which awe originates. Emerson's eye-ball figures a decorporation of self that is also its incorporation into greater circulation. This decorporation begins with an anxiety that not only has familiar Freudian resonances but was common to discourse about the sublime, fear of losing one's eyes. In his well-known image for the sublime—a jagged rock rushing at one's head—Burke registers this feature of awe. Similarly, Thomas Cole, in his notebook narrative, "The Bewilderment," feared when lost in the woods that branches might pierce his eyes. This trope is an appropriate figure for, first, anxiety about the "shrinkage of the me" that James describes but, also, for the fundamental problem of both discourse on the sublime and economic discourse: the nontransparency of value.

At the opening of *Nature,* Emerson goes "into solitude" and first notices the stars, the type of "all natural objects." "One might think the atmosphere were made transparent with [the stars'] design." But we are not in "the perpetual presence of the sublime," and "though always present, [the stars] are inaccessible." Their inaccessibility is the very reason "the stars awaken a certain reverence." Could we feel the perpetual presence of the sublime—that is, were the purpose of natural objects transparent—nature would not be sublime but familiar. And it is excessive familiarity that plagues the narrator of *Nature.* He goes to nature to induce awe, which reminds us of our mere individuality. Rather than Hegel's absolute consciousness or Carnegie's harmonizing, special talent, in nature Emerson's self at first witnesses his own partiality and finitude. For his vision is metonymic rather than synecdochal; he believes that the stars typify a spiritual order, but they fail to body it forth. The poet's self-entitling combinatory capacity compensates for the awful sacrifice and sublation of individual security; it also induces the very feeling of inadequacy that requires and inspires redemption. In Emerson's sublime vision, unlike in Hegel's perfection of individualism, the sense of mere individuality, and of its station in a collective order, survives.

Recall from chapter 1 that Kant conceives sublimity as a utilitarian economy of expenditure and return: the imagination "acquires an extension and a might greater than it sacrifices" (109). Kant takes care to specify that "the ground" of the imagination's extension remains "concealed," and so exaltation retains the feeling of "the sacrifice or the deprivation" that occasioned it; the imagination remembers "the cause to which it is subjected" (109). Emerson's eye-ball exalts the residual feeling of sacrifice in the Kantian sublime; it does not absorb and redeem all difference and loss as would Hegel and Carnegie. When he is part or parcel of God, "the name of the nearest friend sounds then foreign and accidental." Any social or familial relation "is then a trifle and a disturbance" (McQuade 6). Society, whose need for reformation is one reason for the sojourn in solitude, is unrestored, and the very transactions that transcendence ought to facilitate—of "discipline" or the use of commodities—are superfluous. The section ends in melancholy, and Emerson spends the rest of the essay attempting to project at once transcendence and specific action.

In the transparent eye-ball passage, Emerson's perfection of self in its eradication retains the disjunction between mean and universal ego that may thwart its practical aspirations. The poet's combinatory practice cannot in itself teach virtue, which means to use materiality to higher ends. The poetic or sublime goal is "to spend on the higher plane," "to invest," "to take up particulars into generals" (720), and thus give "form and actuality to . . . thought" (702). But the inextricability of the spiritual from the material and egotistical means that Emerson can and will propose no formula for absolutely distinguishing virtuous from degraded practice.

If I was correct to argue in chapter 2 that self-reliance, in its preeminent suspicion of doctrine, is fundamentally amoral, refusing to instruct in any particular, then just as the stars are not transparent, it is not always self-evident whether any act is an exercise of the will or of enslavement to the senses and circumstance. The risk as well as the power of self-reliance is its indifference to present values and conventional duties. Emerson is aware how readily "the bold sensualist will use the name of philosophy to gild his crimes." Emerson's exhilaration crossing the bare common may well be mere whim, and it fortifies him to do what he will. But the caricaturing of Emerson's figure even by followers like Christopher Cranch indicates how little it satisfied his compatriots' desire for prescriptions about the apprehension and application of intuition.

Perfecting Idealism, the Standard

By exploiting the attractions of idealism, the organizational form of the Standard Oil Trust achieved the transcendence of the particular self and its liability to vicissitude that Emerson sought in the figure of the eye-ball and James sought (somewhat differently) in the simple nothing of the "Pure Ego." The trust form was designed to circumvent laws that, in the laissez-faire tradition, prohibited interstate corporate ownership in order to facilitate competition. Yet however desirable in theory, unlimited competition threatened to ruin the fledgling oil industry. Tarbell details how the compulsion of oilmen to drill wherever and whenever there was rumor of oil glutted the market, drove down prices, and drove men out of business.[14] This pattern typified the postbellum era's tendency to overproduce, a crucial factor in the thirty-year decline in prices that fettered farmers and precipitated regular panics. The period's great debates about tariffs, corporate regulation, speculation, and the gold standard sought ways to make commercial conditions inhospitable to fluctuation.

The Rockefeller network of oil holdings was ventured to order the chaotic oil market and protect against shrinkage, or, as Rockefeller put it, "to bring some order out of what was rapidly becoming a state of chaos" (*Random* 59). The signing on 2 January 1882 of the Trust Agreement, drawn up by S.C.T. Dodd, longtime solicitor for the Standard, institutionalized a practice that had been evolving over a number of years (see Nevins 1:382–84). Representatives of the Standard would informally engage as "trustees" of companies in other states, their supervision harmonizing a company's decision making with the Standard family's practices elsewhere. The term *trustee* derived from the tradition of the familiar "trust company," to which people entrust funds for long-term investment, and from the limited trust status the common-law tradition granted certain British trade associations. Formally, the trade associations resembled the cartels common on the Continent, horizontal combinations of corporations operating at the same stage of production. Each member of a cartel retains its separate corporate identity; the cartel sets production levels to regulate market supply. (The British trusts' horizon extended to labor relations as well as production levels.)

The Standard Trust marked a radical development in organizational technology, achieving vertical as well as horizontal integration, the merger of eventually all stages of operation in the oil industry: refining, transportation, storage, marketing, and, finally, drilling. While component corporations remained formally separate, they surrendered functional independence. The trust agreement stipulated that the property and assets of member corporations were "vested in . . . several Standard Oil companies," which then entrusted these by "assignment" to nine trustees.[15] With the trustees "having in their hands the voting power of the separate corporations," Jeremiah Jenks wrote, each corporation was in effect "abdicating its own independent power of self-determination" (122). Corporate self-determination was abdicated by individual stockholders as well. Stocks representing corporate assets were converted to "trust certificates," distributed according to each firm's proportionate value to the whole, and held by the trustees in joint account rather than, as commonly practiced, in individual accounts. The proportions of these assignments were determined not by the affected stockholders but by the trustees, who now held the power to appoint "nominal officers" who were actually, one critic complained, "servants of the Standard Trust."[16] In other words, to earn dividends from the combination, stockholders surrendered both their title to the stock in a particular company and their power of decision over capital assets (see Jones 19–20).[17]

The trustees thus achieved, or were assigned, as they preferred to conceive it, unprecedented authority. The constituent firms were run not by directors or stockholders, whom in theory corporations represent or embody, but by these trustees, empowered to make operating and investment decisions for companies with which they had no formal association. Thus, "complete unification" of the oil industry was achieved,[18] while the consolidating agent remained invisible.

The trust is best understood, I think, as an Emersonian decapitation of the corporation. As the body for Emerson is the transmitter of sensuous representations, the primary form in which the self is represented to consciousness, so the corporation, judicially designated a "legal person," is the institutional form representing the fluid and often intangible capital composing it. Corporations exempt shareholders from individual liability for this artificial person's debts (see Patterson 1). The trust simultaneously perfects and annihilates the corporation. An 1887 column in *The Nation* decried the trust for this reason: it "is the sphinx of corporations, except that it is not a corporation at all" (380).[19] Though controlling enormous assets, the trust represents neither its component corporations nor their stockholders; rather, it is autotelic. Having abdicated self-direction—having given away its head—each corporation is protected as its "mean egotism" is eradicated, since transactions are no longer particular exchanges but manifestations of composite market forces concentrated in the trust. Even William Cook, vocal enemy of Standard Oil, assented in 1888 that because profit "comes from the whole 'Trust,' and not from . . . particular establishments," a member firm's earnings, remarkably, can increase when its production is halted (5–6).[20]

By superseding the agency of corporate persons, the trust accelerated the attenuation of ownership that Veblen identified as a major consequence of the aggregation of capital.[21] It thus posed, a critic from the oil regions pointed out, a threat to the individual "right of property." Managers of corporations control the wealth-making power of capital they only partially own. But the "dogma" of the trust was "that control of property shall be severed from individual ownership" (*Brief* 6, 11). "The Trustmaker never pretends to ownership or possession," wrote another critic (Fleming 56).[22] Since the trustee pretends no ownership, and since, as Cook observes, "the management of the business [is taken] out of the hands of the various discordant elements which constituted the corporation," the trust, Cook concludes, is based "on an absolute right of property, possession, and ownership, vested in the combination itself" (3). Not any person but the trust itself owns the property.

The trust, however, was not a legal person but what Dodd called an "instrumentality" ("Aggregated" 19) that like Emerson's eye-ball is nothing in itself, only the invisible site where economic forces concentrate. Legally, this was true because the trust was not a legal entity. By this I mean not just that it was illegal but that, before 1882 (when it was forced to become a formal entity and Dodd devised the trust agreement), it had, Tarbell observed, "no legal existence" (157). The trust agreement enabled the Standard to operate without a state charter and therefore without filing public reports. Therefore, Cook fumed, "It may carry on any business it desired." Responsible neither to stockholders nor to the states that properly regulate business, it was "neither a corporation nor a well-defined common-law trust" (Cook 53), and it remained, Tarbell noted, "independent of all authority" (151). The Standard achieved this freedom from contingency and mundane authority[23] characteristic of self-reliance by transcending finite boundaries. *The Nation* charged: "We know where to find [corporations]. We know what their powers are." But the trust "is bound by no law. There are no limitations upon it, not even those of time and space." Because the trust has "no fixed abode, no place of meeting, no books of account that anybody can demand access to . . . there is nothing for the state to lay hold of" (380).

In court and during legislative hearings the trustees claimed the same exemption from liability as the trust. Neither initiators of action nor owners of property, they merely actualized operating decisions enjoined by market forces. As in Cowperwood's formulation, by being *mere* agents, they were not really agents. Certainly, responsibility for the Standard's behavior was not traceable to them: no contracts were written, and minutes of meetings were not preserved. With, Cook noted, "no track or trace" of managerial activity (66), it was virtually impossible to find, as under the Sherman Antitrust Act courts needed to find, an intention to restrain trade.

The design of the trust, then, allowed the trustees to deny the charge of unfair practices by denying intention and agency. This design at once observed and transformed the premises of the formalism that governed legal thought of the era. Presided over by Christopher Langdell of Harvard, legal formalism was concerned mainly to protect freedom of contract, and along with it a notion of unconstrained self that entered contracts. Following the liberal tradition, Langdell viewed law as a kind of botany and sought to discover and classify the few universal principles that govern liability in all situations, irrespective of context and purposes (see Grant Gilmore 19–67). Just law, said a framer of the Sherman Act, Senator George Edmunds of Vermont, is law that "work[s] automatically" (quoted in Letwin 100). Dodd's design exploited the formalist premise that social action proceeds by its own (or divine) agency. Even if the Standard's size limited the opportunity of those lacking comparable capital to enter or remain independent in the oil industry, the trustees explained that neither they nor even the trust were responsible. They did nothing; the trust itself did nothing and was nothing: it was sheer form, a symbolic locus of the powers of the market.

In H. H. Rogers's testimony before the 1879 Hepburn Committee investigating the Standard for the New York legislature, we see a remarkable articulation of this formalist conception. Later to become Mark Twain's friend and financial savior, Rogers has just denied that the Standard employed contracts. His questioner wants him to admit that the Standard was an "arrangement," implying contractual obligations that must conform to statutory prescription. Rogers insists that the Standard is composed only of "harmony" (as opposed to the "discordant elements" that Cook said constitute corporations).

> Q: You said that substantially 95 per cent. of the refiners were in the Standard arrangement?

A: I said 90 to 95 per cent. I thought were in harmony.

Q: When you speak of their being in harmony with the Standard, what do you mean by that?

A: I mean just what harmony implies.

Q: Do you mean that they have an arrangement with the Standard?

A: If I am in harmony with my wife, I presume I am at peace with her, and am working with her.

Q: You are married to her, and you have a contract with her?

A: Yes, sir.

Q: Is that what you mean?

A: Well, some people live in harmony without being married.

Q: Without having a contract?

A: Yes; I have heard so.

Q: Now, which do you mean? Do you mean the people who are in the Standard arrangement, and are in harmony with it, are married to the Standard or in a state of freedom—celibacy?

A: Not necessarily, so long as they are happy.

Q: Is it the harmony that arises from a marriage contract?

A: Not necessarily, so long as they are happy.

Q: When you speak of harmony, is it a relation of contract?

A: I mean by harmony that if you and I agree to go on Wall Street and buy a hundred shares of Erie at 33, and we agree to sell it out together at 40, that is harmony. I mean just the same that way—if I go into the Standard Oil office and conclude to buy some oil of them and agree on a fair price to sell it out at, that is harmony.

Q: Is that the harmony that you mean—that you gentlemen have agreed between each other the rate at which you will buy and the rate at which you will sell?

A: Well, not going too far into detail, I would say that the relations are very pleasant.

The questioner is forced to concede that "it is not an abuse to be in harmony" (Tarbell 1904, 362–63). Contending that no particular agency propels stock prices from 33 to 40, Rogers is claiming that the Standard, simply harmony, not quite even form, consists of people simply enjoying the freedom from responsibility that recommends celibacy.

Formally, this denial of agency was not really inaccurate. Its unprecedented scale of operation led the Standard to pioneer the prototype of modern centralized administration, "an organization such as the civilized world has never before seen," as John Moody, well-known publisher of business statistics and critic of the abuses of trust power, wrote in 1904 (131).[24] Alfred Chandler terms the trust form managerial as opposed to entrepreneurial (389), and Sklar contrasts it to proprietary forms of business and property. Entrepreneurial or proprietary management typifies the myth of nineteenth-century individual enterprise, where the outstanding individual accumulates great power by his ingenuity and frontier self-reliance. In the managerial form of organization evolved by the Standard, decision making is delegated to committees supervised by middle-level managers. The Standard's decisions generally were made without the supervision of the trustees. While the trustees occupied themselves with other enterprises and with philanthropic undertakings, Rockefeller's biographer Allan Nevins tells us, the trust virtually ran itself (Nevins 2:18–25).

Conceiving of the trust as self-propelling harmony, Rockefeller was not merely equivocating in his *Random Reminiscences* when he attributed the Standard's success to the actions of friends and market forces. The formalist conception of the Standard enabled him to conceive himself as simply a locus where power concentrated. In this transcendence of agency lies true self-reliance. In "Self-Reliance," exhorting his readers to "self-trust," Emerson asks, "Who will be the Trustee?" of divine intention (155), the powerful agent in whom "all concentrates" (159). An Emersonian "architect" of harmony, "concentrat[ing] . . . the work on one point" (13–14), Rockefeller achieved such trusteeship through what one hostile legal theorist called the "unification of power" and "unification of interests" converged in the Standard (Benjamin 1912, 15, 21).[25] The trust form transcended market fluctuation and justified the virtue of action by ascribing power to transcendent agency rather than mere private interest. "Invented" by necessity and S.C.T. Dodd, the trust, wrote one critic, was "an enormous artifice" "transcend[ing] anything hitherto conceived of in the possibilities of corporate aggregation" (*Brief* 2); its art consisted of converting the self to a transparent trustee of the energy of production, seeing all, concentrating all, seen by none, doing nothing—therefore accountable to no authority, safe from shrinkage, and commanding enormous power and wealth.

Dodd's artifact transcended legal authority through a transcendence of individuality, and in this effect the trust form transmogrifies liberal theory by uniting the spheres of public reason and private interest. The "fiction" of the trust, wrote one judge, "merges in the artificial body and drowns in it the individual rights and liabilities of the members" (quoted in Dodd, "Present Legal Status" 158). As Rockefeller once succinctly put it, "Individualism is gone, never to return." But this disappearance did not signify to Americans, as it does to Alan Trachtenberg, simply the tragic death of the individual.[26] In numerous articles and public lectures, Dodd heralded the drowning of the individual in the greater body as a Christian and potentially democratic ideal. In an 1893 address at Syracuse University he posed himself the question so essential to the voluminous debates about the trust problem: "Has the individual been crushed out [by the trust]? To some extent, undoubtedly, as a solitary individual. But he has found a wider sphere for his efforts through association with other individuals" ("Aggregated" 23). With agency and therefore liability to flux attenuated, the annihilated self transcends its limits to inhabit a larger sphere of action. Dodd's sublime trust perfects and realizes what Poirier has called the "liberation from the boundaries of self and time" that constitutes Emerson's eye-ball metaphor's triumph over "fabricated" boundaries of property (*World* 61). Mere property and physical and conventional limits disappear in the devastation that realizes self and entitlement.[27]

The Attractions of Decorporation

With his transcendental imagination, Dodd could distinguish the Standard's form from the monopolistic effects with which it was popularly equated.[28] We see in its scandalous amorality the final logical link between the trust and Emerson's thought. Dodd always insisted, in an 1894 *Harvard Law Review* essay, for example, that "the term 'trust' . . . embraces only a peculiar form of business association" in which the trustees have "no powers" other than as instruments of the trust's "beneficiaries" ("Legal Status" 157).[29] Because large economies of scale were essential to the nation's health, Dodd felt his combinatory work was consistent with his Puritan, democratic heritage. Faith, he wrote in his private memoirs, transcends "the limitations of man's senses" to perceive the "infinity of existence" behind "phenomena." Like the trust, matter "changes its form only to take new forms." Like Emerson's hierarchy of forms in *Nature* or Thoreau's meditations on "Higher Laws" in *Walden,* Dodd's spiritual architecture traces nature's ineluctable progress "from high to higher forms," perceived by and immanent in the "self-conscious individual." In Emersonian fashion, this spiritual individual "vanish[es] into invisibility almost at the moment of its formation" (*Memoirs* 22–23). Dodd incorporated into his public writing and lectures this interchangeability of the language of theology with the language of the trust, enumerating the ways "aggregated capital" has "penetrated the unseen, discovered the unknown" ("Uses and Abuses" 42). Since combination is business practice "on a higher plane," the trust might readily serve divine and democratic ends (*Memoirs* 19).

If this suggestion seems absurd to us, it did not to disputants occupying all positions of the political spectrum. At the 1899 Chicago Conference on Trusts, reformers and radicals like William Jennings Bryan, Samuel Gompers, and the utopian socialist Laurence Gronlund (whose *Co-operative Commonwealth* helped beget the Socialist Party), hailed the "combination" as "the legitimate development or natural concentration of industry," as an inevitable and necessary device for protecting individuals in an age of expanding capital (Gompers 1900, 330).[30] Gompers and Gronlund drew analogies between the trust and national labor unions: both promoted harmony by combination. Gronlund placed the trust within the same teleology as Dodd: "an irreversible step along the path to universal co-operation," the trust would serve as a model to help prepare boys "to take their place in a perfect democracy" (Gronlund 1900, 571, 573).[31]

Since to many the trust scarcely seemed to promote harmony, it was the scandal of corporate development and "excited," in Henry Demarest Lloyd's words, "an unbroken chorus of protests" (266). Yet labor leaders and socialists could still view the trust as a model of utopian harmony, for they accepted the moral neutrality of the trust form, agreeing that it need not serve only one class, that it was an instrument amenable to myriad interests, that its evils were not intrinsic to its form but results of abuses. In a forum published in *The Independent* in 1897, Daniel De Leon, leader of what was then the Socialist Labor Party,[32] typically argued, like S.C.T. Dodd as well as other radicals, that the trust is an "instrument" or "tool of production." As did Dodd, he extolled its "concentration," which both "raises man to giant's stature over nature" and surmounts the limits of individual action. More important, the trust's sublime "collective" power is the only "remedy" against "the wastefulness of competition," whose cessation De Leon awaits when the trust form structures the entire culture. "The trust is the highest form of collective development," and De Leon ascribes its offenses not to its form but to its private ownership. Thus for De Leon, the trust form exemplifies the contradiction of capitalism, that property is both collective and private. De Leon wants to "improve" the trust through public ownership, purging its contradictory element and expediting its office as both sign and instrument of the historical process, of the collective redemption of individual effort.

If the trust form had no intrinsic effects, its moral versatility was the trust's real scandal. Almost no one, not even Lloyd, ascribed the cause of monopolistic effects per se to the "power to control prices" that "unification gives" (Benjamin 1912, 15). As any evaluation of conduct must, and following the legal treatment of corporations as artificial persons, these debates were conducted in the language of personhood.[33] If many agreed that monopolies were "undesirable persons," "students of the trust problem," as one reformer wrote, distinguished between "the evils of monopoly and the mere power to inflict them," between "abuse of trade *power* and that power itself"—that is, between "*conduct,*" and "mere *capacity or origin*" (Bancroft 2, 6–8).[34] As Theodore Roosevelt insisted in 1903, "the line" distinguishing monopolistic from merely combinatory acts "is drawn as it is between different individuals: that is, it is drawn on conduct" (195).[35]

For this reason the Sherman Antitrust Act lacked specificity in stipulating what conduct constituted monopolistic practice. Ideally, per se or "uniform rules" could identify under all conditions monopolistic behavior. But this ideal of legal formalism was doubly problematic. First, as Richard Ely pointed out in the 1897 forum in *The Independent,* uniform rules cannot anticipate all evolutions of form and conditions available to Dodd's protean conception; their limited range of application could be readily circumvented simply by changes in "forms and names" (268).[36] Their recombinatory capacity would enable trusts to continue, as one jurist put it, to "outstrip the laws of their own time" (Frederiksen 21). Second, formal, per se laws threatened to make criminal all combination, and hence much business, no matter the scale. Casual agreements among, say, bakers in small towns not to engage in cutthroat competition effectively keep prices high. All formal partnerships exclude others from certain benefits of unlimited competition.[37] Finally, many conceded, so do all contracts, conceivably liable to the charge of price fixing. Certainly, all property, an exclusion from possession that constitutes the very conditions of exchange, amounts to a restraint of others' capacities to trade.[38] Consequently, as the Supreme Court later found, a "rule of reason" to determine the intention unreasonably to restrain trade should be applied; yet the attribution of intention was precisely what the Standard's design frustrated, by its claim to be not a legal person but a transcendent harmony of interests.

Sklar has observed that by 1890 virtually all discussants of the trust problem were "procorporate"; that is, they accepted the corporate form as an inevitable stage of cultural evolution. The term *corporate* here is perhaps misleading, suggesting that all discussants favored corporations and their practices. We might more accurately say that all were pro*combinatory,* admiring the protection achieved in the transcendence of agency that combinations offer. The Standard's formal innovation was a crucial catalyst in the cultural transformation from individual to combinatory relations and conceptions of self. The Standard's form was both scandalous and universally attractive because of the security and claim to virtue its combinatory logic of transcendence afforded. This vision of power and personhood perfected the goal of incorporation, to limit liability. The trust's more radical claim that it was merely a conductor of ineffable harmony conformed to the Emersonian ideal that "when we discern justice, . . . we do nothing of ourselves, but allow a passage to its beams" (140). As such a passageway, a corporeal person could command, and justify commanding, vast wealth and power.

Thus we find a conceptual continuity between a metaphor of 1836 and an economic artifice of 1882. One of the "lessons" of the "Discipline" Emerson went to nature to learn was the "combination to one of manifold forces" (21), the "law of one organization" that would enact the "rule of one art" (25). Emerson's thought, often cited by the trust's defenders and critics alike, anticipated and became embedded in America's business imagination as a formal harmony, what Cook called the "masterpiece" of business organization (4), refusing any prescribed relation to conventional morality. The transcendence of agency by which Emerson's logic and the formal elegance of the trust achieved power placed no intrinsic constraints on application.

The logic of the Standard culminates, and was always implicit in, the Protestant tradition of self-reliance and the logic of transparency. Like the transparent eye-ball, and unlike the liberal, individualist ethos promulgated by Carnegie, the trust artifice allowed Rockefeller, Tarbell observes, at once to disclaim agency and "to know every detail of the oil trade, to be able to reach at any moment its remotest point" (110). She is saying of Rockefeller what Emerson said of the "universal" merchant of "Trades and Professions": "His body is in one spot, but his eyes are turned to all regions of the world" (116). Though located in one place and in one body, the trustee incorporated remote events and forces, comprehending "the entire field of production," as Richard Ely wrote (1902, 12); and he did so with the same motive Emerson urged upon his capitalist—"to satisfy" what seemed to Tarbell, and to all acquainted with Rockefeller, "an intellectual necessity"—and with the same proprietary achievement Emerson imagined for his poet: "the oil business belonged to him." Like the universal man of "The American Scholar," Rockefeller's "field was literally the world" (Tarbell 110). By exempting the self from the responsibilities incumbent upon it when it is known through its sensuous or contractual representations, Emerson's transcendentalism and the trust simultaneously kill and transfigure particular interests to inspire an essential egotism through which "the currents of the Universal Being circulate." The trust enacted the sublime "lesson of power" Emerson went to nature to learn: how to be nothing yet incorporate all, how to theologize particular interests as universal, how to perfect the self in the moment of its disappearance.

Theory and the Agency of the Trust

What are we to make of the isomorphism between the logic of Emerson's transcendent self and that of the trust? This question may seem to raise a theoretical question, probing the political or ideological implications of the fact that Emerson's notion of transcendence is, by my account, the prototype of disturbing and still evolving developments in American financial life. Does this mean that a transcendentalist must be a supply-sider or an Ivan Boesky, that even in its most radical moments American literary culture inevitably propagates oligarchy? I do not find the theoretical question especially helpful or relevant. The analysis it solicits, seeking the politics implied by Emerson's thought, conflates Emerson's commitment to the appropriation of nature with its future applications. It equates acts and consequences in ways that ignore the constitutive discrepancy between them. In needing to specify the direction and constraints Emersonian intuition necessarily imposes upon its application, such analysis commits the error that was the ultimate target of Emerson's critique of liberal culture, the error of formalism: the "fetich" or idolatry of forms as essence and prescription, the notion that an idea or symbolic form designates only one use or means the same thing in all contexts. Emerson himself declined the mantle of systematic ethical theorist, I argued in chapter 2, precisely to avoid the idolatry of formalism. The theoretical question, finally, pursues the conventional, moralistic project—attaching moral significance to mere forms—that Emerson, along with his more politically iconoclastic colleagues like Theodore Parker, Margaret Fuller, and Orestes Brownson, repudiated among orthodox theologians.

Emerson's exhortation to use forms according to the needs of the will theologized an extensive (and always theological) heritage of appropriation evolving since the emergence of a handicraft and merchant economy. That is, his understanding of self and action, however peculiarly expressed in the transparent eyeball passage, adapted—more specifically, superseded and perfected—the liberal understanding of self as proprietary. However true, this proposition tells us only about certain of Emerson's beliefs or commitments; it is descriptive rather than theoretical, which would disclose the implied and total political agenda of his thought, the necessary relation between content and formal application.

Nor should we think that because the trust was formally a transcendentalist institution Emerson's ideal of selfhood can serve only monopolistic interests. To catch insiders like Boesky, Boesky's lawyer remarked, the Securities and Exchange Commission must "be everywhere at the same time," must, that is, be more trust-like than the monopolist (*New York Times,* 23 November 1986, sec. 3:5). If Emerson did not disqualify the capitalist as potential idealist, neither did he insist that intuitive action take the historical form of capital accumulation and concentration. Indeed, because idolatry of the commodity is so tempting, he often inventoried the perils of commerce. In the very late essay "The Fortune of the Republic," for example, paragraphs criticizing the culture's "great sensualism" and immanent oligarchy virtually alternate with encomiums to its democratic achievements. Moreover, in the notion of cooperative agency advanced by radical critics of corporate power (among whom we might include Dodd, who refused stock in the Standard and accepted a comparatively limited salary), we find the logic of transcendental selfhood used to resist the liberal conception of social harmony. Oligarchy triumphed because certain actors were better positioned than others to manipulate institutions, better positioned to enforce their will.

The logic of transcendent agency is a historical phenomenon, not a theoretical issue. Cultural currency used for many purposes, it patterned many documents and actions of the era, literary works as well as a crucial development in corporate history. The concluding chapters will examine two other literary examples of this logic. The first is Dreiser's Cowperwood, arguably the preeminent fictional representative of the mechanics and physics of the trust. The second is Alexandra Bergson, the heroine of Cather's *O Pioneers!* who in her redoubtable impersonality resolutely attributes her speculative success to nature. Both characters undertake morally scandalous action. In Cowperwood's hypothecations, we can observe a nearly pure, though finally abandoned, praxis of the trust. Cather's narrative suggests the way in which the logic of transcendent agency underlies the prime element of the liberal regime, the institution of property, though this informing structure was often denied or elided in liberal discourse.

If the logic of transcendent agency fuels the morally problematic achievements of Frank Cowperwood and Alexandra Bergson, Eugene V. Debs's radical unionism, as we will see, and Silas Lapham's more conventional moral rise by the repudiation of big business also display this logic. Silas's prosperity, because it tempts him to venality, and because, ironically, it precipitates anxiety about the "terrible shrinkage in his values," the values of his paint (264), had "nearly stolen from him" his "manhood" (315), a property presumably redeemed by the loss of his paint business. But Silas's business doesn't exactly fail; rather, it is transformed—by a Dodd-like "assignment" (271) of his assets to younger paint farmers—into what he calls "an invulnerable alliance" (279), a paint monopoly no longer forced by competition to overproduce. His business, and thus his manhood, protected by monopoly, the novel concludes with Silas pondering the predictable question, would you do it again? "Well," Silas replies, "it don't always seem as if I done it. . . . Seems sometimes as if it was a hole opened for me, and I crept out of it" (321). His prosperity ensured by monopoly, Silas is, in the best transcendental sense, a real man, not the true agent of his destiny.

In invoking transcendent agency, a transcendentalist, a naturalist, a critic of patriarchal property relations, a Christian socialist realist, and opponents as well as advocates of the trust share a dream of selfhood perfected in the trust. In the tradition of the transcendence of agency, the classic figure of the age of incorporation—the great individual—was conceived to be eradicated in a Foucauldian "ensemble of actions" that constitute power (1982, 786), underwritten in what Derrida calls "a network of effacement" (1977, 182), as impalpable forces circulate and concentrate in the Emersonian poet blessed with sufficient "powers of combination" (McQuade 83). Considering the attraction of the logic of the trust, the virtues of becoming a transparent eye-ball, we may want to rechristen this era the decorporation of America.

CHAPTER 7

Dreiser, Debs, and Deindividualization: Hypothecation, Union, Representation

Failing to Represent; the Realist's Love of Death

"A man, a real man, must never be an agent . . . —acting for himself or for others—he must employ such" (*Financier* 44).[1] When he thinks that to be a real man one should never be an agent, Dreiser's Cowperwood, hailed by critics as "the apotheosis of individualism,"[2] is rehearsing the logic of transcendent agency. Said by Dreiser to aspire to the power of trusts, Cowperwood is clearly meant to be an updating of Emersonian self-reliance.[3] Frequently called self-reliant and self-sufficient, Cowperwood has contempt for conventional morality and law, which he scorns as simply the institutionalization of "mood," and his contempt is authorized by the capacity of his great imagination to sense "that subtle chemistry of things that transcends all written law and makes for the spirit" (*F* 323). Yet Dreiser's paragon of independent manhood achieves his comprehensive power by the paradoxical occulting of individual agency. If I was correct at the end of the previous chapter in suggesting that Howells's Silas Lapham also secures his manhood in a trust-like transcendence of personal agency, then we need to consider why a realist moralist and a naturalist amoralist share the Emersonian, combinatory logic of transcendent agency organizing the Standard Oil Trust. Historical contiguity is not a sufficient answer. What work does this logic accomplish for them?

I point out that the aesthetic projects of Howells and Dreiser were designed to fulfill the conception of agency informing the Standard not just to illustrate the wide currency of combinatory logic but also to probe further the possible—especially, I will argue, protean—political resonances of this conception. Howells and Dreiser employ this logic for different ends but for the same reason: the source of moral stricture cannot be discovered in nature, or in absolute terms; with virtue not palpably available, it is necessary to appeal to some transcendent source of power. Howells, familiarly, like many other authors of the period, such as Garland or Whitman (in *Democratic Vistas*), denounced popular taste as the enemy of mental and emotional discipline and of democracy.[4] This attack is prominent in *Silas Lapham,* mainly through the Reverend Sewell. Most novels, Sewell protests, display "monstrous disproportion" as "representations of life." Imitating such art, people behave equally monstrously, obeying obsolete values imitated in novels. If, on the other hand, novelists "painted life as it is, and human feelings in their true proportion and relation," Sewell exhorts, they "might be the greatest possible help to us" (174–75).

Reprising neoclassical calls for proportion, Howells would reform how we first experience and consequently represent life. As he formulates it through Sewell, realism consists in the proportionality with which it treats feeling; its truth inheres in its moral propriety deriving from geometric proportion. The moral design of Howells's realist aesthetic is unmistakable in the December 1887 "Editor's Study" column in *Harper's* that I briefly considered in chapter 4. He seeks here a "final criterion" of taste—which refers as much to the production as to the appreciation of art—in order to combat the imitative tendencies of both novels and their readers. Howells laments that most artists "have been rather imitators of one another than of nature." Hence, readers absorb from novels a taste for idealized manners and dreams rather than for real emotions and practical necessity. Howells explains his point by discussing the representation of grasshoppers, which, typically, follows a "conventional" standard of "literary-likeness," not "life-likeness." Readers "brought up on the ideal grasshopper" internalize artificial standards; thus "artificial grasshoppers" in fiction engender "romantic cardboard" citizens among the reading public (155).

Howells's alternative aesthetic consists in "simplicity and naturalness and honesty." If the author "held his ear close to nature's lips and caught her very accent," he would be a genuine "authority" of both beauty and truth, or virtue.[5] As nature's authoritative voice, Howells wrote in a review of Ibsen, art may "dispers[e] the conventional acceptations by which men live on easy terms with themselves, and oblig[e] them to examine the grounds of their social and moral opinions" ("Henrik Ibsen" 435). Only in this way may authors actualize Burke's avowal that "'the true standard of the arts is in every man's power.'" In expanding "enfranchisement," not nearly "perfectly accomplished" (EEC, July 1914, 311), realism aims to help inaugurate "the communistic era in taste," when citizens "verify [art] by the standard . . . which we all have in our power, the simple, the natural, and the honest" (ES, December 1887, 154–55).

Howellsian realism is a moral and political enterprise, based on natural propriety in representation. If his formulation of realism is manifestly neoclassical, it is also profoundly romantic. (Howells invokes Keats before he discusses Burke.) Howells's romanticism consists not in thinking of writing as self-expression—as we will see, this is the opposite of Howells's vision. But like the romantic artist, Howells's realist author penetrates conventional and phenomenal forms to divine natural truth, and he transcends his particular interests to comprehend universal ones. As he stresses in other "Editor's Study" columns, especially those propounding Christmas literature, "the sublime truth" the author seeks can be found only in *"the denial of self,"* in the capacity "to love death" (January 1888, 319).

Yet propriety in representing is not so simple, nor simply achieved, as Howells ordains. Neither Sewell nor Howells can enumerate the formal features of the realist aesthetic, except through an economic metaphor, either in Sewell's "economy of pain" or in the retrenchment of Silas's business.[6] Howells is no more specific in his reviews and essays. Beyond the general advice to hold one's ear to nature and to kill the self, Howells does not elaborate any method for hearing or representing natural and honest harmony. Producing without self is, certainly, a difficult achievement. Moreover, the ontological relation between reader and text, self and form, remains inevitably conventional: reader and self experience the world only through representations, and all action proceeds in some cognitive relation to representations. The Howellsian reader is always an imitator; Howells just wants readers provided right models to imitate.[7] Likewise, the writer, in representing actual manners, inevitably imitates other representations.

Howells undertakes to inculcate propriety through art because, much like Emerson, he suspects the authority and efficacy of representation. Since the world is experienced in conventional forms, he sees no guarantee that aesthetic and moral values can be validated according to a standard we know to be real rather than idealized, ours rather than others': "If this should *happen* to be true" (December 1887, 154; my emphasis), then equality of taste and power is possible; but "self-distrust" and hierarchies of taste abound. To preempt the dominion of convention, Howells recommends to authors an aural standard: music. Since art must imitate, music is an ideal model, because it is represented only by purely formal transcription, as form without representational content; thus music may seem to be experienced not as representation but simply as experience.

Howells is indeed more skeptical of the efficacy and stability of vision and conventional representation than Emerson, who tended to extol the capacity of poetic vision to perceive and redeem integrity and harmony. Howells's anxiety about representation charmingly surfaces in *Silas Lapham* when Bromfield Corey discusses what we might call the absolute value of painting:

> "It's very odd . . . that some values should have this peculiarity of shrinking. You never hear of values in a picture shrinking; but rents, stocks, real estate—all those values shrink abominably. Perhaps it might be argued that one should put all his values into pictures; I've got a good many of mine there." (83)

Corey rehearses here the prevalent concern with the shrinkage of values that underlay, we have seen, William James's conception of self and the evolution of the trust.[8] The obvious value of paintings, for Corey, is their tendency to appreciate in the market. But Bromfield, a disciple of Titian's use of color, is punningly observing a formal aspect of paintings that presumably underwrites their financial appreciation. The values that don't shrink in pictures are finally the values of their colors and color relations, which are absolute because they derive from the physical spectrum. These are the intrinsic values that cause the paintings' market value to rise.

The intrinsic values are absolute values, or real values, precisely because they don't represent anything, and thus are not assessed according to whether their representation is, in the terms of the novel, proportionate (realistic) or disproportionate (sentimental). If one did not attempt to represent, if one assayed pure form, one's reputation might not be subject to the shrinkage Howells's reputation was a generation later for being unrealistic, a shrinkage he himself predicted in *Criticism and Fiction,* when discussing the progression of literary movements (from classicism to romanticism to realism). He writes here that each movement arises and aspires "to express [the] meaning" of phenomena in order "to widen the bounds of sympathy." Each seeks "fidelity to experience" in order "to escape the paralysis of tradition"; each proudly supplants the previous tradition, until it, in turn, comes to be disdained as "effete" (15).[9]

Note that Howells emphasizes "fidelity to experience" as the goal of realist representation, which Howells famously (or notoriously) defined as the "truthful treatment of material" (ES, November 1889, 966). Realism, by this account, is an economy, a management of materials, in no way purporting to be mimetic,[10] as many have assumed, like Norris disdaining "copyist realism."[11] Absolute mimesis would, in any case, not be representation but presentation or transcription, and Howells explicitly repudiated transcription. "Verisimilitude" is an "effect" of art, not a formal feature, he wrote in the April 1887 "Editor's Study." To fulfill its moral aspiration, art must be "true . . . to the principles that shape" life, and go "deeper and finer than aspects." "We need not copy" "the red tides of reality," but should "go to the sources of their inspiration" (ES, April 1887, 826). Howells contended that the "great intellectual movement" he was promoting in the United States is but "imperfectly suggested by the name of realism." Genuine realism "is of the spiritual type like that of the Russians, rather than the sensual type, like that of the French" (ES, February 1889, 491). Accordingly, Howells praises Ibsen precisely for not merely copying aspects. Ibsen "finds the reality far below the surface ideality which we see; and he makes us find it if we are capable of so much" (439). Verisimilitude is a moral achievement, by both writer and reader, and this account of realism abates its antithesis to romance. In 1887, Howells cites Carlyle to make his point:

> "Novel writers and such like must . . . record what is true, of which surely there is, and will forever be, a whole infinitude unknown to us of infinite importance to us. Poetry, it will be more understood, is nothing but higher knowledge, and the only genuine Romance (for grown persons) Reality."

Howells enthusiastically inverts Carlyle's terms to say that fiction proper makes "Reality its Romance" (ES, April 1887, 826).

Howellsian realism, then, as Hamlin Garland characterized "veritism," his adaptation of Howells's understanding of art, is not "reproduction"; it represents "the relations of things" and persons (*Crumbling Idols* 42). As such, its morality, or its truthful treatment of material, Howells wrote in his review of *Looking Backward,* inheres in its "allegiance to the waking world," rather than allegiance to mere fancy and artifice (ES, June 1988, 154).[12] Realist representation is not a purely formal problem but an ethical problem, a question of allegiances. But in insisting on the ethical basis of representation, however democratically intentioned, Howells invited, rather than deflected, future derision that he does not

represent accurately but represents according to sentiment. His "hazardous conjecture" of the December 1887 "Editor's Study" on the final criterion of art is too
astute; as long as representation is in any way a question of relations—among
artifacts and things, or among persons—there surely is "no absolutely ugly, no
absolutely beautiful" (153), and art must submit to vicissitudes of fashion, the
very condition Howells drafted realism to combat.

Howells's occasional response to the instability of aesthetic values is to entertain the attractions of a purely formal (although never transcriptive) standard of
criticism and aesthetic practice. What Corey finds valuable in paintings, for example, has nothing to do with morality or an author's choice. An interchange during
the banquet scene illustrates Corey's ironic sense of the superiority of nonrepre-
sentational art. In a discussion of architectural styles, he gently rebukes the
Laphams' architect for his pride.

> ". . . you architects and the musicians are the true and only artistic creators. All
> the rest of us, sculptors, painters, novelists, and tailors, deal with forms that we
> have before us; we try to imitate, we try to represent. But you two sorts of artists
> create form. If you represent, you fail. Somehow or other you do evolve the camel
> out of your inner consciousness." (170)

Corey's irony is twofold: the efforts of the creative artists may result not only in
unfamiliar but in malproportioned forms; moreover, it is doubtful that architects
or musicians do evolve form exclusively out of their inner consciousness, independent of the influence of tradition. At the same time, the effectiveness of Corey's
irony depends upon the attractiveness of the ideal of purely formal or musical
harmony—such art has value because it expresses an author's privileged, because
devoid of ego, access to experience and the universal.

Hypothecation, Surmounting Representation

It is appropriate that Silas's agrarian economy, and his manhood, should actually
be secured by monopoly because Howells's aesthetic, in which "more and more
the individual ceases in importance" (ES, May 1888, 967), is attracted to the formalist ideals that the trust form exploited. Both Howells's aesthetic and the trust
invoke a formal harmony that exempts individuals from vicissitude and accidents
of convention by typifying them as vehicles of universal interests. (Howells has
some difficulty imagining this condition as the expansion, rather than the sheer
death, of self.) Rockefeller and Dodd insisted no less than Howells on the moral
aim of their formal project. In contrast, Dreiser employs the logic of the trust in
the Cowperwood narrative not to ground moral convention in the universal but
as a metaphor for the impossibility of doing so.

The discussions in *The Financier* and *The Titan* of the squid and lobster, the
black grouper, and the equation inevitable are meant to classify financiering as a
type of the natural process. Dreiser's 1920 essay, "The American Financier," corroborates this point. "The financial type" is "plainly a highly specialized machine

for the accomplishment of some end which Nature has in view."[13] Dreiser observes that financial genius or the "organizing mind," "blazing with an impulse to get some one new thing done," encounters law and morality as obstacles, and is generally occupied with "driving a wagon" through laws, rights, constitutions, or declarations of independence (AF 81–82). Dreiser refuses to entertain the obvious question of whether organizing genius has "ethically the right" to do as it does. Reminiscent of Emerson, and explicitly following Schopenhauer and Nietzsche, he responds that "in spite of all the so-called laws and prophets, there is apparently in Nature no such thing as the right to do or the right not to do" (AF 87).

Charles Yerkes was an appropriate choice as the model for Cowperwood. Yerkes, the ruthless Chicago gas and rail magnate, whose machinations very much instigated the development of the modern loop and elevated system, suits well Dreiser's vision of the American financial type as avatar of the lawless law of nature. Matthiessen tells us that in determining to trace "'the drift of the nation to monopoly,'" Dreiser found Yerkes "the most interesting" of the twenty capitalists he studied (129–30). Matthiessen conjectures that Dreiser's choice was probably less deliberate than he claimed since Yerkes's career was most notorious during Dreiser's Chicago days. Yerkes's career was undoubtedly memorable, but Dreiser was fascinated by the American financier as an ethical type (or antitype), fondly recalling in *A Hoosier Holiday* that "most lawless" period of "the great financiers" (171). Yerkes was distinctive because he flouted the conventional moral justifications sounded by such as Carnegie and Rockefeller. It is Cowperwood's amorality rather than merely his financial cunning that marks him as special.[14] For example, Cowperwood probably could have avoided being tried for embezzling from the city of Philadelphia had Edward Butler not discovered that Frank, a married man, is having an affair with his daughter Aileen, a fact Frank does not rigorously hide.

Cowperwood's defense during his trial is that as merely an agent of the city he was not embezzling because he was not really doing anything. In one way, Frank's defense is semantic equivocation, but more central to Dreiser's concerns, it is by this effacement of agency that Cowperwood consummates individualism. Frank designs his entire career to protect himself from the dangers of vicissitude and shrinkage, whose effects underscore the frailty of individuality as ordinarily conceived. Cowperwood is early on persuaded of the volatility and unpredictability of market movements (*F* 40–41), and he repeats that he wants to get out of stock gambling and stockjobbing (*F* 48, 197). The market's adventitiousness is exemplified by the panic set off by the 1871 Chicago fire, without which Cowperwood would have been able to meet his obligations and hence avoid trial despite Butler's vengeance.

Vicissitudes of valuation symbolize to Cowperwood the vulnerability of individuality. Ordinarily,

> We think we are individual, separate, above houses and material objects generally; but there is a subtle connection which makes them reflect us quite as much as we reflect them [in a relation like weaving on a loom]. . . . Cut the thread, separate a man from that which is rightfully his own, characteristic of him, and you have a

peculiar figure, . . . much as a spider without its web, which will never be its whole self again until all its dignities and emoluments are restored. (107–8)

The passage concisely presents the dilemma of the liberal conception of self we have encountered in many forms: self is independent of its "emoluments" and material properties but also a cognitive function of them. The representational identification between self and property that underlies "individuality" here subjects it to Jamesian shrinkages in the me. If self represents and is represented by its properties, the destabilization of value caused by vicissitude is also an undermining of self.

Dreiser narrates the psychological effects of the self's subjection to vicissitude in a particular economic order in *An Amateur Laborer,* the account of his bout with unemployment, poverty, despondency, and finally neurasthenia in 1903, composed, though never completed, in the year after his recovery. Returning to New York from Philadelphia, he looks at the social order of the city, and at his faltering place in it, and feels that individuals are in the ideal "an integral, though, of course, infinitesimal part of it," like a "microscopic portion" of a whole body (13). But after extended unemployment, Dreiser cannot "bring myself into correct alignment with something" (26), and he begins to feel primarily the infinitesimal quality of individuality rather than its integration into a whole body. He starts to experience the social body, and his sense of self, as discrete rather than coordinated, and therefore no longer a body. In this condition of mere individuality, he "lose[s] consciousness of that old, single individuality" (25), feels more like two persons (one thinking, the other unable to act), and eventually must enter a sanitarium. Dreiser's representation of Cowperwood is in many ways a response to the threat of individuation in an economy of fluctuating, or failing, fortunes (especially for the underclass). Like the protectionists, Emerson, Twain's pilot, Basil March, Howells's Christmas author, Norris and Curtis Jadwin, and the Standard, Cowperwood undertakes a protectionist economy of self. His financial schemes, exercises in hypothecation, are his version of exercising power by taking himself off the market, out of the constraints and risks of representation, virtually out of the sphere of action.[15]

In order to become a real man without being an agent, and thus to surmount "the strange vicissitudes of life" (*F* 500), Cowperwood tries by his extraordinarily subtle speculative dealings essentially to obtain a monopoly on the commodity exchanges. To succeed, he must abandon speculation. Some contemporary reviews and most modern criticism have found the complexity of Cowperwood's scheme confounding, even superfluous, and significant only insofar as it indicates his ruthless cunning and enormous acquisitiveness. In the best discussion of *The Financier,* Walter Benn Michaels has understood how the novel's concern with the relation between speculation, the ceaseless production of desire central to capitalism, and Cowperwood's need to have affairs is representative of late-nineteenth-century anxiety about overproduction.[16] Yet the *specific* properties of Cowperwood's subtle and finally antispeculative scheme in *The Financier,* and of his turn to streetcar systems in *The Titan,* are central to Dreiser's notions of aesthetics and the self.

Cowperwood is the "new . . . type of financier" in "the day of the trust" (*T* 90, 399). His sinking-fund plan in *The Financier* is a study in hypothecation, pledging security for a debt without transferring possession or title, nor without necessarily having title in the first place. (Contemporary arbitraging has evolved from this principle.) His arrangement with George Stener, the Philadelphia city treasurer, enables him both to offer city bonds on the exchange as a secret agent of the city, and to buy them through his own financial organization. As secret manager of the city's sinking fund he can release bonds at will; hence, his own company always buys or sells at a profit and its actions help ensure that the city's bonds will rise to par and thus pay off the city debt. Cowperwood profits on both ends of the transaction, and because as Stener's agent he can borrow on the city's funds and hypothecate these for credit, he profits in enormous disproportion both to his assets and to the appreciation of the city's funds. Constant hypothecation of borrowed (and previously hypothecated) funds, a method Cowperwood calls "pyramiding," unfetters his resources from referring to any tangible quantity, and thus "there was no limit to the resources of which he now found himself possessed" (*F* 110).

His power derives from his success in manipulating the "endless chain" (*F* 112) of intangible relations that putatively but in fact no longer reflect or represent him in the social world. It is this reiterative pyramiding that he calls the "art of finance" (134). His hypothecatory art secures the self by exploding—at once proliferating and annihilating—the representationality of value and property. In the ordinary account of individuality Cowperwood rehearses, self depends on its property because valuation is likewise tangibly founded: value refers to materiality (labor or products). It is precisely the visibility and materiality of value's lineage that subjects it and its attendant self to vicissitude. Hypothecation so complicates and attenuates—subtilizes, in Cowperwood's term—the representational lineage of values that the basis of value can never be finally identified in order to be vulnerable to change; consequently, values may surmount local fluctuation. It is only appropriate, then, that the narrator ascribes to the trust, whose techniques Cowperwood anticipates, "watery magnificence" (*T* 399). Cowperwood's art of finance is designed to make his person and resources fluid, endlessly plastic, finally unaccountable. Self and resources become like the winds, with "no tracing to their ultimate sources" (*T* 188).

In this watery and gusty aesthetic, the self and identity are not dialectical, arising in the dynamic between consciousness and possessions that Cowperwood is trying to surmount. For example, when deflecting one of the attacks on his solvency in *The Titan,* Cowperwood feels his "self-sufficiency" precisely in his capacity to hypothecate (*T* 432). His hypothecatory self-sufficiency, his very subjectivity, consists of the "subtlety" of his diffusion of resources: spiderlike, "he had surrounded and entangled himself in a splendid network of connections, and he was watching all the details" (*F* 157). In this trust-like comprehensive network of vision, he has control without "actual ownership" (*T* 473). Like a trustee in the trust form, he enjoys in hypothecation the power of possession without its public obligations and duties. Having "spread himself . . . thin" (*F* 151), he can observe all the relationships among fluctuating conditions and moral values without identifying himself with or conforming to any particular conditions or relations. Vision, here, as with Emerson's transparent eye-ball, Standard Oil's design, or Howells's Christian aesthetic, effects formal power *without* representation. His power is protected because it is *not* fungible with its material effects. No one can "trace out exactly where he stands" (*T* 401). He thus capitalizes on what we now call conflict of interest without being identifiable with any interest. None of his ventures or hypothecations adequately represents Cowperwood; if they did— if his resources, and thus he, were visible and thus graspable—he would no longer be the strong individual. He is individual and powerful by virtue of his ineffability.

Unlike Howells, more like the Standard, and especially like Emerson, Dreiser forges a hero whose identity is secured not by modifying but by inhabiting vicissitude. Cowperwood's scheme structures a selfhood adequate to what he takes to be the distinctive feature of personhood: "all individuals are a bundle of contradictions" that "alter at varying ratios" (*F* 90, 302). That persons are contradictory ratios of change is, for Dreiser, exactly what makes them natural. As Dreiser emphasized in the essay "Change" (*Hey* 19–23), it is vicissitude, constant change, that defines the self-regulatory character of nature, of which the financier is the type. To adapt to natural conditions, Dreiser, following Nietzsche, dispenses the following counsel: "Learn to revalue your values" (*Hey* 19). He is exhorting pure adaptation, irrespective of moral convention.

Thus hypothecation composes the self in concord with nature's definitive principle of change. What we might call Cowperwood's hypothecated self is a ceaseless revaluation that transcends speculation. He hopes to command vicissitude because he absolutely inhabits and is constituted by it, for his goal is to reside, as he designs the sinking-fund escapade, "on both sides of [the] market" at once (*F* 199). The sinking-fund plan is not a corner on the market, which cannot sustain a crash; rather, it comprehends all fluctuation and profits whether the market goes up or down. Thus he achieves self-reliance without resorting to the claim of "inherent virtue" common (perhaps necessary) to morality and law (*F* 329), a claim tied to the claim that values represent tangibles. Hypothecated selfhood fits an Emersonian model insofar as it has disappeared and comprehends universal forces; at the same time, Dreiser shares Howells's more radical skepticism of the adequacy of representation to signify the universal; unlike Howells, and also unlike Emerson, he disdains the delusory impulse to justify action in universal law.

Dreiser's conceptualization of the self makes for a paradoxical aesthetic of transcription without representation. Cowperwood's "financial imagination" (*F* 84), like, he believes, the artistic imagination, makes one able to "grasp" the "boundless" and often contradictory relations of nature that law, politics, and ordinary notions of art habitually reduce to fixed categories (84, 64). Art and action are thus formulated in terms of appropriation. But the agency of history does not originate with, although it resides in, the self-sufficient "personality" (64) of the artist or financial titan. Law and morality need to ascribe definite agency; otherwise, they have no rationale. Cowperwood is convicted of embezzlement in the name of this limited "orderly and artistic whole" that juries seek (363). But in Dreiser's financier aesthetic, art does not actually represent; instead, although it necessarily takes the form of linguistic or visual representation, its forms do not represent its subject.

One of the most transcriptive novelistic enterprises in American literature (rivaling Norman Mailer's docu-novel *The Executioner's Song,* and providing by far the fullest account of Yerkes's life and career), the *Trilogy of Desire* nevertheless displays and thematizes a profound anxiety about representation. It thus doubly epitomizes the nineteenth-century tradition I have been tracing, as well as an older Christian legacy. Dreiser rejects representation as the office of art or any formal activity. Art, rather, manipulates representations and is an instrument for realizing and producing desire. The site of consciousness, which both represents and is represented, is protected from the limitations and constraints of convention by its interminable networking, which is figured in Cowperwood's "subtle" hypothecation and insatiable sexuality and acquisitiveness. Hence, the *Trilogy of Desire's* title.

Personality and [De]Individuation

Because the hypothecated self is insubstantial and insatiable, Cowperwood's motto, "I satisfy myself," acquires both a paradoxical and emblematic significance. It is emblematic because selfhood emerges in webs or networks of power and attraction, and desire is the definitive quality of this structure of self. It is paradoxical for two reasons. First, with self as a drama of attraction (Cowperwood frequently imagines his life in terms of drama), satisfaction is, as Veblen might say, in principle out of the question. Hence, Cowperwood's need for mistresses, a tendency all the more conspicuous and ruthless in *The Titan,* where he has fourteen affairs, often with the wives of his rivals.[17] Second, since the self is a matrix of relations—or, more precisely, an intangible principle of organization structuring, though not entirely determining, this matrix—the self seeking satisfaction, albeit a powerful individual, is hardly sufficiently stable to be capable of satisfaction. This hypothecated self is, as we saw Emerson write of nature, a phenomenon not a substance.

The individual here is at once agent and sublated. At one point Cowperwood reflects that the atoms composing the "form" of experience "represent an order, a wisdom, a willing that is not of us" (*F* 364). The financial and artistic personality, if irreducible, does not wholly originate the order or will that it represents; it is the expressive site where they become manifest and it exploits the forms in which they appear. In "Personality," an essay in *Hey Rub-A-Dub-Dub,* Dreiser elaborated this logic in order to distinguish the emerging ethos of "personality" from a more traditional notion of character, an Aristotelian sort of selfhood that in chapter 4 we saw Basil March, Howells, and most American economists trying to identify and cultivate.[18]

"Character" is what Washington, Lincoln, and Grant (Grant?) had, a quality we "develop for ourselves." Dreiser remarks that we hear less of character nowadays and more of personality. Personality is what traditionally we should eschew, a capacity "in many realms and forms [that] comes without volition on our part, fate or circumstance causing it to blaze for us whether we will or no." Dreiser's notion of personality, an instance of blazing intuition and capacity, concedes more to fate and circumstance than Emerson would prefer, but the sublime exaltation of personality is comparable to the affect of the transparent eye-ball. Personality is a unique aspiration bequeathed by nature. It is "a sense of power resting on a feeling of capability or usefulness, and hence a right to be." But the source of the capacity and right is not "inherent." "The man of personality . . . realizes the guidance, enmity or favor of not necessarily higher, we will say, but different powers" (*Hey* 107–9). As the paragon of personality, Cowperwood both consummates individualism and also explodes "the myth of individuality," as Dreiser titled a late essay, and individual agency (*Notes* 92–95).

Cowperwood's financial success results from his recognition and exploitation of the conventional desire and legal propensity to see individuals as discrete agents. Cowperwood watches and tries to anticipate the entire complex of actions. His business plan in *The Titan* consummates this principle of self and finance, and indeed, though its details have generally been found even more supererogatory than those of Frank's sinking-fund scam, Cowperwood's Chicago activities transcend his methods on the Philadelphia stock market. The Chicago novel begins with Cowperwood "sick of the stock-exchange," yet determined to exploit "the boundless opportunities of the far West" (*T* 1). He decides that "street-cars . . . were his natural vocation" (*T* 5). His desire to enter (that is, to control absolutely) the streetcar industry, by first consolidating the gas industry, illustrates the differences between his protectionist enterprise here and in the first novel. Gas and street railways are public services tied to cities, especially to "the expansion of the city" (*T* 22).[19] Thus the fortunes of the self are invested in an entity less volatile than the market, and inexorably expansive.

Concomitantly, streetcar and gas lines are less individualistically based than stockjobbing. Whatever the problems tracing ownership in the period, and however much ownership and control were diverging, ordinary speculation and investment always retain the risk of liability, since they point to assets (or the lack thereof) and persons. Albeit "fictitious persons," corporations still point to persons, fictitious or otherwise. This fact, after all, is what returned Cowperwood to fortune after his prison term. During the panic of 1878, Jay Cooke's investment house closes. The problem with Cooke's house, in Cowperwood's view, is that it is "dependent upon . . . one man" (*F* 491). It is such dependence and traceability—that is, such conventional individuality—that Frank seeks to eschew in his obsession with surpassing speculation. In urban utilities he sees an opportunity to disappear entirely, to become an element of the city's inexorable expansion and thus endlessly satisfy his insatiable, because unlocalized, self.

The gas and railway lines symbolize the expanding city, and are crucial to its transindividual circulation. To control them is to control a vast law of nature, "totality" rather than "individuality and separateness" (*T* 70). This is why Frank thinks streetcars exemplify far more than stockjobbing "the vast manipulative life" (*T* 5). In so doing, they fulfill his aesthetic urge. Cowperwood inveterately associates financial achievement with art, and he imagines that as "sole master of street-railway traffic in Chicago" (*T* 186) he will be "laureate" of the city (*T* 6). The streets are at once a piece of art and a subject to be grasped and formed. He "studied these streets as in the matter of art he would have studied a picture" (*T* 4). His artistry consists in exploiting the expansive opportunities of the streets. His method is not just appropriative (which it of course is), but combinatory, in the manner of the trusts. His objective is to merge the disparate lines, to "combine and control them all!" He "busied himself with various aspects of the scene quite as a poet might have concerned himself with rocks and rills. To own these street-railways! . . . So rang the song of his mind." Combining the railways would effec-tively combine "the divisions of the city" into one great metropolis. The thought "was lovely, human, natural" (*T* 167, 169).

Like Emerson's poet in *Nature,* Cowperwood's art integrates "various aspects." In this combinatory achievement, Cowperwood is himself by *being* the city, as a person, in Emerson's view, is self-reliant by being nature. Cowperwood's fascination with the map of the city typifies his notion of the self, diffused through-out the city's circulatory system (underground as well as overground), but never, unlike in stock manipulation, discoverable in and limited to one spot. Like Rockefeller absorbed and emboldened in the matrix of the Standard, Cowper-wood would be everywhere at once; whatever location he occupies on the street map of the city, he can experience his extension to every other part. If stock manipulation and even the corporation are finally all too material, the achieve-ment here is at once idealist and materialist, ideal since the materialism is so com-prehensive. Individuality is secured by inhabiting the "totality" in which "indi-viduality is nothing" (*T* 70).

The ontological economy of street-railway combination would surpass that of mere financiering, achieving the combinatory self-reliance of "personality" through the deindividualization of hypothecation. The East District Penitentiary of Pennsylvania, where Cowperwood serves his sentence in *The Financier,* is designed to deprive inmates of such power. The prison is conceptually, though not architecturally, based on the Benthamite model of the panopticon so dramat-ically analyzed by Foucault. "The 'Pennsylvania System' of regulation" enforces "solitary confinement," "a life of absolute silence and separate labor in separate cells" (*F* 425). The inmate experiences only conformity to "custom." If Cowper-wood's resources are untraceable outside prison, in prison he is subject to metic-ulous inspection of his history, possessions, and body parts. The cells are window-less, and, worst, when inmates move from one wing of the prison to another, they wear a hood "to prevent a sense of location and direction." This hood "almost cost [Cowperwood] his sense of self-possession" (435), which he does lose later when Aileen is permitted to visit him: his mere individuality apparent to the sex-ual force that on the outside satisfies and reinvigorates him but on the inside can do neither, he sobs, "completely unmanned" (462).

In the technology of Cowperwood's unmanning, Dreiser captures the essence of panopticism as Foucault characterizes it. Foucault writes that "each actor" in the theater of a cell (Cowperwood views himself as in a play) feels "perfectly indi-vidualized and constantly visible," in "a state of conscious and permanent visi-bility." "He is seen but he does not see; he is the object of information, never a subject of communication." The agency of power, on the other hand, is invisible and transindividual, virtually agentless. In the Benthamite regulatory system, "the perfection of power should tend to render its actual exercise unnecessary; . . . this architectural apparatus should be a machine for sustaining a power relation independent of the person who exercises it" (Foucault 1979, 200–201).

Foucault is describing how the subjectification and production of individuals take place while the agency of power is displaced and deindividualized. There are strong historical reasons that Foucault's account of panopticism illuminates so well the logic of agency informing Cowperwood's machinations, and those of the trust as well. But the logic of power and personhood I've been examining registers the "automatic functioning of power" with a different emphasis from Foucault's in *Discipline and Punish,* and resembles more his altered emphasis in some later work. The various fictional and historical reiterations of the logic of the trust involve a double gesture that, it seems to me, revises the force of Foucault's description of the carceral vision, and especially the force of many critical applications of Foucault's analysis of the carceral system.

The transcendental imagination of power and the subject I have been delineating differs from Foucault's theorization of power as agentless, wholly "saturating" subjectivity, and hence always "carceral," in the sense that all acts of purported freedom are contained. For Foucault, in *Discipline and Punish,* "the carceral texture of society" (304) and its production and normativization of knowledge regulates the formation of the body and mind of all subjects in a society. Subjects are constituted as individuals by a network of power that "deindividualizes . . . in a certain concerted distribution of bodies, surfaces, lights, gazes" (202). Individuation is the triumph of deindividuation; subjectification is objectification. In the trust era's paradigm of organization and agency, however, designed with varying emphases but for similar reasons and according to a morphologically related logic, deindividuation marks the possibility of individuation and, as the condition of subjectivity, is the condition of power as well as subjection.

Foucault's later work, like *The History of Sexuality* and "The Subject and Power" emphasizes precisely this point. Here, as for Dreiser, Emerson, or the Standard, power is not "something that one holds," yet it is "both intentional and nonsubjective" (*Sexuality* 94). Power does "structure the possible field of action" and subjectivity. But though it is the condition of action and subjectivity, power— or rather the relations constituting it—does not utterly determine or "saturate the whole." Its exercise presupposes resistance—that is, that subjects' actions are not systematically determined; albeit a physics, power is not a "physical determination." In short, freedom is not what power constrains but is the very "condition for the exercise of power," and has meaning only in a network of power ("The Subject and Power" 790).[20]

The work that we call Foucauldian generally emphasizes or exposes the carceral, saturating, containing force of agentless systems. Foucauldian or other poststructuralist work that propounds the primarily policing (or hegemonic) function of social and aesthetic formations at times forgets to ask the question, carceral or policing as opposed to what? If all action and subjectivity take place within cultural networks as their limiting conditions, then all action and thought are, structurally speaking, disciplinary. If there is nothing other than disciplinary (that is, discursive) practices, then to underscore the disciplinary function of a practice is redundant. Such a project retains a residual, romantic impulse, to find a moment or site outside structure and practice whose irruption upon practice might precipitate true change.

What I have been calling the logic of transcendent agency does not imagine or enact a fall from freedom but engenders a sensation of subjectivity and freedom that we should call combinatory. Yet the transindividual logic of power circulating in the period I'm exploring is not power itself, but a trope employed for moral justification. It is one mechanism for claiming authority. If agency was denied, interest was not; the goal was to persuade others of the superior morality of any particular interest, although this interest was often posed as the absence of particular interest. The notion of transcendent agency was at once historically specific and the inheritance of intertwined older traditions, Protestantism, liberalism, and the sublime. Skeptical (or exploiting skepticism) of the adequacy of representations to body forth the forces generating them, the powerful being of transcendent and transparent agency sees all but is individually invisible. More specifically, the agent appears as emblem of transcendent forces. Moreover, power claimed through the logic of transcendent agency can be (and was) deployed to defy and perhaps transform, as well as serve and exploit, the extant disposition and distribution of power. The term *carceral* resonates constraint and political suppression. The trust era's combinatory conception of power and the self, although inspired by the image of subjection Foucault so powerfully analyzes, maps out an appropriative stance toward the deindividualized disposition of power. The fact that it underwrites Eugene V. Debs's conception of political agitation suggests that this combinatory logic of subjectification does not prescribe univocal (political) effects.

Transcending Labor: Debs's Unionism

Combinatory logic performs anti-capitalist work in the figure and career of Eugene V. Debs, generally regarded as the most radical, and also charismatic, labor figure of this period (at least) of American history.[21] Debs's logic and rhetoric resemble that of his abominable enemy, the Standard, even more than Cowperwood's. This historical affiliation does not annul his radicalism but, rather, elucidates its specificity. Debs advocated industrial or revolutionary unionism rather than the craft or trade unionism, which he called reactionary, pursued by Samuel Gompers, Paul Arthur, and others. The difference between craft and industrial unionism is the difference between a combinatory and a merely corporate imagination, and this difference, in Debs's account of labor organizing, forms the difference between radicalism and reform.

The International Workers of the World and the Socialist Party of America, both products of the first five years of this century, evolved from the conflicts within and among the "brotherhoods" and emergent unions of the 1880s and 1890s, conflicts that hampered the 1894 Pullman strike and the corollary Great Northern strike by the American Railway Union, catalyzed by Debs. Debs believed that the refusal by independent labor organizations, like Gompers's

American Federation of Labor, to meld their discrete interests with others, and finally always to support strikers from other brotherhoods or unions, thwarted the labor movement.[22] While serving a six-month prison term for conspiracy after the Pullman episode, Debs, already tending toward the principle of "fraternization," studied the work of Edward Bellamy, Robert Blatchford, and especially Marx and his American popularizers, Laurence Gronlund *(The Co-operative Common-wealth)* and Karl Kautsky.[23] This course of study completed what he called his baptism in socialism ("How I Became a Socialist" 46), and led him to call for a global unification of laborers, most dramatically at the Unity Convention of 1901, where the Socialist Party of America was formed.

Debs believed the political differences between craft and industrial unionism derived from their different principles of organization. In his view, craft unionism, the organizational pattern the American labor movement has since followed, keeps individual crafts, and finally individual laborers, individuated. Individuation (of unions) generates wasteful competiton: each "craft union seeks to establish its own petty supremacy." So, "to organize along craft lines means to divide the working class and make it the prey of the capitalist class,"[24] for craft division organizationally reproduces the bourgeois commitment to contract, to private property, and to individualism and individual agency. The distinct interests of each union often conflict with those of other crafts. Unions may as easily compete as cooperate. Moreover, the goal of craft unionism is increased individual property, greater individual access to the means of production and to the middle class.[25] It pursues prosperity "in [laborers'] individual pursuits" (*Walls and Bars* 269). Thus craft unionism posits and seeks to realize an "identity of interests of the exploited and exploiting classes" (IU 125–26). It is reactionary because it imagines proprietorship and power in individualistic terms, the very terms underpinning the exploitation of labor.

Debs sees this structural liberalism informing important aspects of craft-unionist policy. First, it means that trade unions respect "the sanctity of the contract they have made with the employers" (IU 132). Consequently, unions are reluctant to strike and, more important, reluctant to respect other unions' picket lines, not to mention venturing sympathetic strikes. Second, as Mother Jones also observed, it means a willingness to cooperate—hobnob, in Debs's preferred word—with progressive industrialists who, seeing the virtues of avoiding strikes and violence, established the National Civic Federation, led by Senator Marcus Hanna and championed by such as Carnegie.[26] This organization, which in 1908 sponsored the second, less critical, Chicago Conference on Trusts, was designed expressly to negotiate conflicts between management and labor. Debs objected that this organization "guides [the labor movement] into harmless channels,"[27] but in his view craft unionism's individualistic principles of organization already channeled it comfortably.

The consequences Debs ascribes to craft unionism resemble the amputation at the trunk that Emerson censures in "The American Scholar." Just as the craft movement divides laborers as a class by individuating their interests, it also divides laborers: each laborer, motivated by self-interest rather than class interest, is divided into parts: mind/body, body/limbs, especially hands. Capitalists and craft unions both "contrive to keep you divided" (IU 134). Because the parts of the self and body are not synecdoche for the whole, the contrived self-division of individuated action effectively enslaves labor to capital and technology; laborers are mastered by rather than "masters of the tools" of production (IU 132). Thus in the regime of individual action and property, the burden of labor is finally (predictably) a depersonalization; workers become "merchandise" and "machines" (IU 122; "The Socialist Party's Appeal (1908)" 167). What is foremost sacrificed, Debs repeats throughout his writing, in language familiar from Cowperwood, is not labor time, or even self-respect, but "manhood."

Only through global unionism can you "stand erect in the majesty of your own manhood" (IU 143). If this exhortation invokes a masculinist iconography, it is nevertheless not finally invidious. Long an admirer of Susan B. Anthony, Debs was in the minority in insisting on the inclusion of women as full citizens in unions; in addition, he was virtually alone in denouncing not only the segregation of blacks into discrete unions but their often wholesale exclusion from the labor movement. The goal of industrial unionism was the *universal* supersession of the capitalist mode of production and its attendant system of private property. In the language common to the period, Debs formulated this goal as the redemption of personhood. Union is the condition for becoming "a real *man* or *woman*," as Debs put it in his famous Canton speech (1918), which denounced the war effort and for which he served (at the age of sixty-three) nearly three years in prison for violating the Espionage Act.

Debs foresees the advent of personhood through the submission of individuality to ineluctable historical processes and to the collective will. The "worldwide mission" of the IWW could be realized only through "organization," not by individual action (IU 129). He decried "'direct actionists,'" "anarchist individualists," those who espoused "the 'propaganda of the deed.'" Specifically, he was warning, in this case prior to the 1912 nominating convention, against violent acts. But if "the collective reason of the workers repels the idea of individual violence," it is because any "artifice of direct action" is itself a form of "sabotage."[28] Undeniable artifice, by which he means manufacture, invalidates the collectivity of individual action.

Debs's logic will be by now familiar. When he speaks of socialism as "a collective system of industry," he does not mean just that ownership will be collective; he means to emphasize that industry (labor) must itself be, or rather feel, collective. Collective agency is expressed through an aesthetic that is nearly antiaesthetic. Like individual agency, contrivance or artifice—the visibility of making—will be transcended in the sublation, but also empowerment, of self within the collective will.[29] The collective, global mission will redeem the expenditure and pain of labor. Nor can the collective will be achieved by action, even though Debs recommends tactics to facilitate it, specifically education and discipline. Global union is an "irresistible force" "of evolution" (165, 170). The onset of collective industry is a "world-wide organic social change" (*Walls and Bars* 228) like the rising sun of a new day ("Canton" 279), or like gravity.[30] As individuals, we can "hasten . . . or retard" organic evolution, and of course Debs wants us to "get on the right side of it" (IU 142, 134), since the movement stalls without individual action. But as mere, discrete individuals we are rapidly becoming obsolete, and we will be overwhelmed in any event. It makes sense, then, that Debs exhorts education and discipline, both forms of subjecting the self to greater principles. Debs views persons as natural entities always (and properly) subject to devastating natural force:

> Come as the winds come, when
> Forests are rended;
> Come as the waves come, when
> Navies are stranded.
> ("The Outlook for Socialism" 67)

Only adequate organization can redress the imminent annihilation of individuals, and accordingly Debs occasionally charged craft unionism and direct actionism with leaving laborers "unorganized" ("Canton" 267). Joining the global mission prompts our "redemption" from the enthrallment and nullity of labor and from the annihilating power of natural history ("The Western Labor Movement" 88). It is in these terms—both sublime and Christian—that Debs characterized the fate of the American Railway Union after the traumatic Pullman strike: the union was "overwhelmed but not destroyed," transmuted (into socialism) rather than merely annihilated ("How I Became a Socialist" 49). Debs describes the condition of subjection to collective industry in terms readily identifiable as sublime and transcendentalist. When discrete entities are overwhelmed, they ascend to "an exalted plane" and "feel the ecstasy of a newborn aspiration" (IU 140, 143). In their "absorption" "within one organization," divisions of self, body, and labor are healed, integrated "within one compact body," "the whole body of organized labor."[31] The body of industrial unionism materializes the ideal of Christian unity, as self is annihilated as a singular entity and transformed into the body immanent in the eucharistic wafer.

Union absorption heralds the "rising" of a "disenthralling, emancipating sun" ("Outlook" 60) because it is a typically transcendentalist double movement, of individuation through deindividuation. Subjection of self is also the end of "subjection" ("Working Class Politics" 173). Absorption by the evolutionary organization is itself arousal, "stimulation" to "cultivate self-reliance and think and act for [oneself]" (IU 135). Typically, this self-reliance spurns the "inviolate . . . traditions of the past" that craft unionism reveres ("Unionism and Socialism," *Speeches* 129). Nonetheless, it does submit to new, wider powers. Debs concisely captures the double economy of self-reliance in his essay "The Socialist Party's Appeal," written during the 1908 election campaign: in collective industry, "each individual will be his own economic master, and all will be servants of the collectivity" (168).[32]

Here is the language and economy of sublimity. The burdens of labor, subjection, and finite individuation are, once overwhelmed, converted to ecstasy and the "boundless enthusiasm" of "the Socialist spirit" ("A Plea for Solidarity" 214). Global union achieves for Debs the utilitarian economy of the Kantian sublime we observed in chapter 1: a narrative of subjection and expansion. In learning "how to serve," he declared in the Canton speech, he learned how "to multiply myself over and over again." In "service I can feel myself expand" (258), because the self becomes the site of the inevitable agitation of the American Movement.[33] Each day "the area of its activity widens." "It has no limitations but the walls of the universe." This expansion confers value on the self and on action. "Life has no value" under the regime of individual profit taking (IU 138); only in service to the organization, Debs says over and over in his work, can one feel "worthy" or "worthwhile." The subsumption of individual form and artifice gives value to that form, and is the condition of freedom. Union, then, is a transcendent appropriation of self and particular forms that perfects the idealist component of the Lockean fable of the creation of value.

Debs's paean to the dynamic and emancipatory virtues of "the all-embracing bonds" of one great organization distinctly rehearses the central tenets of the trust, the magnificent artifice whose goal is to annihilate artifice by perfecting it. Debs was explicit that if laborers "acted as one" they would usurp the techniques of corporate action ("How I Became a Socialist" 46). In his article on the 1908 campaign, he magnified the scale of his vision, comparing "the whole industry" to "a giant corporation" ("Appeal" 167). During the 1912 election campaign, he formulated this vision in yet more encompassing terms: "Each industry must be organized in its entirety, embracing all the workers, and all working together in the interest of all," until union controls all "productive forces" ("Danger Ahead" 181). When two years later Debs speaks of "the merging of the hitherto conflicting elements into a great industrial organization" ("Plea for Solidarity" 214), he exactly echoes H. H. Rogers defending the Standard's practices or William Cook decrying them.

He is most explicit of all when lauding the organic nature of the great organization in a 1911 essay in the left-wing organ of the party, the *International Socialist Review.* Much like Veblen, Debs emphasizes the need to evolve institutions according to the forms of technological evolution, and he recommends that unionism pursue "the same improvement and enlargement, the same high modern efficiency . . . that there is in machinery and production." Thus, "the industrial union corresponds to the locomotive, the steamship, the railway and telegraph, and the trust which controls them" ("Labor's Struggle," *Review* 142). In short, Debs was in no way an agrarian utopianist, nostalgic for small-scale individual enterprise and its attendant notion of autonomy. Although the vehement enemy of trusts and combinations, he was hostile not to the form of the trusts per se but to the particular policies and tactics they pursued.[34]

This strategic and finally aesthetic intersection does not betray Debs's political ideals. Rather, to him, as to S.C.T. Dodd, combinatory agency provided the very conditions of democracy, freedom, and finally "moral worth" ("Appeal" 165). In his view, of course, industrial unionism would supersede the trust, because it perfected the combinatory ideals the trustees abused for private gain. Debs's unionism would consummate the, for Debs, still unperfected historical movement toward collective agency, the only condition in which individuals have true worth. In the transcendentalist and protectionist tradition of the trust form, Debs envisions an expansionist organism/organization that achieves "mutual protection" of individuals against "masters" by "tak[ing] possession" of more and more persons and institutions, in turn expanding the power and redeeming the manhood or womanhood of those whom it absorbs.[35] When "the dividing lines" among "the forces of labor" "become imperceptible," he once declared, "Such an army would be impregnable. No corporation would assail it. . . . An era of good will and peace would dawn" because all discord would be absorbed in the whole body of the organization (quoted in Ginger 1962, 130).

Industrial unionism is a "mission of global emancipation" through universal absorption and appropriation. It "open[s] new avenues of vision" and "spread[s] out glorious vistas," he declared in his Canton speech (258). Debs is articulating here his version of Twain's feudal fantasy of piloting and of Huck and Jim on the raft, of the unrefracted scope of Emerson's transparent eye-ball and of the monarchial moment of self-reliance at the end of "The American Scholar," of Curtis Jadwin's corner, of Cowperwood's railway network. "When you unite and act together, the world is yours," Debs foretold (IU 134). The Standard sought no less and employed the same conceptions of self and agency—though distinctly not the same political tactics—to achieve its goal.[36]

The Standard's conception was also called imperialism, or rather the term preferred by executives and diplomats, "missionary diplomacy" (Sklar 83). Industrial unionism is clearly such a missionary political economy. Debs equated the form of craft unionism with its effects, but in the case of combinatory action, he distinguished mere form from its content and affect. Debs saw combinatory agency *among labor* as the only possible challenge to the hated trusts and to the imperialism that aided their extension.[37] Global unionism would be emancipatory because its unity of action pursued a different ethos, that of complete absorption by organic inevitability rather than, in his view, the merely economic and formal absorption that the trust practiced in the service of individual property. Debs envisioned, we might say, kind rather than cutthroat absorption—or more exactly, complete absorption *as* kind absorption—but this is precisely absorption's "great moral worth" ("Appeal" 165).

Debs's combinatory and transcendental conception explains three idiosyncrasies of the American Movement and his participation in it: the visibility and at times prominence of clergy and of religious rhetoric; Debs's withdrawal from internal policy debates; and his devout refusal to call himself a "leader" of the movement. The first two matters have understandably troubled commentators on Debs's career. The third has not, and its reasoning helps to elucidate the others.

Debs frequently derided the leadership of the trade unions, and not just for socializing with the Civic Federation. "The natural tendency of officials is to become bosses. They come to imagine that they are indispensable" ("Sound Socialist Tactics" 196). The problem was "officialism," "self-importance," thinking of one's representative function as constituting rank-and-file interests. Opposing the syndrome of the "'great man,'" where "principle is subordinated to personality," he admonished members not to look "for some mythical Moses to lead them into the promised land."[38] "I am not a Labor Leader," he always insisted.[39] He called himself, instead, an agitator and organizer; he was an instrument of organization and change, catalyst in "the alchemy of social order" ("Outlook" 67). As he said in accepting the party's first nomination for President of the United States in 1904, he was honored "not in an individual capacity, but as your representative," as servant of the "supreme" "collective will" of the membership and of history.[40] "I have simply been commanded . . . to perform a certain duty," he wrote in accepting the 1916 nomination (234).

The aesthetics of the unionist's representativeness is, I have suggested, antiaesthetic, resisting "artifice," any individual or direct action. It is instead idealist and teleological (something akin to Howells's vision), not transcriptive but expressive and affective. Debs felt he was the "congealed, tangible expression" of the Socialist Party ("On Race Prejudice" 98), just as the party was only the congealed expression of the American Movement. For this reason, Debs never attended nominating conventions and always withdrew his name from consideration well before each convention. To accept the inevitable command, to be worthy of the arousal that accords him individual worth (see *Review* 1904, 692), he must withdraw from participation.

This conception of agitation explains as well why Debs absented himself from most internal party conflict. He could not be representative if he initiated action. Moreover, if the party consisted in "the merging of the hitherto conflicting elements into a great industrial organization" ("Plea" 210), its disputes must seem to resolve according to their own dynamic rather than by individual action. He who is the congelation of the supreme will of history must not immediately influence party action, and can intervene only as the expression or vehicle of the future will.

Debs's vision is in some ways more Augustinian (that is, permitting even less individual intervention) than the Protestantism of the Kantian sublime, Emerson's eye-ball, and Dodd's trust. It is nonetheless identifiably Christian, and somewhat notoriously, the American Movement, unlike its European counterparts, welcomed clerical input. Its apocalyptic version of historical materialism ("Come / as the winds come, when / Forests are rended") did not, in the view of its apostles, betray historical materialism but elicit its already Christian basis: old forms, bearing their own contradictions and the principles of their supersession, are not destroyed when overwhelmed; they are transmuted and redeemed in another form. Debs compared the American Movement to "the early days of Christianity" ("Appeal" 165), and applauds the likelihood that members of this insurgent group will have to withstand persecution (IU 142). The sense that individual sovereignty will also be absolute service to an inexorable and global will, he says, was "taught by Christ nineteen centuries ago" ("Appeal" 168). When he denies he is a leader, and warns against seeking one, Debs's words, like Emerson's, recall Christ's refusal to specify, as Exodus had endlessly specified, formal rules obedience to which constitutes righteousness. Emerson insists that the true teacher (Christ) does not teach but provokes. So Debs: "I do not want you to follow me or anyone else; if you are looking for a Moses to lead you out of this capitalist wilderness, you will stay right where you are. . . . if I could lead you in, someone else would lead you out" (quoted in Reynolds 71). Like the unionist, and unlike the leader and lawgiver Moses, Jesus was, in overturning the moneylenders' tables in particular and the Pharisees' authority in general, a "vital agitator" (quoted in Ginger 1962, 419).

In the narrative prefacing Debs's *Writings and Speeches,* Debs's first biographer aptly intuits that "to teach, to serve is [Debs's] mental and moral mission" (Reynolds 71). More specifically, his mission was to inspire, especially to inspire agitation. Ordaining and legislating were too rigid to be Christian, or rather Christlike. Given Debs's sublimation of self in the movement, his willingness to suffer for it—serving a jail sentence in his sixties, repeatedly damaging his health by a brutal appearance schedule—and his refusal to act as leader or ordain policy, it is perhaps not entirely blasphemous that Debs occasionally imagined himself as Christ. Usually such blasphemy was indirect. In recounting a story about his incarceration after the Canton speech, for example, he does not reveal that his nickname in the Atlanta penitentiary was "Little Jesus" (Ginger 1962, 411), but he does profess his legendary effect on other prisoners, his kindness inducing reciprocal kindness in lifelong criminals, his cooperative disposition inducing small-scale democracy in cells he shared. One counsel he regularly dispensed was to cleanse speech. One black recidivist answered, when asked how he had "cleansed his vocabulary, " "Mr. Debs jest asked me to. . . . He is the only Christ I know anything about" (*Walls and Bars* 97).

Debs's theological bent was not in any way unique among the most radical and militant agitators, nor was it an embarrassment to them. The career of Mother Jones offers an interesting comparison. Jones took the principle of "ceaseless" and "continual agitation" to such a degree that she held no permanent post in a national or local office (153). Moreover, as consummate agitator and servant of labor, she denied that she effected any action. For example, she once organized a group of women to compel their own arrest at a protest, and then to attract public sympathy by singing ceaselessly in jail. A sheriff tells Mother Jones he wishes she had brought him many men rather than these few, singing women. Resisting the baited remark about gender, Jones promptly denies that *she* had brought the women: the mining company and its purchased judge did! These two aspects of her imagination of union—as agitation that involves the effacement of agency—culminate in a baptismal organization ritual that she administered spontaneously for an audience. After a rousing speech, the audience begins chanting, "Organize us!" But official organizers refuse to enlist them because they lack dues. Renouncing the officials, Mother Jones invents a "ritual," and the protesters "raised their hands and took the obligation to the Union" (155–56). Like Debs, whose speeches often were said to empty the churches,[41] Mother Jones felt her work to be salvific in the deepest religious sense.

Mother Jones inflects her agitation differently from Debs, insofar as she often imagines her office as maternal, whereas Debs's sense of his office is more conventionally priestly. Mother Jones's tactics often exploit stereotypes and pieties of feminine gentility. Armed with brooms, aprons, and female sweetness, the striking women she led radicalized domesticity against genteel politics. Debs's devotion and tactics are more customary. Moreover, if Debs could occasionally see himself as Christ, Mary Harris Jones did not, although even in her nineties she led march-ers through snow and sleet. She was always "Mother" to the union people. But if the two agitators had differently gendered rhetorical and strategic arsenals at their disposal, and if Mother Jones could pursue more iconoclastic tactics, partly because she had no official connection with a union, the two shared basic, and in their view essential, notions of agency and organization. Individual action, however distinguished and charismatic, was not the work of the individual; and union was at base a conversionary ritual, enabling the self by obligating it within a vast network.

Bars, Stars, and the Residue of Individualism

Both Debs and Cowperwood pursue the self-protection and self-perfection of combinatory logic, but I do not mean to suggest that their undertakings, however isomorphic, are identical. The differences between them do not lie in their general notions of organization and self. Early in *The Titan,* for example, Peter Laughlin, an old-time speculator whom Cowperwood enlists to help him acquire gas holdings, recognizes that he is "only a trader by instinct," not an "organizer" like Cowperwood (*T* 25). Like Debs, Cowperwood scorns the random particularity of individual trades. Nor, perhaps more surprisingly, do Debs and Cowperwood differ on the significance of desire to the powerful, deindividualized self. The Debs legend celebrates his supreme sympathy; Frank's ruthless self-sufficiency displays this capacity as well: he could "fit himself in with the odd psychology of almost any individual" (*T* 21). Howellsian liberalism would identify a sympathetic disposition, and the self in general, with curbed desire. For Debs as well as Dreiser, however, sympathy and the self dispersed in webs of relations entail the magnification of desire. Cowperwood, supreme practitioner of combinatory economics, quite obviously behaves by proliferating desire. Debs was a staunch opponent of the capital accumulation Cowperwood endlessly pursues, yet he too urged desire (of a particular sort) as a political imperative. Remaining "unorganized and content" reinforces the thralldom of labor ("Canton" 267), and is "treason to your manhood." Instead, "increase your discontent" (IU 133), which both inspires and is inspired by organization.

But though they share a notion of conglomerate action and self, Debs and Dreiser emphasize different starting- and end-points. Unlike Dreiser, Debs never calls the personhood achieved in union "personality" or "individuality," and by attending to this difference we can discern in Dreiser's vision the residue of liberal individualism. Throughout this study, I have tried to show that the conceptual tension of liberalism is the ideal distinction it would maintain between inevitably interdependent variables: reason/desire, private/public, freedom/constraint, self/other. One term is appealed to in order to establish the value and validity, but also the independence, of the other. The source of value is held to be the rational self, for example, even though desire is a central component of value. If not resolved, this typical contradiction jeopardizes claims to have and know value. The combinatory logic we have observed in various forms, most prominently in transcendentalism and the trust, traverses (without resolving) the antinomies of liberalism.

Debs's industrial unionism is a supreme example of the dissolution of individual differences, which, in Debs's view, effectively secures individuality and deliberateness. But Dreiser relieves the conflicts of liberalism without fully embracing the transcendental paradox of the transparent eye-ball, the trust, or industrial unionism. In Dreiser's vision, desire overwhelms reason and, correlatively, self overwhelms society.

Cowperwood's hypothecated self, protected by its diffusion in inevitable or natural structures (the market or, better yet, urban growth), finally stands singly, as personality. Emerson's transparent eye-ball can feel like harmony because the relation between the Me and the Not Me approaches (though does not *absolutely* achieve) mutual synecdoche. But in Dreiser's vision the distinction between self and society is allayed only when the Not Me (here, the city) becomes synecdoche for the Me; and, as the legal realism emerging in the period posited, all principles of reason and law become synecdoche for desire and personal mood. With his Schopenhauerian and Nietzschean notion of desire and will, Cowperwood is heir not so much to Emerson's exaltation as to Ahab's monomania, which is clearly a radical defense against brute shrinkage in the Me.

Cowperwood's understanding of self finally retains the residue of the conventional desire for an untrammeled self that is the generative center of experience and value. For Frank, the goal of amalgamation is individuality in the absolute sense: his "conception of individuality" impelled him to meticulous and exhaustive tactics "that would carry him finally to the gorgeous throne of his own construction." When Cowperwood is enthroned, not only would the vast combination be ascribed to him, he would achieve equilibrium. Cowperwood seeks finally the "spiritual equipoise" he thinks he beholds in Berenice. "My ideal has become fixed," a "pole-star." This development in his vision "spoils other matters for me" (*T* 485, 463, 465), and he begins to value art not as emblem of the appropriative and conglomerate nature of self and relations but as the ideal expression of absolute beauty. His antirepresentational, hypothecatory aesthetic becomes neoclassical.

This shift in Cowperwood's aesthetic principles manifests his residual liberal commitment to a notion of self, independence, and ideality that hypothecation annuls. In this conventional individualism, desire is the mark not of the self's supersession but of its irreducible singularity, and desire feels like the lack that, in brute form, it is: "life at its topmost toss irks and pains. Beyond is ever the unattainable, the lure of the infinite ache with its infinite ache" (*T* 20). Dreiser is expressing what Derrida in another context calls "the teleological lure" of consciousness and essence, which amounts to the fantasy that one's intention could absolutely govern events and experience (1977, 192). Cowperwood's residual ideal of individuality literally desires to possess infinitude, and therefore he continually encounters its sheer unattainability.

This feature of Cowperwood's conflicted psychology partly explains his desperation in prison. Ironically, the individuating discipline of the penitentiary reminds him of the condition he most seeks because it is most impossible. He feels his singularity most when he wears the hood and also when he stargazes. Cowperwood's wealth obtains him a cell permitting study of the heavens—a standard trope in Western writing at least since Dante for the relation between self and larger forces. The "fancy" he takes in astronomy indicates his remoteness from Emerson's comprehensive vision. He wonders whether "the peculiar mathematical relation" of the stars

> could possibly have any intellectual significance. The nebulous conglomeration of the suns in Pleiades suggested a soundless depth of space, and he thought of the earth floating like a little ball in immeasurable reaches of ether. His own life appeared very trivial in view of these, and he found himself asking whether it was all really of any significance or importance. (*F* 468)

Cowperwood can temporarily shake off these rather hackneyed "moods" because of his vital and material sense of self. But this scene indicates his adherence to his singular individuality. Unable to apprehend the "conglomeration" of relations that constitutes the heavens, he does not, unlike Emerson, feel power in the sublation of self that the remoteness of the stars portends. Finally, Frank's vision is carceral in the sense Foucault formulates in *Discipline and Punish*; more specifically, it displays an anxiety about self similar to Twain's: he feels that any compromise or subjection of the self is imprisonment.

If Cowperwood succumbs to the individuating discipline of the penitentiary and to the stars' sublation of his personality, in contrast, during the three years he served for the Canton speech, Debs felt empowered by his absorption in the body of the prison. The posthumous *Walls and Bars* (1927), Debs's only full-length piece of writing, narrates his prison term while analyzing crime and the prison system as epiphenomena of the wage system. Debs describes the process of incarceration in ways similar to Dreiser's and Foucault's. So that each prisoner is imprisoned in "his own loneliness and isolation," every movement and body part are scrutinized and "fraternity" is punished (132, 65). For Debs, as for Dreiser and Foucault, the rigorous individuation of prison life is also a deindividuation and unmanning. Just as workingmen are mere "hands," prisoners "lose their identity" and become "cases" (83). The individuated prisoner is, in Debs's view as in Cowperwood's, "stripped of his manhood" (73).

Debs is saved from such a fate by his faith in union, which he readily confesses is also a religious faith in "the triumph of the spirit over the material environment under any possible circumstances." His prison sentence is a service to his "loving comrades" (50). As a result, "As my cell became my world, and I understood its limitations, it began to expand and I so adapted myself to my prison situation that the steel bars and gray walls melted away" (103). His "ideals" of duty "sustained me inviolate in every hour of darkness and trial" (208). Outside prison, laborers are so enthralled to capital that "no star sheds a ray of hope for them" (*Debs Speaks* 139); inside Debs's cell, however, although he cannot see the sky, Debs experiences the march of heavens and of time, symbolic of his salvation. "Often at night in my narrow prison quarters when all about me was quiet I beheld . . . the majestic march of events in the transformation of the world" (*Walls and Bars* 229).

Debs envisions the sublime march of events without the material sensation

Cowperwood at once needs and bemoans. If visionary, however, his redemptive vision is not fancy but metamorphosis. His presence in the prison inspires, we saw earlier, democracy and kindness among the prisoners. Factiousness, crime, and drug use virtually disappear, replaced by fraternity. But Debs, typically, "arrogate[s] to myself no importance whatever" for the transformation of his fellow inmates. "They did vastly more for me than I was able to do for them" (271). His vision and the "redemptive influence" (269) of his presence are to Debs "prophecy fulfilled" of "the intellectual and spiritual estate to which all human beings are heirs who live in accordance with the higher laws of their being" (229). Socialism, socialist, and agitator have become what they always at base were— faith, convert, and preacher.

Debs's narrative concludes with a ritual exodus and epiphany. When he is released,

> 2,300 prison victims . . . spontaneously burst their bonds, as it were, rushed to the fore of the prison on all three of its floors and crowded all the barred window spaces with their eager faces, cheering while the tears trickled down their cheeks. . . .
>
> In that brief moment prison rules were stripped of their restraining power. . . . Men and women on the prison reservation, including the officials who bore witness to that unusual scene, stood mute in their bewilderment. (270)

Here is a convention-rending outburst to do Emerson proud. Typically, Debs attributes the prisoners' "unparalleled demonstration" not to anything "of a personal nature" but to their treatment according to humane and "scientific" precepts (271). His rhetoric combines principles of universal incorporation and Taylorism to epiphanic purpose, the liberation of selves in the service of the one great organization.

Cowperwood cannot surmount an ethos of contract. At the end of *The Titan,* he vainly ventures to obtain a fifty-year streetcar franchise, a sanctioned monopoly on the system. His ability to secure self-reliance in the inexhaustible city depends upon the tangible, and hence traceable, representation of contract and institutional authority. He remains, finally, imprisoned by the conventionality from which he would stand out as exceptional. In contrast, Debs's term in prison culminates his career as servant/agitator. Organizational bonds, spontaneously absorbing inmates, transcend customary regulations. Unlike Cowperwood, who needs the city bureaucracy, Debs did not need the Socialist Party to license his agitation. Indeed, his visionary organization in prison arose precisely as the Socialist Party was disintegrating into factions.

Debs's prison gospel thematizes the combinatory logic of transcendent agency structuring Emerson's vision and the trust form. It also exemplifies the prophetic foundation of this logic. In the submission of self to greater powers, customary bonds melt and greater individual authority issues. We might call Debs himself the embodiment and apotheosis of the trust form, surrendering personal agency while ascending to folk hero. Of course, Debs was trustee of labor, not the Standard. It is Debs's differently applied sense of self and agency—his dramatically antithetical political agenda—that distinguishes his apocalyptic narrative and folklore career from Rockefeller's and from what is finally Cowperwood's tragedy. If Cowperwood's fate and genre manifest his residual commitment to a conventional ideal of individualism, Debs's redemption from the divisions of the liberal order more closely resembles the transcendental organization animating the trust. Yet Debs's narrative of personal relinquishment, by which he transcends the conflicts informing the liberal vision of value, self, and representation, may appear to a postmodern, secular world too visionary, too redemptive. Willa Cather's *O Pioneers!,* a novel about owning the land, is a less epiphanic appropriation of the liberal habitus. As directly as any other figure of the period, Cather comprehends the sublime relinquishment informing proprietorship, self, and representation in the regime of liberalism.

CHAPTER 8

The Natural Law of Property: O Pioneers! and Cather's Aesthetics of Divestment

Marking the Land

O Pioneers! begins with the wind, which in 1883 is threatening to blow away Hanover, Nebraska, a little town serving homestead settlements. Returning to the Bergson homestead, Alexandra, her brother Emil, and their neighbor Carl Linstrum pass homesteads whose sod houses, low-slung shelters made of the land itself, crouch in a hollow to survive the elements, their very materials marking their subjection to nature. "The great fact was the land itself" (15), "a wild thing that had its ugly moods; and no one knew when they were likely to come, or why" (20), only that "the land wanted . . . to preserve its own fierce strength." Nature's unpredictable impulses to self-preservation "overwhelm[ed] the little beginnings of human society that struggled in its somber wastes" (15). In this novel, then, Willa Cather introduces the relation between nature and culture within one part of the sublime tradition, in which the initial experience of nature discloses the limits of human faculties and threatens their health and continuance.

The thwarted human attempts to claim the sublime are figured as inscription. "Men were too weak to make a mark here" (15). The sod houses common on homesteads finally "were only the unescapable ground in another form," the roads "but faint tracks in the grass." "The record of the plow was insignificant, like the feeble scratches on stone left by prehistoric races, so indeterminate that they may, after all, be only the markings of glaciers, and not a record of human strivings" (19–20). Claims on nature are writing, records of labor, marks of the difference between self and world; "the landscape," Judith Fryer has written, "is for Cather where representation begins" (249). But human marks here fail to signify because the unfriendly "Genius" of nature so naturally assimilates them to itself. Cather's figuring of labor as writing is not just an exaltation of her craft. It exemplifies the condition of homesteaders, faced with claim-filling procedures so complicated as to perplex land registry officers as well.[1] In homesteading and on Cather's prairie, only when labor is embodied in writing does proprietorship begin. Writing is the true mark of labor, of self.

What is so "depressing and disheartening" about the prairie's erasure of "human landmarks" (19) is that their illegibility extends to the men and women who scratch them. In *My Ántonia* the fact that the land is "outside man's jurisdiction" causes Jim Burden to feel "erased, blotted out" (7–8), an "erasure of personality" Cather herself experienced when transplanted from Virginia to Nebraska in her tenth year, 1883, the year *O Pioneers!* opens.[2] Here, Alexandra's father, dying now that he has at last struggled out of the debt so typical of the homesteader's lot, spends his time "trying to count up what he is leaving" behind (16). He "looked up at the roof beams he himself had hewn, or out at [his] cattle" and "counted" them "over and over" (22), as if his labor had given birth to them. John Bergson's conflation of property and generation personalizes the lineage of property, labor, and identity so famously embedded in the Lockean justification of property underwriting the liberal, natural-rights tradition: property is created by labor because labor is the expression of one's inalienable property in oneself. But counting up the marks he leaves behind indicates to John Bergson the very opposite of self-expression, extension, or continuance: "He knew where it all went to, what it all became"—dust (25). On the prairie your remainder is not integral to you, your ratification and regeneration, but your negation, unmistakably separable from you, its endurance a reminder of your impermanence and subjection. This separability, the inevitable alienation between self and property, is what Cather figures by inserting writing into the natural-rights equation of identity that joins self and property through the copula of labor. For Cather, that one's labor, and hence one's identity, is experienced only through acts of representation emphasizes the difference rather than the identity between the terms.

The novel thus begins by at once linking and sundering writing, identity, and proprietorship. This double action implicitly challenges the Enlightenment premises that underpin the liberal social order and conception of identity. Liberalism is known for conceiving the self as interiority, a space of intrinsic and hence inalienable qualities, faculties, and properties. With property the direct expression and extension of these qualities, the right to property may be viewed as absolute. This liberal vision of self is epitomized in two reform movements Cather's novel intersects, homesteading and the women's rights movement. The compelling claim of nineteenth-century agrarian agitation lies in the Enlightenment premise that labor is the mark of and thus belongs to the self. This was the founding principle as well of women's rights advocacy of the period, as it is of contemporary court and legislative challenges to discriminatory employment, financial, and educational practices. Both movements contemplated that political amelioration—a less competitive and exploitive society—would result from the natural-rights reforms they pursued.

Recent discussions of women settlers' writing, foremost by Annette Kolodny, make clear the common foundations of agrarianism and women's proprietorship. In this view, a feminine apprehension of nature, sympathetic and cooperative, would engender a less conflictual social practice because it would be more respectful of—more continuous with, an extension of—its resources (nature and persons). Female culture on the frontier would bring to fruition the liberal vision. In *O Pioneers!,* however, the prairie resists the natural-rights vision of ownership as self-extension so basic to the Constitution's inalienable rights. Moreover, because nature is so inhospitable to human interests, proprietorship, even by a woman, remains an act of appropriation—affiliated, that is, with the terms of the patriarchal tradition of the Founding Fathers and western expansionism. The virtues of natural rights and feminist reform seem absent from this materially agrarian depiction of female proprietorship.

This disowning of natural-rights and women's-rights premises seems rather curious in a novel about a successful female landholder. Moreover, as is well known, *O Pioneers!,* though not Cather's first published novel, is the first she considered her own achievement (preface to *Alexander's Bridge* v–ix). Both narratively and artistically, then, as Sharon O'Brien has recently argued, the novel registers an ascension to identity and authority. The close of Cather's career seems to complete this ascent. Having discouraged the preservation of her letters to friends, Cather prohibited in her will the publication or citation of those surviving, and succeeded more than most authors in controlling her future persona. The novel and Cather's career appear, then, to commemorate the natural-rights conception of ownership and attendant conceptions of identity and independence. Instead, I believe, the novel and Cather's testamentary performance exhibit a different sense of what it means to own, neither endorsing classical patriarchal proprietorship nor, as Alexandra's well-known "love" for the land might suggest, supplanting it with a domestic, feminine conception. Foregrounding and exploring the antinomy between autonomy and authority inherent (but elided) in the familiar liberal notion of proprietorship, the novel is a meditation on the interior structure and limits of property.[3]

Two Accounts of Property

Alexandra Bergson is unique among capitalists in American literature between 1880 and 1915, and not just because she is a woman who accumulates property. Her business practice is unmarred by the unscrupulous behavior or melodramatic catastrophes of a Silas Lapham or Frank Cowperwood, of Hetty Green, "Witch of Wall Street," of Joseph Kirkland's Zury, or the characters of H. B. Fuller, Robert Herrick, Hamlin Garland, or even Wharton. To understand Cather's achievement, we must understand the phenomenon of Alexandra's success. Her relation to the land is characterized in a frequently cited passage as "love": "For the first time, perhaps, since that land emerged from the waters of geologic ages, a human face was set toward it with love and yearning" (65). Inheriting her father's "Old-World belief that land, in itself, is desirable" (21), Alexandra "expresses herself best in the soil" (84), and this quality explains much of Carl Linstrum's attraction to her: "You belong to the land," he tells her when they decide to marry at the end of the novel (307).

Alexandra's love for the land has been thought to bespeak a true agrarianism, which we may contrast with the attitude of Howells's Lapham. Silas, we will recall, defends his controversial, although common, advertising method—painting signs on rocks and hillsides—by mouthing his vernacular version of the Lockean glorification of appropriating the land: the landscape "was made for any man that knows how to use it" (14). Emerson's remark that nature "is made to serve," "meekly" receiving "the dominion of man . . . until the world becomes at last only a realized will—the double of the man," expresses, however more elegantly, the same appropriative vision (*Nature,* McQuade 22). Numerous texts of the domestic tradition posed their domain as an alternative to this appropriative attitude and the competitive culture it underlies. For example, in *The Country of the Pointed Firs,* by Cather's supporter Sarah Orne Jewett, to whom *O Pioneers!* is dedicated, the town of Dunnet Landing, centering on Almira Todd's herb practice, affords the novel's narrator relief from the nervousness she feels in Boston's commercial culture.

Alexandra's sympathy with the land displays without reconciling the contradictory impulses of both Lapham and Mrs. Todd. With her father and Jewett she values the land in itself, but unlike Jewett's characters she engages in commerce and expands her holdings; yet her abiding faith in the soil does not issue in the comfort that Silas and her father desire from owning property and that Dunnet Landing's primitive herb economy fosters. If Silas and her homesteading father represent a patriarchal agrarianism, and Dunnet Landing an alternative feminine agrarianism, Alexandra's practice is nonagrarian. She loves the land not for its service but for its resistance to human will. Her love for the land is thoroughly Eurocentric, indicated by the omission in the very passage about her love for the land of the Indians, who certainly inhabited the land sometime after the geologic ages. As Blanche Gelfant writes, this "hyperbolic sentence" bespeaks a "metaphor for appropriation" (viii–ix). Nor does Alexandra's love for the land inhibit her from trying to acquire as much of it as possible. In 1820, Colonel William H. Jackson of Athens, Georgia, tried in his will to bequeath to one of his trees itself and its plot of land.[4] No such sentiment touches Alexandra. She wants to acquire as much land as possible in order to free her family from service to the land's whim, and she pursues this goal by speculating in land.

And Alexandra is an accomplished land speculator. She prefers not to work, and her goal is to make the entire family "independent landowners, not struggling farmers" (67). Even as a young girl she "followed the markets" (23), and she believes one should watch the "shrewd ones" who buy "up other people's land [and] don't try to farm it" (68). The passage describing her love for the land inaugurates her first speculative venture, mortgaging the family's 640 acres (acquired through homesteading), buying farms surrendered during the drought and panic of 1885–1886, waiting for values to rise, and then selling off "a garden patch anywhere" to redeem the rest (67).

This plan's success ultimately dissolves the family unit, after a dispute over whether Alexandra can do as she pleases with land she has acquired beyond that of the original homestead. In this confrontation, her mulelike brothers employ a confused series of arguments opposing their proprietary rights to speculative and feminine interests, and their confusion illustrates the contradiction informing the homestead movement and common understandings of property. "The rest of your land," cries brother Lou. "Did n't all the land come out of the homestead? It was bought with money borrowed on the homestead, and Oscar and me worked ourselves to the bone paying interest on it" (167–68). Oscar and Lou claim title to the expanded holdings because their bodies, incised by their labor, have seeded it. As John Bergson had imagined his cows were generated by his labor, here the new land "come out" of their work, and this alluvium belongs to them. The brothers' agrarian premise, that manual labor begets value and property, epitomizes the Lockean premise that property is created by mixing with nature one's labor, an inalienable part of the self—a property in one's own person—because it is the expression of divinely granted reason. As Oliver Wendell Holmes wrote, this view holds property to be "an extension of the *ego*" (*Formative Essays* 180).

When Alexandra will not yield, the brothers intensify the genetic aspect of their argument. "Everything you've made has come out of the original land that us boys worked for, has n't it? The farms and all that comes out of them belongs to us as a family" (168). Tracing title back to the original homestead recovers their original investment of labor, subject to no alienation by contract, and accruing with every addition, however acquired. This move elevates proprietary rights of labor to a divine right, indivisible and indissoluble, their feudal and aristocratic basis indicated in the insistence that property belongs only to, and is transmitted only through, families.

If this strategy perfects natural-rights logic, whose goal is clear and unchallenged title, Alexandra, impatiently waving her hand, rejects it. "Go to the county clerk and ask him who owns my land, and whether my titles are good" (168). When the sons married, the three children signed agreements individuating their titles, and Alexandra is reminding her brothers that property is distinguishable from congealed, ego- and blood-conveying labor. Alexandra is voicing what legal theorists called at the time a "legal theory of property" (Ely 1:121–26), also known as conventionalist. This conception of property, controversial for reasons that I will examine, derives from Jeremy Bentham, who argued "that there is no such thing as natural property . . . [property] is entirely the work of law" (137). Alexandra's titles are protected because the force of law excludes others from attaching her property at will.

Alexandra does invoke natural-rights logic, noting that her financial management is labor.[5] Yet she entwines this appeal with an appeal to sovereign authority because her brothers deny that management is labor and, finally, that women can own. Only fieldwork qualifies as "real work," they say, and "the property of a family really belongs to the men of the family, no matter about the title . . . because they do all the work." "Good advice is all right, but it don't get the weeds out of the corn" (169–70). Management is effete, and labor's production of property is a male parthenogenesis. These sentiments are integral to the brothers' patrimony. Recall their father's dying visions of his labor's giving birth to his cows as he had his children. Clearly he did not beget his cows, nor is it clear that he, rather than his wife, gave birth to his children. Indeed, a misogynist impulse inspired the entire dispute: Alexandra is "keeping" the unproductive artist Carl (166), and Oscar and Lou worry she will marry and parcel out the "family" land.[6] The brothers' clearly unfair denial of property to Alexandra reveals their appeal to natural law to be an appeal to a hypostatized sovereign authority; for them the words on a title deed always read "the men own." Experiencing the natural-rights justification of property, heralded for its egalitarian tendency, as authoritarian and invidious, Alexandra concludes with a final Benthamite parry: "Go to town and ask your lawyers what you can do to restrain me from disposing of my own property. . . . the authority you can exert by law is the only influence you will ever have over me again" (172).

The dispute between Alexandra and her brothers reprises a debate about the nature of property taking place in the decades before the First World War (its structure perdures in contemporary debates about rights).[7] Authors like Justice Holmes, Thorstein Veblen, and Richard Ely, first organizer of the American Economic Association, sought to dispel what was called the personality theory of property embedded in the natural-rights vision. In their view, confusions like those informing the Bergson brothers' arguments result not just from paternalistic intransigence but from an antinomy inherent in the classical liberal notion of proprietorship. It is important to remember that, technically, property is not things or the possession of things (as popular usage has it), but the right to make enforceable claims to the use or benefit of things.[8] Property has classically served as the paradigm of rights.[9] And "like other rights," wrote one contributor to the influential collection *Property, Its Duties and Rights* (Gore, ed.), private property "is a creation of society, yet in the institution of private property society seems to be denying its nature and insisting not on the social but on the individual and exclusive nature of its members." The contradiction in property is that it is produced and protected by tradition, convention, and positive law, yet celebrated as a pre-contractual quality inhering in individual character.[10]

Instead, Holmes held, "legal duties are logically antecedent to legal rights"; rights are a limit on action, unimaginable outside society and convention (*Common Law* 173; *Formative Essays* 181). This is not to say that property (or any other) rights are sheer constraints. But in a double movement we now associate with Foucault, independence and rights, another jurist wrote, "are meaningless except as the converse of duties" (Rashdall 44). As Wesley Hohfeld influentially argued, rights and duties are "correlative."[11] Yet these authors find most writers on the subject insisting on absolute freedom even as they assume the sovereign protection of the institution of property. Long after the conditions of its emergence have passed, the tradition C. B. Macpherson has called "possessive individualism" still posits that an individual's authority over possessions is absolute because property is "an objective realization of the will" (Holmes *Common Law* 164; *Formative Essays* 180).[12]

Why does the natural-rights, personality theory of property perdure? Partly out of habit, of course, but also because of the ideal, or idealist, relation of representation it promises between self and property. Property comes to figure as an analogue to subjectivity, more precisely, as a structure of representation and reproduction in which self is experienced. The natural-rights, personality theory of property imagines property as a simple, linear artifactual relation: self produces its representation and owns that self-reproduction. The personality theory of property, however, only uneasily accommodates compromise of the agent's will (putatively sovereign), though clearly various abuses of property are proscribed. As Ely writes, "the right to regulate" is "part of the institution" (1:177).

If Alexandra's relations with Carl offend the town's moralists, her complication of natural-rights claims to property threatens the social fabric. Her brothers' contentions, albeit confused, at least claim a foundation that transcends written title, without which it might be difficult to justify unusual practices, in this case dissolving family property and "keeping" a male friend. One jurist explained that the natural-rights theory of property, however outmoded and internally contradictory, at least ministers to the "sense that any existent legal enactment on property [must] justify itself at a higher bar. It [is] never, in itself, ultimate" (Holland 173). If there is no natural law, there is no bar of higher appeal.[13] Perhaps worse, the higher law that ideally lies behind and generates its conventional inscription cannot necessarily be distinguished from fallible positive law, because it is only through determinate form (in title or statute) that natural law is known.

Holmes's common-law analysis clarifies the correlative, perhaps circular relation between rights and positive law. "A right is nothing but a permission to exercise certain natural powers in certain circumstances." A right, that is, is "a fact," "a consequence attached by the law to a group of facts which the law defines. When . . . such a group of facts exists," a person "is said to be entitled to . . . corresponding rights" (*Formative Essays* 177). "Hence," Holmes writes, "it is almost tautologous to say that the protection" attaching to property is a right (*Common Law* 170).[14] That's what rights *are*, an inference of "practical cohesion" from the fact of enforceable possession (*Formative Essays* 88).[15]

At one point in *O Pioneers!*, Oscar expresses a "dread" of signing his name to deeds and contracts (69). His dread typifies natural-rights anxiety that rights are, in Holmes's phrase, "almost tautologous" with facts of property and law. That is, rights are known only in terms of practical facts though revered as their prior foundation. Laws signify the inadequacy of natural law to enforce itself. As Hobbes observed, the unwrittenness of the law of nature means that not only does it, like all laws, "have need of interpretation," but also it is perhaps hopelessly "obscure" (322). To have effect, natural law or the principle of law must be embodied in positive form. But embodiments of transcendent law embody simultaneously its irreducible disembodiment, its remoteness from the sphere of action.[16] This dynamic is incarnate in (and is the reason for) the alienability of property,[17] the difference between representations and the self they ideally represent.

The alienability of representations from their proprietor imperils the ideal artifactual and cognitive relation premised and promised in the personality theory of property. In her study of the Western imagination of creation, Elaine Scarry has analyzed an analogous promise in Marx's vision of labor as "a prolongation of the body," a rescue and redemption of pain expended: disembodied labor, embodied in material products, expands personhood (244–51). Marx understood the idealist base, "the *subjective* essence," of materialist logic (*Early Writings* 147): property and rights represent and extend inalienable, intangible qualities of self transmitted in labor. But since the personhood and rights conditioning property are themselves inferred from concrete dispositions of property, the artifactual, generative relation is reciprocal. To paraphrase William James, what is most "us" turns out to be "ours,"[18] experienced and structured like property, a historically specific product, alienable like any other. It is finally this hazardous relation between self and its cognitive representations that Oscar's dread of signing deeds signifies. The novel's use of writing to figure the alienability of property figures as well the spectre of the legal or conventionalist theory of property.

Homesteading and Domiciliation: Legislating Property

It is the structural conventionality and alienability of property—acceptance of which might make inequitable and invidious distribution seem legitimate—that radical agrarianism would remedy. Popular mythology has viewed the Homestead Act of 1862, the means by which the Bergsons first acquired their land, as an attempt to enact the fundamental agrarian impulse to base property in labor and thus obviate the difficulties of a legal theory of property: after five years of residence on and cultivation of a quarter section of land (160 acres, one-quarter of a square mile), a patent of ownership was granted. Since she makes Alexandra a land speculator and entwines natural-rights logic with Benthamite premises, Cather is perhaps guilty of profaning the historical and political provenance of her novel. Careful scrutiny of homesteading policy, however, suggests that it, too, exhibits the conflicts evident in the Bergson brothers' natural-rights logic.

The Homestead Act climaxed thirty-five years of agitation against speculation and concentrations in landholding. Especially after the Panic of 1837, advocates of land reform—eastern labor leaders, Andrew Johnson of Tennessee, Congressman Galusha Grow, Free Soiler from Pennsylvania, Horace Greeley—rhetorically pitted farm and immigrant labor against land monopoly, yeoman independence against subjection to vested interests. Tilling the soil was lauded as "the original foundation of title."[19] The greater independence afforded by homesteading would, it was frequently proclaimed, "elevate" the cultivator/landowner. With land grants to the railroads, and as the creation of territories, especially Kansas and Nebraska, intensified debate over the scope of slavery, the debate over homesteading increasingly marshaled the rhetoric of natural rights to one's own person and the product of one's labor.

A substantial percentage of homestead patents ended up in the control of merchants, railroads, and speculators rather than cultivators. The homestead movement's mixed results had two sources: it belonged to what Paul W. Gates has called an incongruous land policy, an unsystematic aggregation of 3500 laws (422); and its claim procedures were liable to abuse. The Homestead Act followed upon, among others, various Pre-emption Acts (especially the Pre-emption Act of 1841), the Graduation Bill of 1854, and railroad and state land grant bills. Similar in principle to homesteading, preemption required only one year's residence and unspecified "improvement" of the quarter section, which could then be purchased for $1.25/acre. The brief residence requirement, lack of definition of improvements, and required payment favored single, middle-class settlers, who could then resell the land at market price to "paper townships" or corporate interests. The Graduation Bill auctioned unsold surveyed land, the price declining in proportion to the length of time the land had remained unsold or unclaimed. The result of these bills and the land grant acts was that much of the more desirable land was privately owned by 1862, and landseekers arriving in prairie towns were often greeted by advertisements offering the choicest plots, closest to facilities, for sale by the railroad or even the state agricultural college.[20]

Labyrinthine procedures for "proving up"—filing an affidavit avowing fulfillment of the law—could be exploited to use written claims, as it were, to improve on "improvements." Three techniques were most prominent. (1) "Commutation"—at any time after six months the land could be bought at the government $1.25/acre rate. Railroads often took advantage of this option, buying from settlers or sending agents to build a shack; many other commutations were by professionals. Nearly one-quarter of all claims were commuted. (2) Parts or all of claims could be "relinquished," that is, sold to another party. Some claims passed through the hands of five or six parties. At the least, a settler could partially relinquish to commute the remainder. (3) Because one could homestead and preempt simultaneously, one type of claim could be relinquished or mortgaged to commute the other.

Future laws increased potential for exploitation, laws like the Desert Land Act of 1877 or the Timber Culture Act of 1873, intended to adapt the homesteading principle to conditions other than those of the black-soiled midwestern states. The Timber Culture Act, for example, passed in hopes of treeing the prairie and thereby increasing rainfall, abbreviated patent time on an extra quarter section if one acre in sixteen were timbered "in a good, thrifty condition" (Copp 1875, 28). The difficulty of monitoring claims meant that planting only a few fledgling trees was often more than was necessary to prove up.[21]

Homesteading rarely took place in its pure form. The difficulty of cultivation and the allure of profit in land induced many, including the Bergson family (21), to accumulate multiple sections through preemption or the Timber Culture Act, or to claim land in the name of a family member, or to reside only sporadically on the land. The ease with which the law could be manipulated is an index of the separability between the will, action, and their written representations. Agents of corporations and also settlers or their friends might "swear with a loose conscience," and "professional swearers" sold their wares about land offices (*Public Lands Commission* 98, 586). Manuals like Henry Copp's *American Settler's Guide* clarified intricacies of the law and cited official holdings suggesting its elasticity; for example, "Residence is largely a question of intent" (Copp *Guide* 57).[22] "Proving up" was so easily manipulable that cartoonists in local newspapers caricatured the discrepancy between the representation and the act (figure 20).

Loopholes in the law thus helped keep homesteaded land within the circumference of conventional market practices. Paul W. Gates's criticisms of Congress's failure to devise a more systematic public-land policy have been widely accepted as the cause of this "ignoble failure" of representative government, as one legal historian has called it (Friedman 418).[23] As historians have amply demonstrated, however, the laws were passed over decades by different congressional bodies and by shifting alliances among East and West, North and South, old state and new state, industry and agriculture.[24] The possibility for discrepancy between will, act, and their representation is a condition of a republican legislative system, which

A BONA FIDE RESIDENCE.

A HOUSE 'TWELVE BY FOURTEEN.'

A HABITABLE DWELLING.

FIGURE 20. Three public domain sketches of land claims fraud. Titled in illustration. Albert D. Richardson, *Beyond the Mississippi* (Hartford: The American Publishing Company, 1867).

works by accretion rather than systematically, not the sign of its failure. Moreover, the point of homesteading was not, as we might casually think, to shelter land from market practices.[25] The language of land reform suggests that the point was to extend the market.

Three overlapping principles fueled homesteading advocacy: (1) increase of national wealth; (2) yeoman independence; (3) greater security for the republic. The ultimate goal of land reform was expansion of the American empire, indicated in the frequent comparisons of treatment of America's "imperial possession" to land-reform policy in Ancient Rome, and in the rallying cry "Westward Ho! . . . Westward the star of empire wings its way."[26] Formerly disaffected laborers now eager to defend their property against foreign or Indian attack would compose "a voluntary army worth all the conscriptions of Europe"; for property, proclaimed the militant Galusha Grow, "fasten[s settlers] to the country by a tie stronger than the oath of allegiance."[27] The identification between self and property obliges freeholders to the nation by obliging them to their property.

The empire would be secured also because bringing "waste lands into market" would stabilize it (*Globe,* 29th Cong., 1st sess., Congressman Bowlin of Mo., 9 July 1846, 1059). Older states objected that what were called donation or eleemosynary laws robbed them of land-sales revenue to pay off debts. Reformers countered that land donations, inalienable to prior debt and untaxable until patent was taken (an event settlers could delay for years), would cultivate markets for eastern goods and later generate a tax base from what was now "waste land." Donation policy, that is, would generate greater and more regular revenue than a one-time sale.[28] It would stabilize the market also because speculators and railroads could less easily "keep [land] out of market," "fettering its free passage from man to man."[29] Donation laws, therefore, although explicitly hostile to speculative interests, belonged to market and speculative culture; they sought to democratize entry to the land market. "Widening the market," in Andrew Johnson's phrase, would not "terminate" but "restrain" speculation (*Globe,* 32d Cong., 1st sess., Appendix, Congressman Hall of Mo., 20 April 1852, 439). Bolstered independent proprietorship was viewed as a redistribution and stabilization of the market, not an attack on it.

This point is substantiated by the sudden support that donation policy enjoyed in the 1840s. The initial proposals of Senator Thomas Hart Benton of Missouri, dating from 1826, were routinely tabled. Unable to dislodge squatters from unsurveyed government land, however, the older states conceded a few ad hoc preemption laws to generate at least some revenue from land de facto already occupied. Partisanship shifted drastically after the Panic of 1837, at which point Greeley, formerly an opponent of settlement legislation, began to modify his views (Robbins 96–116, 152–59). Foreclosures, failures, and unemployment populated urban streets amid fear that the economic structure would collapse. What was called the safety-valve theory arose at this time, widely editorialized by Greeley and George Henry Evans, labor leader and publisher of labor newspapers: cheap western lands could absorb "the unfortunate poor man" and thus situate the unemployed and stimulate markets for eastern products (*Globe,* 29th Cong., 1st sess., Congressman Thompson of Miss., 9 July 1846, 777). But even proponents of this prototypical welfare plan, first enacted in the Pre-emption Act of 1841, acknowledged that preemption did not salve the economic crisis: many poor could not afford to relocate.[30]

The especially severe winter of 1845–1846 intensified clamor for liberal settlement laws, led by Evans's National Land Reformers and promulgated by the now fully converted Greeley. It is in this year that the word *homesteading* was first applied to donation policy,[31] thus marrying land policy to the cult of domesticity. Future discussion routinely sung the benefits of home and family, the sites where citizens "learn their rights" and "practice virtue and independence" (*Globe,* 29th Cong., 1st sess., Appendix, Congressman Thompson of Miss., 9 July 1846, 780). This commitment informed the language of the numerous versions of the Homestead Act Andrew Johnson submitted (the 1862 bill listed the "single person" as eligible third after "the head of a family" and "widow"). By the Republican triumph in 1860 the slogan "Homes for the Homeless" was heard nearly as often as "Land for the Landless" (see Republican Association pamphlet).

The celebration of the independence and elevation resulting from the "purifying" "influences of the domestic fireside"[32] suggests that just as freeholding was a form of proprietary obligation, domestic independence entailed customary obligation. Opponents muttered that donation policy's stimulus to mobility would loosen the social fabric. Johnson countered that "domiciliation" "would create the strongest tie" between citizen, family, and state: eager new taxpayers would tend their property more assiduously than the single man for having a family to support. Another legislator extolled the measure as "the best possible to create a great conservative interest of everything valuable in our institutions."[33] The independence of domiciliation, that is, reinforced rather than loosened social constraints.

I point out that the rhetoric of independence and natural rights was also a rhetoric of constraint and socialization not to expose hegemonic conspiracy but to suggest that homesteading should not be overromanticized, as it often has been in the American imagination—in the moralism of the Bergson brothers, or in works like Oscar Micheaux's *The Homesteader* (1917) or Mary J. Holmes's *The Homestead on the Hillside* (1867), in which love and the homestead are secured simultaneously;[34] or even negatively in stories by Hamlin Garland, where threats to the homestead occlude the avenues of sympathy. Homesteading and the rhetoric that advertised it were part of a hybrid and multiply motivated policy—most policies are in America—broadening access to land in an attempt to manage land and labor markets.[35] For example, the old states relaxed opposition to homestead proposals only in the 1850s, when developments in transportation evaporated anxiety about "depopulation" of the labor supply and the loss of land sales revenue,[36] and inspired enthusiasm for the markets promised by the new states (Hough 1921, 85). Moreover, homesteading came to seem a mechanism for inhibiting slaveholding (Republican Association, "Lands for the Landless" 1).

Domestic impulses, partisan concerns, and fee-simple principles are neither discrete nor scandalous in their conjunction. Johnson unabashedly proclaimed that his freeholder would be both "an independent man and efficient citizen" (*Globe,* 31st Cong., 2d sess., 23 January 1851, 313). And in his labor organ, *The*

Radical, George Evans similarly pressed for free access to land "on the ground of *expediency* as well as *right*" (8). Natural-rights rhetoric dwelled comfortably with the economics of settlement, though it was not the exclusive property of donation factions. Indeed, the Lockean assumptions of donation policy were precisely those of the interests opposing it (Young 106). Why, eastern congressmen continually asked, should land be granted to some, "deprecating" real estate values in the old states, when others—their constituents—had borne hardships to pay off market-value mortgages?[37] This bald unfairness, an indirect taxation of "old settlers," was the reason President Buchanan gave for vetoing the first Homestead Act in 1860 (Richardson 5:611–14).

A few proponents conceded the potential inequity of homesteading.[38] This is "class legislation," Congressman Dawson of Pennsylvania concedes, but not the first example of it, and no previous enactments favored labor (*Globe,* 33d Cong., 1st sess., Appendix, 14 February 1854, 182). Dawson assumes that the state inevitably redistributes benefits in light of its greater interests and principles. His admission about government power seems discordant with the natural-rights principles of agrarianism, but not if we consider the political meaning the term had at the time. All proponents of homesteading, including Johnson, Grow, and Evans—even the later Grangers!—denied that they were "agrarians" (or, from the period of the English Reformation, "levellers" or "destructors"), those who would arbitrarily divide up property and declare land permanently inalienable if cultivated. Such policies absolutizing natural rights were renounced even by radicals as abridgements of rights.[39]

In the homestead movement, natural-rights rhetoric, domestic rhetoric, and various accounts of proper market development met in a complex coalition of radical natural-rights and Benthamite logic. On one hand, law and property originate in and express labor and citizenship; on the other, a citizen's property in labor is constituted by positive law. As in *O Pioneers!,* the entanglement of natural law with statute and title is figured by writing, in homesteading's obsession with labyrinthine claim-filing and proving-up procedures. Cultivation may be the natural law of property claims, but these, finally, possess authority by virtue of written title and the institutions protecting it. In homesteading, the authority of law and disposition of property are at once natural and sovereign.

Gender and Identifying with Property

Is there a way to resolve the internal antinomy between a priori rights and sovereign authority that informs the Bergson brothers' stance, Alexandra's counter-argument, the homesteading movement, and natural-rights justifications in general? Clearly, both settlement policy and the Bergsons' contentions belong to a patriarchal culture, and recent critics, like Annette Kolodny, have studied American women's fantasy of "an idealized domesticity" as a salutary alternative to patriarchal "erotic mastery" of the land (Kolodny xiii). Virtually every pamphlet and congressional speech invokes the biblical command to Adam to "subdue" nature. In contrast to this appropriative stance, Kolodny has compellingly argued, the New World Eve regarded the land sympathetically and cooperatively as an abundant garden to be stewarded (cf. 178–226 passim). This thesis is shared by recent editors of women's diaries and letters and by many critics of Cather,[40] and carries a clear moral and political implication: the feminine ontology of nature fosters an alternative, communitarian politics less exploitive of the land and fellow citizens.

I believe this position is historically inaccurate, imposing on the writings of women pioneers idealizations of the gentler sex that underlay the cult of domesticity and the enduring disenfranchisement of women. Instead, women settlers' and homesteaders' conception of the ontology and purpose of the land is consonant with men's. I mean to suggest neither that there are no alternatives to Adamic patriarchy and capitalism nor that women's experience of the frontier is identical to men's. Yet while subduing the prairie was perhaps harder for women, they shared the classical liberal impulse to own and manage it.

In the decades after Turner announced the closing of the frontier, frontier nostalgia seems to have combined with the emergence of the "New Woman," with women's suffrage agitation, and with the last great land rush, the first to advertise "Homesteads for Women,"[41] to produce a small boom in the subgenre of the woman settler and homesteader. *O Pioneers!* belongs to this literary movement. One widely read woman homesteader was Elinore Pruitt Stewart, whose 1913 *Letters of a Woman Homesteader* served as the basis for the movie *Heartland.* Pruitt[42] writes to inspire "the troops of tired, worried women, sometimes even cold and hungry, scared to death of losing their places to work" (216). "Any woman who can stand her own company . . . and is willing to put in as much time at careful labor as she does over the washtub will . . . have independence . . . and a home of her own in the end" (215). Typically melding the ideals of domesticity and independence, Pruitt's appreciation of the beauty of her Wyoming surroundings takes for granted that her labor and title afford her profit in the landscape. Having completed a mountain ascent whose rigors make her wish she had stayed home, she finds the vista "beautiful, and the views many times repaid us for any hardship we had suffered" (196).

Pruitt's comparatively innocent proprietary stance toward nature is evident in other accounts of women pioneers. When Mollie Dorsey arrives in Nebraska in 1857, she interprets the "glorious" sunrise, flowers, and chorus of animals as a prayer commemorating her family's "safely . . . coming to a home / of our own once more" (Sanford 32). The sunrise and nature's song seem to celebrate her family's proprietary achievement. Sometimes this natural achievement takes more violent form. Christiana Tillson celebrates the possibly disturbing slaughter of sheep to relieve hunger during severe weather and strenuous labor, "for one of the most urgent demands of our nature was in full force" (41–42). Her labor and hardship have earned her digestive nature the right to sacrifice and consume nature.

One of the marks of the New World Eve that Kolodny expounds is communal cooperation, but women's writing about pioneering sees no intrinsic conflict between cooperation and possessive individualism. For example, in 1909 Luella Shaw can praise simultaneously the "unselfish consideration" of "the fearless frontiersmen" (266) and their "determin[ation] to conquer the West" (iii). In her 1886 memoirs of frontier life Sarah Royce has larger ambitions than just to compose a resource for her son Josiah's history of California. She hopes to persuade him to a faith in and obedience to God that will underpin his later, famous "Idea of the Community" (140, 144). The manifestation of communitarian faith to Sarah Royce is the material prosperity Californians have achieved by collectively overcoming natural disasters and domesticating nature.

Women's concern with domestic details complements rather than challenges the classical ambition to secure property in land. Women pioneers' writings regularly contemplate the formalities of claim filing, the proximity of land to the railroad, whether or not to extend holdings by means of the Desert Land Act or Timber Culture Act, or how to meet residence requirements even while "working out" for others.[43] The proprietary basis of interest is evident in a letter by a Norwegian homesteader whom Elizabeth Hampsten compares with Cather's Ántonia Shimerda for her distinctly feminine concerns. Let me test the relation of her concerns to possessive individualism. This writer supports herself and her mother mainly by crocheting and sewing:

> . . . I cannot take care of the land like a man. All I can do is live there as much as I can and to plow what I can so that no one can take it away from me because I hope in time to sell and maybe get a couple of thousand dollars for it if we get railway over here. (Hampsten 34)

Exploiting the possibilities of the Homestead Act, this woman is superficially breaking ground to prevent contests to her claim till she can sell to the railroad.

Her concerns belong to the tradition of possessive individualism. Insofar as any citizen was eligible to homestead, the act extended possessive possibilities to both genders. Such extension was, of course, the essential goal of the nineteenth-century women's movement, which challenged patriarchy as a contravention of natural law and divine entitlement.[44] The Seneca Falls Declaration of 1848 famously and ironically emulated the Declaration of Independence. Elizabeth Cady Stanton lobbied for the 1860 New York Married Woman's Property Act to secure woman the "best right to her own person" (Scheir, ed. 118). And Charlotte Perkins Gilman traced social ills to the "discord . . . between will and action" inaugurated by the deprivation in women and other subjected classes of the innate and definitive human impulse "to express one's . . . spirit" in labor (333, 67).

By seeking harmony between the will and its productions, women's challenge to patriarchal law marshals patriarchy's possessive logic in order to exorcise its misogyny and perfect its implementation. The promised identification between persons and property was fundamental to reform and radical analysis. It fuels Marx's analysis of alienation under capitalism. He understood the appeal to *Recht* as a tactic of the bourgeois class for justifying inequities in the disposition of property, and sought what he called the "supersession" of the internal contradictions that the language of *Recht* equivocates.[45] Labor and land-reform leaders like George Evans and Thomas Skidmore argued that there should be *no* property in land because property in land inevitably leads to tenantry and thus abridges natural rights to one's labor.[46] Critiques of capital accumulation affirm the same account of the origin and basis of property as classical capitalist defenses. Whether seeking to eliminate private property, or, like Henry George's "Single Tax," to perfect market operation by melding land to labor and eliminating land markets, they are attempts to purify the identification of self and self-productions that the actual operation of markets has corrupted—to recuperate one's inalienable right to one's labor and one's self.

The Antinomy and Aesthetics of Property

Alexandra's business conduct foregrounds the antinomy in natural-rights premises that natural-rights movements aspired to resolve. These movements (conservative as well as reform) imagined or sought the identification and ratification of self/independence through the "extended personality" of property (Ely 2:536). But the distinctive triumph of the liberal reformation was its foregrounding of the alienability of property. Property, as a structure of representation in which self is experienced, is by definition contradictory because never wholly continuous with personality. To adapt a phrase from Holmes, property is "*almost* tautologous" with its owner (my emphasis). Property delineates the self's inscription within external authority. Irreducibly different (alienable) from the identity it identifies, property is simultaneously us and not us, marking both our extension and our limit, and finally the historical contingency of all rights and properties.

Property, then, like the various inquiries into the ground of value this book has examined, is fundamentally sublime, just as the Enlightenment discourse of the sublime, as we saw in chapter 1, was at base an inquiry into the grounds of self-possession. Like Kant's transcendental sublime, one's property tells the story of both the self's empowerment and circumscription within an order not of its creation. We may think of property as Jacques Derrida and Ralph Cohen have described genre, as a liminal scene of classification or description. Genres or properties (or rights) are known in terms of and conditioned by what differs from and exceeds the class of identifying traits. Traits represent a subject only by differing from it.[47] In signifying us, properties circumscribe us; they simultaneously belong and do not belong to the proprietor/subject/genre. The generic structure of property means that, as I suggested earlier, the self is experienced as reflection and loss, as something severed from itself, and as an inference from social facts.[48]

In its very structure property exemplifies the discrepancy between self and its representation that Gilman hoped to bridge when she sought to resolve the "discord" between the will and its productions; and such resolution or "harmony" (through the universalization of rights or political agency) was the goal of nineteenth-century reform and radical movements. The need to justify a disposition of rights inclined both apologists and critics to idealize property as the plenary production and sign of intrinsic qualities of self.[49] Natural-rights logic, radical and conservative alike, tends to elide, or seeks to elide, the antinomy in property, and thus to naturalize or make necessary relations that are political.[50]

In the actual year *O Pioneers!* opens, the political and conflictual nature of these relations was painfully evident. The economic panics of the early 1880s "demoralized" (47) farmers witnessing the easy alienability of their titles. The panics incited, a decade before Turner announced his frontier thesis, a lively debate, in the *North American Review,* for example, on whether "the public domain is quite exhausted." The author of this conclusion, Thomas Gill, is joined by Henry George in his assessment that this event is rapidly "manufactur[ing] . . . a tenant-farming class."[51] The disposition of public lands underscored the power of written title and its difference from labor and its fruits.

This difference is thematized in Cather's novel by nature's sublime resistance to human management. And the sublimity of nature incites sublimation in Alexandra. What Alexandra loves about nature is its capacity always to reclaim itself. We own only "for a little while" (308), she says, and we can "lose everything" "in a single day" (183). Nor does Alexandra believe herself the true owner or creator of her property. Neither labor nor ingenuity in applying new technology causes prosperity; nature is the agent in property: "We had n't any of us much to do with it. . . . The land did it . . . [;] it worked itself. It woke up out of its sleep and stretched itself, and it was so big, so rich, that we suddenly found we were rich, just from sitting still" (116). In this sublimation of human agency, we own nature only by permission of its "Genius."

Alexandra's sublimation of agency is also a sublimation of desire. She has never been in love (154, 205), and her relations with Carl are unfailingly dispassionate. When she and Carl part early in the novel, for example, their recollection of their attachment ("Yes, that's it; we've liked the same things") lacks any typical adolescent curiosity about sex (52); their later marriage is motivated not by passion but by the shared understanding that they will someday be vanquished by the wind with which the novel opens: "We come and go, but the land is always here" (308). Like all sublimation—and all property is sublimation—Alexandra's is a redirection of libidinal urges. And when exhausted, she dreams of ravishment by a potent nature. After these dreams, Alexandra "prosecutes" a cold bath (206), but her response, if neurotic, is not morally based; rather, the economy of sublimation is symptomatic of her rejection of the natural-rights wish-fulfillment fantasy. She leaves "Rapture" (255) and "sharp desire" (v) for those who, like Marie Shabata and Emil, "cannot feel that the heart lives at all unless it is still at the mercy of storms; unless its strings can scream to the touch of pain" (226).

Why sublimate agency and "sharp" desire? In the natural-rights narrative of property, desire underlies independence. In 1854, Congressman Dawson of Pennsylvania declared that when "the Almighty" "created man after his own image," he first "filled him with desires" and only then "endowed him with reason" (33d Cong., 1st sess., Appendix, 14 February 1854, 185). Desire, here, is the precondition for compensatory reason. This sense of the primacy of desire is elaborated in the myth, recounted by all participants in land debates, that our instinct to acquisition symbolizes and responds to the expulsion from the harmony of Eden. But as the condition expressed in acquisition, desire is the supreme contradiction in natural-rights logic: desire, the condition of need and of incongruity with present conditions, cannot be the sign of self-sufficiency or the cause of harmony.

In the fate of Marie Shabata, Cather depicts this classic narrative of desire as sheer suicide. At one point Marie gazes at "the remote, inaccessible evening star."

The star's remoteness—typically symbolizing nature's transcendence of human claims on it—stirs Marie to mourn her own remoteness from others (specifically from her illicit lover, Emil), and to yearn for union: "always the same yearning, the same pulling at the chain—until the instinct to live had torn itself and bled and weakened for the last time, until the chain secured a dead woman, who might cautiously be released" (248). Desire here, the mark and result of difference from others and the world, suffers a self-consuming career: at once seeking gratification in identification with other persons and objects, yet testifying to the impossibility of its ideal, desire finally desires its eradication, release from itself. In this novel where self-extension is always incomplete, classic "yearning," exemplified in the lovers' passion, "secure[s] a dead woman" (Marie) rather than self-fulfillment.

Alexandra also longs to eradicate longing, "to be free from her own body, which ached and was so heavy. And longing itself was heavy: she yearned to be free of that" (282). But her longing has different results, and finally a different basis, evident in her contrasting fascination with the stars. Her stargazing rehearses the mood of Emerson, whom Cather much admired, at the opening of *Nature:* "She always loved to watch them, to think of their vastness and distance, and of their ordered march. It fortified her to reflect upon the great operations of nature, and when she thought of the law that lay behind them, she felt a sense of personal security" (70–71). Experiencing her sublation in the stars' remoteness fortifies Alexandra, suggesting that for her longing and personhood are not, as Marie wishes, absolute, or absolutely individuated. Not only does Alexandra, through her sublimated, speculative practice, accumulate land without viewing it as self-extension, but she feels her productions have significance precisely because nature exceeds them: "If the world were no wider than my cornfields, if there were not something beside this, I would n't feel that it was much worthwhile to work" (124). Her love for the land is ultimately a love without passion or specificity. The famous early passage I discussed earlier says that Alexandra's face is "set toward [the land] with love and yearning" (65). The force of the narrative is to disentangle love from yearning, from identificatory desire, from viewing possessions as monuments to the self.

Nineteenth-century natural-rights defenses or critiques of capital distribution ignore or would resolve the antinomies of property. Both within and against the classic liberal vision, Cather explores in *O Pioneers!* the extension of property holding to a subaltern class (women) but challenges the liberal premises about property usually underlying such a politics. Property for Cather is not a linear self-extension, a narrative linking the will and its productions. She insists on the discrepancy between property and the self, self and its representations. Cather suggests property is always both a sublimation and an appropriation of the other term in the discordant equation between self and its properties that compose identity. Property for her marks the self's constitutive division and sublation. In this sense, Cather emphasizes more definitively than even Kant that in expanding, the self continues to feel the "deprivation" that incited its appropriation of phenomena. Alexandra's growing power and freedom from convention remember what Kant called "the cause to which it is subjected" (109), which includes both the physical and social forces external to her.

By not identifying herself passionately or absolutely with her property, Alexandra can conceive both her inscription in a greater order and her nominal independence. She feels at once submerged in the "impersonality" of the soil and blessed with "so much personality to put into her enterprises" (203). Her memories, hence, are both "impersonal" and "very personal" (205). We might call Alexandra's practice a minimalist capitalism, sensing loss as impending and refusing to justify property or its distribution in ontological or moral terms. But this minimalism questions what liberal, natural-rights advocacy too quickly assumes, that tying property to selfhood, or to gender, would have specific and necessary political effects. Her proprietary practice, sublimating self and gender, resists the romanticism of the liberal impulse to purify the lineage of land, value, and self.

Alexandra's minimalist economy produces an unusual psychological response to the inevitable alienability of property. Elizabeth Cady Stanton's 1892 address to the Senate titled "Solitude of Self" sounds the usual, tragic tone. To deny woman's "birthright to self-sovereignty" is to dismember her, put out her eyes, paralyze her reason, and thus incarcerate her irremediably in the originary "isolation of every human soul." The promised self-extension of property rights finally promises only partial compensation for the anomie caused by our exile from Edenic unity; ironically, it underscores the "deeper tragedies" of solitude (325–27)—that individual rights exist by social authority rather than by sheer acts of will, and that subjectivity is always subjectification. Stanton's tragic tone stems from a romantic conception of property. The property relation can never fulfill the conditions of satisfaction implied by natural-rights logic: an inalienable expression of intrinsic qualities so fundamental as to constitute an ontological identity between material, rights, and owner.

In contrast, Alexandra achieves what Cather calls an "impervious calm" (226), leading her to forgive Frank Shabata his double murder because she locates her independence precisely in the irreducible difference rather than identity between self and property.[52] It is instructive here to recall the very different attempt of Colonel Jackson of Athens, Georgia, to bequeath a tree to itself. This provision of his will was held invalid; the tree was declared a city park—owned by the public. As we might expect, trees can't own; it is their inability to experience a difference between themselves and property that disqualifies trees from ownership. Only in this disparity does ownership make sense. What Alexandra finds beautiful about property, what she calmly loves about the land, is its difference from you, its transcendence of personality.

In this, she embodies what Fleda Vetch understands, at least momentarily, in James's *The Spoils of Poynton*. Mrs. Gereth cannot accept the loss of any piece of Poynton, now controlled by Mona, but for Fleda

> the array of them, miles away, was complete; each piece, in its turn, was perfect to her. . . . Thus again she lived with them, and she thought of them without a question of any personal right. . . . They were nobody's at all. . . . It was Poynton that was theirs; they had simply recovered their own. (162–63)

When Fleda later speaks of the "beauty" and "poetry" of "something . . . relinquished," she articulates (though her tone retains traces of tragedy) the aesthetics of property that Alexandra calmly loves. Alexandra's response to the fact that we can know ourselves only in determinate form alienated from the essential qualities ideally inspiriting these forms is not lament or despair. She regards the transcendence of property as the basis of, rather than a threat to, identity. Property's difference from us, like the remoteness Alexandra feels when she gazes at the stars, is fortifying; in the incompleteness of ownership she feels the reciprocity of her subjection and empowerment within a political order.[53]

The novel's aesthetics share the sensibility of Alexandra's economy. Alexandra, as many have observed, is the artist of the novel. Her "pioneer" imagination displays a speculative structure, "able to enjoy the idea of things more than the things themselves" (48). Carl, an engraver, contrasts his imitative work to her creativity. While he "engrav[es] other men's pictures," "you've stayed at home and made your own" (116). Alexandra, however, as she does in her land dealings, denies agency for her imaginative productions. She concurs with Carl's remark that "there are only two or three human stories, and they go on repeating themselves as fiercely as if they had never happened before" (119). Stories, like the wind, "fiercely" follow a course independent of human actors.[54] The narratives of ourselves are not our own.

Cather often characterized her episodic "narratives" in terms of this fundamentally sublime paradox, in order to distinguish them from "novels," which have elaborate plots and a strong authorial voice.[55] Rather than "dramatic treatment," she wrote about *Death Comes for the Archbishop,* she sought "the style of legend," which finds "all human experience, measured against one supreme spiritual experience," to be "of about the same importance" (*OW* 9). She aspires to the "simplification" of Gothic sculptors who "told the same stories (with infinite variety and fresh invention) over and over on the faces of" cathedrals (40, 27). In a sense, Cather seeks not two or three narratives but one: the achievement of uniqueness and authority in the divestment of self.

This unresolved paradox will be familiar to us by now as the one structuring the Emersonian vision of transcendence and the organizational principle of the trust form and Debs's unionism. In Emerson and the trust (more so in Emerson), the antinomies of self and world, particular and universal, matter and spirit are at once transcended and maintained. The operative economy, as we have encountered it since Kant, Cole, Allston, and Henry Carey, is protectionist, an expansion or even perfection of self to manage vicissitude by inhabiting it. Self-possession begins in dispossession or abandonment, the conditions Twain and Norris can scarcely bear and that Howells grows accustomed to. In the sublime economy, dispossession is often redeemed, best of all in comprehensive vision and power. Cather, however, lingers in relinquishment, savoring rather than rapidly redeeming the sublation that is constitutive of personhood and the dispossession that is the formative antinomy of property.

This dynamic informs the language of self-possession with which Cather famously defined artistic authority. In her remarks on writing (for example, her preface to the 1922 edition of *Alexander's Bridge,* preface to Jewett's stories, "The Novel *Démeublé*"), she demands in art "a quality of voice that is exclusively the writer's own, individual, unique" (*OW* 50). Yet Cather's vocabulary of possessive individualism points toward dispossession, toward the condition of *démeublé* (being unfurnished), the critical term in the title of one of her best-known essays.

She seeks to advance "the revolt against individualism," a special burden on the artist because "of all men the most individual" (26). She praises Jewett for "not [writing] about exceptional individuals" (55). Concomitantly, the artist achieves his own voice, his "own material" (*Alexander's Bridge* v) only in "giving himself absolutely to his material." The uniqueness of the artist emerges as "he fades away into the land and people of his heart, [as] he dies of love to be born again." Loving and writing sympathetically are self-investiture by means of suicide, a Christian pattern Cather again describes as a gothic divestiture. The artist writes what "haunts" him, his possessions that possess him (*OW* 51).

Cather's sense of narrative and authorship as divestment contrasts with Judeo-Christian aspirations in general and specifically with those of women settlers, for whom writing is urgently a form of self-preservation, an attempt to "will and bequeath" (Sanford xii) a "record" of the self (Tillson 3). According to Cather, the young writer feels similarly: he thinks he is learning to express "his one really precious possession." But maturity persuades him "he has less and less power of choice" about his experience and about what he writes (*Alexander's Bridge* vii–viii). As owning the land does in *O Pioneers!*, writing signifies to Cather the circumscription of possession and self-possession. A detail of her publishing practice recounted by her companion Edith Lewis illustrates this point. Though Cather was adept at arranging the marketing of her product (33), and even risked changing publishers from the Brahmin Houghton Mifflin to the fledgling Knopf to gain greater product control (108–15), she nevertheless "had an extremely impersonal attitude toward her writing." Unlike most of us, "she did not cherish her words and phrases," and she readily, though dispassionately, revised proofs, no matter the cost (106). Her unthrifty galley-reading practice suggests how conditional was her investment in particular embodiments of her unique voice.

The paradox of property structures Cather's last accomplishment, the prohibition in her will against citation of her remaining correspondence. This interdiction has been troubling to Cather studies, especially to biographers, because it obstructs attempts to link a person's productions to that person. Her will thus appropriated some control over her surviving persona. The accomplishment of a will would seem the paramount example of writing as an extension of self, a disposition of property, all that remains of a deceased. But as Susan Stewart has remarked, destroying letters is a form of suicide (14),[56] and the troublesome force of Cather's will is finally to stress the conventionality of the link between the person and its phenomenal disposition, for she controlled the assigning and production of her persona by erasing herself.

Wills are paradoxical. In wills we proclaim property over representations of ourselves in the very moment that the independence of these representations is unmistakable. I have been arguing that Cather knew this, having recognized that the alienability of representation that structures writing also structures property and identity. Wills, in this view, exemplify all acts of writing and proprietorship. Cather's last testimony to this intuition, her last will and testament, thus culminated a narrative of coming into possession by dispossession that she first recounted in *O Pioneers!*.

Afterword

In his ardent *History of the Grange Movement* (1873), James McCabe explains why the American farmer deserves legislative protection against land-grabbing monopolists: the farmer "is faithful to the obligations imposed upon him by the laws of the land" (286–87). McCabe's is a fairly dense appropriation of the strategy that is central to the artifacts and actions this book has studied: invoking the law of nature as the ground of value. McCabe's law of the land here is civil law, and farmers' fidelity to this law makes them industrious citizens, morally superior to European farmers. In effect, McCabe's remark naturalizes the morality of civil obedience. But as part of his defense of the Grange, this remark is also unmistakeably ironic and critical. Throughout, McCabe stresses the limitations on farmers: they possess, of course, less capital than monopolists, but also their economic elasticity and capacities for capital growth are directly limited by weather patterns and by the fertility and location of the land they work. The laws of the land, then, should be reformed to conform to the refractory laws of the physical land. America both incarnates and transgresses the law of nature, and McCabe's invocation of the law of the land intimates simultaneously the nation's election and corruption.

Other examples that I have presented of this strategy are more dense than McCabe's, often by several degrees, and it is the particular density and texture of artifacts that I have tried to explore. One question I have not programmatically emphasized is why the law of nature was so regularly invoked. I have not, because the genesis or cause of this gesture is so irreducibly, and so particularly, historical. The law of nature was ubiquitously invoked and sought partly because, as Enlightenment philosophy judged, especially the work of Hobbes and Rousseau (less so in Locke, much less so in Montesquieu), it was remote, if not altogether inaccessible. Though many imagined the law of nature to be immanent, all knew, following Hobbes and others, that it was in need of interpretation, and they tried to supply an authoritative interpretation. The law of nature was invoked, that is, because this strategy was fundamental to the liberal tradition deeply imbuing the intellectual and affective inheritance of most Americans, including Afro-Americans. But if invoking nature is a reflex of the liberal tradition, liberalism itself belonged to an older moral tradition, and it emerged as already a form of religious appeal. The strategic appeal to nature, then, does not originate wholly with liberalism.

Nonetheless, nineteenth-century appeals to nature were specifically inflected by liberalism in that they routinely invoked nature's authority in order to appropriate nature; self and action were axiomatically proprietary. Still, even liberalism is not a sufficient explanation. The liberal appeal took particular shape in different contexts and in the hands of different actors with varying needs, interests, psychologies, positions in the culture, and skills. I have tried, as students of culture must, to discern some patterns, both historical and conceptual, in these hybrid, protean, and multifarious appeals. But I hope I have also attended to the particularity of each instance.

One might note that I have not completed a theoretical analysis of these historical materials. What is the theory of, in this case, acting or justifying action by the law of nature? It is today nearly a critical imperative to pose the theoretical question, but I am not sure what this question means and accomplishes. If it means that critics must establish the causal relation between the metaphysics or class position or gender position, and so on, of agents and the agendas they pursue, then I find the question deeply essentialist. I reject it for that reason, not so much because essentialism is abhorrent but because it is wrong. No doubt, many of the authors, speakers, and actors I have discussed were convinced of the harmony— or sought harmony—between their view of nature and God and their aesthetic, economic, and political values. But this historical fact does not make the relation between logical premises and events linear or causal; indeed, we have seen that the relation between paradigm and act often contradicted authors' self-imagination, and it was always profoundly artificial, inflected by specific historical factors. For an explanation to be theoretical, in the strongest, and only meaningful, sense of the term, it must seek more than the contingent factors in artifacts and acts, the patterns they assume, and disjunctions they display. In other words, if a "theory" of action or of a historical problem is not idealist, it is simply a hypothesis or thesis. A hypothesis may be correct, and it may be ambitious or wide ranging, even meant to predict other phenomena or legislate future conduct; nevertheless, it has the same status as any other description, proposition, or explanation: it expresses an empirical conviction about phenomena.

I have indeed ventured a thesis about aesthetic and social practice over several decades: the appeal to nature as the ground of values, however hybrid and protean, often takes the general form of what I call transcendent or combinatory agency, whereby individual identity and agency are perfected in their occulting, absorption, or sublation by wider forces, usually called natural or universal. The strategy seems to have been routine, evident, for example, in a remark by Charles Peirce recently cited by Cornell West. Peirce is defining vocation as "a generalized conception of duty which completes personality by melting it into the neighboring parts of the cosmos. If this sounds unintelligible," he apologizes to his readers, he offers motherhood as the type of vocation (cited in West 48–49). I am concerned not just with the ubiquity of the logic Peirce employs, in this case to characterize the philosophical and domestic arenas. What I find striking is the way this justificatory paradigm was employed to conflicting ends by parties directly disputing policy or aesthetic values. Samuel Gompers, for example, repudiated industrial unionism, and parried charges by radical unionists that he and his trade-union model were individualistic and self-serving, by declaring that craft unionism is the "natural form of associated effort of the working people" (1971, 49). Industrial unionism, he countered, despite its own claim to be the natural form of organization, was coercive, that is, manufactured. Similarly, when the Grangers underscore their allegiance to the law of the land, they are claiming a natural basis for their legislative proposals to override the land monopolists' corollary claims.

I have generally clustered a variety of positions around particular problems, in order both to evolve and to test my hypothesis, and also to ascertain the range and diversity of its instantiations. Transcendentalists and protectionists, like Debs, Dreiser, Cather, and the Standard Trust, articulated an affiliated vision of agency and representation. In no way do I mean to suggest that these persons held the same beliefs or pursued the same social vision. Rather, I have tried to indicate the variety of actions, programs, and aesthetic visions that employed, explored, and transformed the combinatory impulse. This paradigm of thought, feeling, and action appeared in many, often conflicting permutations. We may think it disingenuous or hypocritical, or plain insidious, when Andrew Carnegie, extolling the "Gospel of Wealth," salutes the economic age in which "the man of wealth [is] becoming the mere trustee and agent for his poorer brethren" (1962, 15). Carnegie could well afford this philanthropic sentiment, earned at the expense of laborers whose union association he aspired to crush. Nevertheless, Carnegie believed this sentiment, and attempted to fulfill, through philanthropy (not to mention union-smashing), its paradigm of action or identity, the same one that Debs dedicated himself to while fighting concentrations of capital. People with as different aesthetic visions as Bryant and Cole (perhaps close friends, but not really kindred artistic spirits) believed that art engaged in combination—of forms, genres, traditions—and pondered the best way to accomplish this goal. When Henry Clay lobbied in 1820 for a systematic protectionist policy, he argued that under the regime of free trade, citizens (industrialist, capitalist, labor, and farmer alike) could achieve only "partial combination" (225) with other economic agents, and would thus be enthralled to Europe, laws, and habit. William Cullen Bryant, in contrast, a leader of the Free Trade League, denounced protectionism because *it* obstructed the complete combination of citizens.

In addressing the economic contractions of the late 1830s and early 1840s, Orestes Brownson powerfully articulates the imperative to combination: self is already combinatory. "The Laboring Classes" tries to dissuade citizens from the popular myth that "each man is his own centre," "a whole in himself" (379). Instead, individuality is a confluence; indeed, Brownson writes in the sequel, "Our Future Policy," since any act of labor has value only in exchange, all labor is already *"associated labor"* (109). This social principle may be best realized by abrogating banks' monopoly on credit authorization and paper issue, for banking policies have victimized and vitiated labor. Nevertheless, forbearing the period's widespread anxiety about incorporation, Brownson insists that "corporations are not strictly anti-democratic" (109); any community, he explains, since it is "the combination of individuals," is in principle a "sacerdotal corporation" ("Laboring Classes" 379–80). When bank monopoly is arrested, corporations may consist of "operatives and employers in the same persons," and will then perform their true sacerdotal function. Such corporations will be "blessings" since in them individual action will palpably be the "collective action" it always in principle is ("Future Policy" 110, 83).

I do not adduce this example in order to celebrate, like Brownson, the democratic potential of corporations. (Brownson's proposals for fulfilling this potential in corporations differed vastly from those of conservatives.) I merely wish to observe the connection between the imagination and (often) justification of corporations and the spiritual mission of antebellum importers and adapters of Continental idealism. The corporation developed in this transcendentalist tradition, I have argued, culminating in the trust form and also its enemy, industrial unionism. These organizations attempted to traverse or supersede antinomies animating the liberal notion of property. We would do well, following Pocock (1985, 52), to think of liberalism's notorious possessive individualism as also a spiritual materialism, for articles count as property only when invested with human spirit, and spirit is palpable only when invested.

It is today a critical commonplace to be suspicious, if not censorious, of artifacts that manifest what we should call the liberal habitus—experiencing identity through property—as if this fact assures their complicity with political evils. One could hardly study the materials, literary and otherwise, treated in this book without feeling the limits and often dreadful effects of this habitus. But if I have occasionally registered my disapproval of, and sometimes bewilderment at, the strategies and logic employed in the search for, and service of, natural law, I nevertheless think that, overall, condescension or scorn is an inappropriate tone to take toward one's subjects. There are other ways to conceive the self and social organization than as realized and organized through property, but few in the period I've studied perceived them, even people, like Thoreau, who explicitly sought alternatives.

William James's understanding of the cognition of self exemplifies the way in which people conceived identity as a problem of belonging. What is most us? he wonders in *The Principles of Psychology;* at least, what is most intimate to us? Even "our bodies themselves, are they simply ours, or are they *us?*" It turns out that our bodies are ours, but they are also something more since they are not "mere vestures" (1:291); moreover, the self can be experienced only as a property. For James, identity emerges in a dynamic between the "I" that cognizes its collection of attributes, and the "Me" that consists of those attributes and properties. The cognitive "I" should be the most intimate part of the self; it knows the self as "my historic Me, a collection of objective facts"—"powers, possessions, and public functions, sensibilities, duties, and purposes" (1:322). The "I," then, is a collection agency. Identity, he wrote two years later in the *Briefer Course,* is an appropriation of these material and corporeal properties: "I" is " *'that to which'* all the concrete determinations of the Me belong and are known," a mental state "which at every moment goes out and knowingly appropriates the *Me* of the past" (180). James sees the self as proprietary, but the proprietary self is not static, for it interminably appropriates and invests itself in its properties.

In James, as in Locke, this idea that the self is realized through properties is axiomatic, but the relation is not simple, and it thus exemplifies the discrepant dynamic of the liberal (at least the Lockean) narrative of identity. The cognitive "I" achieves positive existence only through the appropriation of attributes. The collecting agent "I" must remain more intimate than the "Me," however, because the "Me" consists of dispensable attributes: physical, neurological, emotional, rec-ollective. These elements vary over time and are subject to variable conditions (*Briefer Course* 180), but vicissitude precipitates the psychological phenomenon of the "shrinkage of our personality" (*Psychology* 1:293); hence, the need or desire for an "I" independent of its attributes. At the same time, however, cognition can apprehend phenomena only insofar as they are different from the cognizing agent. Therefore the "I" cannot be directly cognized. One can only experience "I" as an "historic Me," as a collection of past properties and mental states. The most central element of the self, then, can be experienced only as any other attribute or complex of attributes; the inalienable is available only in alienable and contingent form. The "I"'s very independence is an attribute and effect. Identity, then, is fashioned (and refashioned) over time, and is inseparable from the malleable, appropriated attributes composing it; yet it must also be different from them, feeling, if not more independent of its components, at least more intimate than they, because it is a principle of coordination, which James calls spiritual.[1]

I argued in the previous chapter that in the natural-rights tradition property and other rights display this unresolved paradox. For rights to be precedent to positive laws, they cannot be merely conventional, though they are always, in John R. Commons's term, "acquired" under sanction of law (111). That which is natural and inalienable acquires existence and value through the alienable and conventional. Natural rights and spiritual identity are cultural achievements, though they claim a prior and authoritative provenance. Let me further clarify the dynamic of the liberal self and of proprietorship that James, among others, articulates, by comparing it with Lacan's sense that the subject is known and indeed emerges as loss. Lacan's conception of self and the proprietary self both dramatize the structural alienation of identity. "Nothing contains everything," Lacan writes, and the difference between everything and any local situation is "the gap that constitutes the subject. The subject is the introduction of a loss in reality." The Lacanian self exemplifies particularity: being less than, because not identical with, totality, and therefore unable to control the refractory. Relatedly, the proprietary self, in both Locke's and James's formulations, is compensation for and protection against particularity, which here means the fact that one is not nature. But because this self emerges through continuous appropriation, endless self-investment in and through property, it is available only through attributes that are other than itself, which cannot embody or instantiate the self in its entirety. Thus the proprietary self must experience itself as reduced from its ideal animateness and integration. In both the Lacanian and liberal visions, then, self is experienced through an expenditure (of labor or cognition) irredeemable in identical terms.[2]

I have deliberately assayed a comparison that appears ahistorical. I do not probe this intersection to impose modern theory on an earlier period, or simply to proclaim that William James, or Locke for that matter, anticipated Lacan. Rather, Lacan's vision of self resembles the liberal dynamic of self because it is forged within it and is part of its inheritance. The notion of self often associated with liberalism, the notorious autonomous self—a precept currently enjoying a resurgence in American political discourse—was by the time of the trust, as Rockefeller said, dead, at least in the view of many. It had been dying for a long time, partly because its supersession was already one pole of the liberal vision of value and self. When the trust form was finally ousted, between 1911 and 1920, by legislation and jurisprudence, it was not entirely deposed, for business practice now looked more like that of the trust than like anything else. Criticisms that trustbusting did more to legitimate than eliminate trust-like practices were, then, well taken. But though we may deplore the finally legitimating effects of trust-busting, we should understand trustbusting to be not really a failure; from the start it was an attempt to adjust law and other institutional practices to the kinds of cultural transformations that the trust both fulfilled and catalyzed. The trust's pro-tean operation, its capacity to reform itself and recombine, continued and advanced the concern with self and its relation to institutions that occupied the central figures treated in this book, from Henry Carey, to Emerson, to Debs, to Alexandra Bergson. Since the self is experienced as proprietary—which is to say as a complex of rights and obligations, freedoms and limitations—how best to protect that self?

The proprietary notion of self organized substantial parameters of action but not inflexible ones. Nor were the politics of invoking natural law entirely predict-able or internally consistent. McCabe, for example, published his history of the Grange under a pseudonym, Edward Winslow Martin, because he was a well-known apologist for Confederate political principles.[3] In his view Grangers' rights and slaveholders' rights warranted equal and analogous protection. At the same time, however, the proprietary habitus also provided the terms for contesting the regime of property—underwriting the women's rights movement, or abolition, or industrial unionism—and hence this mode of consciousness and self-conception was a central propellant of change. Mother Jones, for example, exploits the pro-prietary conception of self when she criticizes journalists' failure to report the plight of strikers at a Pennsylvania mine. The mill owners hold stock in the news-papers, the journalists explain. "Well, I've got stock in these children," Mother Jones retorts (71). Mother Jones is clearly challenging the mill owners' and news-paper owners' notion of investment; yet her point is not to eliminate investment but to invest the self in different forms and bodies.

How does it feel to be (or want to be) a combination, to be more than one, and to be yourself by relinquishing yourself? Many of the aesthetic and organi-zational projects I have considered contemplate this complex of questions, or for-mulate their enterprises in terms of this formal, historical, and affective problem. They routinely stage this condition as a problem of (self-)representation, and often illustrate and try to realize it by conceiving representation as attenuated or oblique, not as simply mimetic. Criticism tries to explain how it feels to inhabit a construct or to engage in a practice at a particular moment. Criticism also explores the effects of these practices and what those effects felt like. This study has tried to understand the combinatory impulse, its formal properties and some of its effects and affects. I hope the foregoing inquiry has been resonant, because we live still in the regime of property, combinations, and a liberalism that (often idealis-tically) aspires to its own perfection.

Yet while the propositions of this inquiry may resonate in arenas we call political, the truth or validity of my characterizations is not a political question.

Literary and historical analysis may, surely, have effects that are identifiably political. Just as surely, these effects are mediated institutionally, just as they are institutional effects. But historical or literary inquiry, if I may speculate somewhat, asks questions inevitably about the formal and affective properties of artifacts and actions. They are empirical questions about the shape and content of data, and about their effects as well. I am concerned that the recent politicization of criticism, with its ambition to explain theoretically the relations of cognitive models and formal strategies to hegemony, risks reprising an Arnoldian ethos, albeit with a different political bent, and reprising also a fallacy of identity and action that, we have seen in various forms, is one component of liberalism. I have tried to delineate the relations—the differences as well as the affiliations—among a variety of artifacts and acts addressing related issues in the regime of property. Believing they are valid, I hope my characterizations are provocative.

Notes

Introduction

1. *Congressional Globe,* 29th Cong., 1st sess., Appendix, 306. Hereafter the *Congressional Globe* will be cited as *Globe.*

2. In "Fate," McQuade 674. Unless specified, and wherever possible and convenient, subsequent references to Emerson's work will come from this widely available edition. References to the Harvard *Collected Works* will be cited as *CW.* For a thorough discussion of the way in which the concept of natural law was "a flexible instrument" in revolutionary and early-national political thought and culture (327), see Benjamin Wright's *American Interpretations of Natural Law* (1931). Wright traces the appropriation of the concept by widely competing parties. For example, he details its various applications in debates about slavery and government restrictions on property holding (191–209). Wright also analyzes the indefiniteness and confusions in Americans' understanding of natural law (327–41).

3. *Second Treatise* 145. Unless otherwise specified, page references to Locke are to the *Second Treatise.*

4. Locke did not address the paradox of why readily available necessities, like air or water, have no or little market value (whereas wild berries may have considerable value), and this question was in general a conundrum for mercantilists. Adam Smith explained the apparent paradox by introducing the factor of scarcity into the composition of value. He thus could, as mercantilists (including Locke) could not, begin to contemplate the complex interrelation between production and consumption, supply and demand, and could see that in a market society, value in general and the value of labor in particular are bound up with institutional policies and market forces. Smith's recognition of the problem of scarcity and his stress that value is inflected by market forces, and not simply a function of production, are two of the defining characteristics distinguishing classical (liberal) economic thought from physiocracy and mercantilism.

5. Pocock writes: "Property was the foundation of personality" (1985, 70). This point is unaffected by debate that has surrounded Macpherson's elaboration of possessive individualism. The debate concerns what Locke's vision implied. Macpherson argued that the possessive nature of the Lockean self implies both (1) that this self is inclined interminably to accumulate property and, more importantly, (2) that Locke sanctioned interminable accumulation, which justifies and even necessitates invidious distinctions in the distribution of wealth. Thus Locke's vision negates the universal proprietorship and enfranchisement— nature given to mankind in common—that was its starting point and that it was partly intended to recover. See Tully and, more recently, Rapaczynski, who, disputing Macpherson and defending the moral grounding of Lockean liberalism, reviews the debate clearly (177–80; 194–204). Pocock has criticized Macpherson's interpretation from a slightly broader perspective. Macpherson too quickly reduces his subjects to "economic man,"

ignores the component of "virtue" informing their conceptions of governance, and over-looks the fact that liberalism had sometimes positive, prescriptive force and sometimes neg-ative force, limiting the scope of government (70–71, 48–50, 40).

6. Pocock calls the conceptual complex conjoining self and property a "hylozoistic spir-itual materialism" (although he coins this phrase to argue that not all Enlightenment thought follows this pattern) (1985, 50).

7. For a helpful discussion of Hobbes's view of the nontransparency of signs and hearts, and of the way this conviction shapes his moral philosophy, see Kay's *Political Construc-tions* (19–44).

8. This is the basic thesis of Walter Benn Michaels's essay on the nervousness of *Wal-den* (1977), often cited as an exemplar of deconstructive criticism. Fred See's recent and dramatically deconstructive work on the realist and naturalist periods emphasizes desire essentially as a trope for the remoteness of a substantial cognitive ground of action. John Irwin's study of the impact of hieroglyphics in the antebellum period, and Michael Davitt Bell's more psychological study of the way American romanticism depended on a "sacrifice of relation" are other examples of work one might place in this category.

9. Published in 1955 in the *Harvard Theological Review*, and reprinted in *Nature's Nation* and, under the title "Nature and the National Ego," in *Errand into the Wilderness.*

10. Moretti calls this "universalizing immodesty" the *"Zeitgeist* fallacy" (25).

11. Moreover, as we have seen in the case of Senator Chestnut, and as we will see in chapter 3 in the work of George Fitzhugh *(Cannibals All!),* some premises of liberalism were deployed in defenses of slavery. Locke's own formulation of his premises has been criti-cized, by Macpherson for example, because it justifies the inevitable disenfranchisement of some subscribers to the social contract.

12. "Afterword" 433; "Problem of Ideology" 645. Bercovitch evolved versions of this thesis in *The American Jeremiad,* where he argues that American millennial discourse "denies divisiveness" and "bypasses" conflict (1978, 17, 20). See especially 9–25, 184–93. His recent essays have developed the point *American Jeremiad* urged into an argument that the American millenial imagination insidiously "represses alternative or oppositional forms" by its capacity to "deflect dissent, or actually to transmute it into a vehicle of social-ization" ("Afterword" 433). This thesis inverts the force of, say, Fredric Jameson's argu-ment, in the closing chapter of *The Political Unconscious,* about the utopian potential of popular discourse, and takes very seriously Jameson's sense of ideology as a "strategy of containment" (1981, 193). Bercovitch's thesis about the American Jeremiad may be use-fully viewed as an Americanist counterpart to the notorious containment thesis of some Renaissance New Historicist critics.

13. Indeed, American religious institutions, Puritan and (generally) Congregationalist, promoted disaffection from British rule.

14. Hartz 5. Hartz takes this phrase from Santayana.

15. See chapter 1, "Allan Melvill and Some Versions of Romance," in *Subversive Genealogy.* But *Fathers and Children,* as well, cites Hartz, and thanks him as a scholarly model.

16. By what Jehlen terms an "ideal-material fusion" (11), conflating mind and matter, American thought imagines that nature (or any phenomenon) incarnates human values. Therefore, people could take for granted that "the ethos of liberal individualism inheres in the American continent" *(Incarnation* 43).

17. Dimock elaborates a similar point when she speaks of the antebellum discourse of empire as "allegorical" (21ff.), one dimension of which is the claim of atemporality.

18. The two critics' theoretical approaches differ vastly, but, still, compare Jehlen at pp. 5, 71–74, and 147–50 with Hartz's discussions of America's ambivalence toward history

and of the difference between Europe's and America's relation to liberalism (achieved through conflict in Europe, inherited in America). Compare also Hartz's remark that American liberalism—a "pride of inheritance, not a pride of achievement" (37)—develops "an absolute and irrational attachment" to Locke's ideas that "has within it, as it were, a kind of self-completing mechanism, which insures the universality of the liberal idea" (6). Like Jehlen, Hartz explores the American fantasy of universality in order to explain the culture's durable fear of political stimuli—socialist or aristocratic—not devoutly middle-class.

19. Earlier, I associated this view, the third general thesis about the status of nature, with a pragmatist tradition. I do not mean to ascribe to Jehlen (or Bercovitch) the premises about cognition and the purportedly political agenda for which the so-called neopragmatism (of Stanley Fish, Walter Michaels, and Steven Knapp) has been decried (see Knapp and Michaels, "Against Theory," and Mitchell, ed., *Against Theory: Literary Studies and the New Pragmatism;* and see Fish 1985). The goal of ideological criticism to develop a method that will disclose the political consequences of form and induce change through a chosen form of analysis is expressly counter to neopragmatist arguments. These deny a necessary connection between forms of action and effects, and more specifically, between critical method and consequences. In contrast, Bercovitch's explicit aspiration is that awareness of ideology and "suspicion of the . . . 'natural'" will help critics "open up interpretation" and thus "make use of the categories of culture" rather than "be used by them" ("Afterword" 438–39; "Problem of Ideology" 649–50).

20. Pocock is dissenting from what he regards to be Macpherson's excessively "monolithic" understanding of Enlightenment possessive individualism, or rather from excessively monolithic applications of Macpherson's work. Whether or not Pocock is correct in his field, Leo Marx has recently registered a similar dissent in the field of American studies. He suggests that some Americanist "studies presume to generalize about American society and culture as a whole," and neglect the culture's "multilayered, fragmented character" (1988, xi). Relatedly, in his recent review of Jehlen, Rogin warns of the dangers of trying to use the Hartz thesis to achieve "an Archimedean point" from which to judge America (1989, 412–13).

21. Hartz does not so directly assert that American liberalism preempted conflict and dissent, merely that it circumscribed and restricted its shape and boundaries. Such circumscription is surely an inevitable result of *any* historical and conceptual position.

22. 1988, 8, 12. Greenblatt's work on what he calls cultural poetics is in the tradition of Althusser's point about the relative autonomy of instances or arenas of cultural activity. See Greenblatt's introduction essay to *Shakespearean Negotiations,* "The Circulation of Social Energy." My argument here about the nonuniversal and nonunified nature of culture, and its corollary point about the impossibility of a universalizing and exhaustive analysis is influenced by these lines of argument.

23. Adapting a point Raymond Williams, among others, has urged, Moretti nicely terms this project in the subtitle to his book, "the sociology of literary forms."

24. Bercovitch provides a concise description of the ways in which aesthetic work and other individual acts are historically embedded and resonant (Preface, 1986).

25. Jane Tompkins characterizes this project very well in the conclusion to *Sensational Designs,* whose subtitle alludes to "the cultural work of American fiction." While her thesis about the kind of work done by sentimental literature differs from Tompkins's, and is explicitly criticized by Tompkins, Ann Douglas's *The Feminization of American Culture* is, of course, a ground-breaking example of historicist criticism. Kaplan, typically, phrases this aspect of recent critical concerns powerfully: we must "ask *what* [literary and aesthetic acts] do accomplish and *how* they work as a cultural practice" (8).

26. This approach should indicate that I find neither literary nor nonliterary artifacts

and events, neither the cultural nor the economic sphere, to be primary and determinative. The present study, then, like much recent scholarship, rejects the classical Marxist base-superstructure model. Raymond Williams, for example (1977), questions what Althusser called the Hegelian and Leibnizian idea that an immanent "expressive causality" determines any particular phenomenon (*Reading Capital* 187). Instead, Althusser argued, determination and effectivity are structural, specifically metonymic, by which he means two things. First, effective cause is never uniform and unilateral—phenomena are vitally and densely enmeshed, at once causes and effects. Second—a point consistent with pragmatism—cause or structure exists and comes into being only *"in its effects"* (188). Throughout, we will encounter literary work as explicit social practice, intended as a conceptual and sometimes practical intervention in the social. And aesthetic categories will be shown not just to inhabit the same world as the economic but indeed often to structure the economic. Althusser's notorious concession to previous Marxist analysis notwithstanding, there is *no* determination in the last instance. Although all entities are conditioned and exist within (and have existence only by virtue of being within) a matrix of relations, determination is always local and contingent.

27. We might contrast the impulse to an ideological grammar with Kenneth Burke's notion of a grammar. Burke cannot conceive of grammar that is not action, nor of action that is not bound up with motive. Hence, his *Grammar of Motives* seeks the effective relations between scene, act, agent, agency, and purpose. Such a grammar would consist of a series of attributes. An ideological grammar, on the other hand, as currently proposed and practiced, risks reducing action to Burke's categories of scene and agency, that is, to mechanical instruments and determination.

28. I am alluding here to one of Foucault's criticisms of the notion of ideology, made in the interview "Truth and Power." He finds the concept unhelpful because it "refers, necessarily, to something of the order of a subject" (*Power/Knowledge* 118). Foucault is arguing, I think, that the notion of ideology implies that resistance must originate with a subject somehow at least partially exterior to ideological structures. The whole point of *The History of Sexuality* (and also a primary insight of deconstruction) is that nothing originates outside contexts and matrices of relations. It is because Foucault believes that Marxist criticism tends to posit an individual subject, and because he thinks it committed to positivist notions of truth and falsity, that he finds much Marxist analysis finally consistent with "the liberal university tradition" (*Power/Knowledge* 110). For an elaboration of Foucault's critique of the notion of ideology as subjectivist, specifically as this point pertains to analyzing economic action, see Michaels (1987, 176–80).

29. See Bourdieu's chapter 2, "Structures and the Habitus," in *Outline of a Theory of Practice* (72–95).

30. For detailed arguments for this point, see Fish (1985; 1988), Maslan, and Horwitz (1988), which also challenge the widely made claim that critical self-consciousness enhances critical work: this assertion, I argue, risks licensing a new eruption of moral criticism, for it finally exhorts evaluating literature and criticism primarily in terms of values they seem to embody. For a related argument, see Michaels, "Is There a Politics of Interpretation?" (1983). On the particular topic of discussion here, Wright argues that if natural law was regularly invoked, no "particular form . . . or content [was] essential" to it (1931, 342).

31. This book's historicist objective belongs to a certain poststructuralist account of the relation between literature and social practices. Recent discussions of sociological criticism have taken, generally, two positions on this question. Both dispute the classic formalist distinction between literature and political, social, or popular discourse, or between unique individual expression and the conditioned and conventional. One position can be characterized as more totalizing, or as holding totalization as a utopian ideal. Jameson is the most

eminent representative of this position in the United States. He sees a political unconscious and total History (the Real) as the absent cause of literary articulation, and his express utopian goal is to overcome the repression of, if not make conscious, political and aesthetic engagement. Against Jameson's rather Lukácsian ideal, critics more inclined to Althusserian and Foucauldian paradigms tend to stress the historically specific, relative autonomy of various arenas or instances of cultural production. The present book generally tries to observe this view, in which the primary goal of study is to apprehend the specific nature of differentiation, and thus the relations of articulation, between arenas of discourse and action. For a discussion of this debate, see Arac 264–70, 305–08. See also Jameson's "Interview" (1982). For Althusser on the relative autonomy of instances of culture rather than the Hegelian expressive or spiritual unity of all levels of the social totality, see *Reading Capital* 97, and his chapter, "The Errors of Classical Economics." Chapter 13 of Martin Jay's *Marxism and Totality* elucidates these issues very clearly. For Foucault's espousal of the "specific" rather than universal intellectual, see "Truth and Power," in *Power/Knowledge* 109–33. For a challenge to the theoretical and utopian attempt to coordinate and harmonize discrete arenas of discourse and cultural production, see Greenblatt, "Capitalist Culture and the Circulatory System."

32. The exact nature of ideology—the relation between consciousness and the real or history—has been a classic problem in materialist criticism, occupying Marx and Engels *(The German Ideology)*, Althusser, Pierre Macherey, Terry Eagleton, Jameson, and many others. Jehlen's introduction to the collection *Ideology and Classic American Literature* nicely recounts transformations in the definition of the term *ideology*. See also McLellan.

33. Althusser means "imaginary" in the Lacanian sense. The term doesn't mean fantastical but denotes a presymbolic relation, experienced as if it were immediate and total rather than mediated.

34. This understanding of the term informs Bercovitch's recent definition of it: the network "of interlinked ideas, symbols, and beliefs by which a culture . . . seeks to justify and perpetuate itself" ("Problem of Ideology" 635). Ideology, here, is an agent enlisting us, the alienated, to reproduce it and its values.

35. Following Williams (115–27), Howard and Louis Montrose are sensitive to the "monolithic" (Montrose 10) and "monologic" (Howard 30) tendency in uses of the term *ideology*. They thus emphasize the heterogeneous and shifting nature of the cultural fabric and cultural production. Howard then suggests that since "there are no inherent laws governing the functioning of those texts we call literature," we should study not the ideolog*y* of literature, but the many "ways in which literature is traversed by—and produces—the ideolog*ies* of its time" (30; emphasis mine). Howard's terminological revision is useful, but it suggests as well the inherent redundancy of the term *ideology* and the limits to its usefulness. It has positive content only when imbued in advance with an agenda, political or moral, usually both. I have suggested above that perhaps "paradigm" or Bourdieu's "habitus" better expresses the descriptive or empirical usefulness of "ideology," without the baggage this term generally carries.

36. In either case, as Porter argues, following Williams, the term is too inflexible to "understand the dynamic interplay between social and cultural formations" (15). See Williams 108–12. I am indebted to Edward Hutchinson for discussion of these and related issues. In certain institutional contexts, of course, emphasizing the study of ideological structures and formations may have impact, helping to contest scholarship that denies that subjectivity emerges historically. If few literary scholars still pursue such work, nevertheless dissenting from classic humanist (positivist) premises is valuable (because they are incorrect). Yet beyond this rather broad professional and institutional agenda, disclosing the ideological nature of thought, action, and consciousness in itself accomplishes little analytical

or political work. More generally, let me suggest that the spatial vocabulary so prominent in discussions of culture and cultural criticism is mistaken and perhaps misleading. The question is not whether one is inside or outside culture—in this formulation, there are no real options—nor even where one is in a culture. Instead, one must examine what opportunities, options, alternatives, strategies are available (or impermissible) at particular moments. The key analytical question concerns agency, not space.

37. Moretti cautions that "true isomorphisms never occur," even in physiology, and he urges us to attend to the "categorical discrepancy" in the organization of cultural forms and "to assess *each instance* carefully" (9, 19). I take his point; in the view I am expressing here, emphasizing the morphological aspect of the term, no isomorphisms exhibit true identity.

In discussing the dissemination of identity in a "relational matrix" and "network of associations," Dimock also adopts the language of kinship to describe the connections among diverse acts. Following, like many of us, the insights of anthropology, especially as found in the work of Clifford Geertz (1973), Dimock proposes to "map forth a kinship in difference" (4–5). For historical reasons, however, I would differ with Dimock's emphasis on the imperial imagination. In her discussion, which she attests parallels Jehlen's, the differences among agents and contexts finally do not matter, since all display the impulse to empire, domination, and atemporality.

Chapter 1

1. The term "Hudson River School" was hurled scornfully by young, foreign-trained artists at the academicians of the National Academy of Design, some of whom (John F. Kensett and Frederic E. Church, for example) served on the first Board of Trustees of the Metropolitan Museum of Art, elected in January 1870. For a history of the term and of the Hudson River School's changing reputation, see Avery 1987, 3–10. I retain the name American School here because it registers the nationalist designs of its participants.

2. Instead, Trumbull tried to enlist Jefferson's support for a program of government art patronage to "diffuse over the world" the nation's glory. Trumbull's letter to Jefferson, like other materials I cite later by Asher B. Durand and Washington Allston, is collected in John W. McCoubrey, *American Art, 1700–1960: Sources and Documents,* 40–43.

3. Quoted in Merritt 12. Verplanck was a prominent New York political and cultural figure. He served in the State Senate, in the House of Representatives, as a judge, and on boards of education; he published political satires and narratives, and was an editor of Shakespeare. See William Cullen Bryant's commemorative address to the New York State Historical Society (*Prose* 1:394–431).

4. Quoted in Merritt 17, from *The American Landscape,* an 1830 collection of essays and engravings Bryant edited with Asher B. Durand, at the time an engraver.

5. For discussions of associationism and its influence on American landscapists, especially on Thomas Cole, see Merritt, Baigell *(Cole),* and Roque.

6. Clinton is cited at Roque 22. Led by Samuel F. B. Morse (painter of *The Gallery of the Louvre* and *The Old House of Representatives*), those who thought the directors of the art academy should be artists rather than patron stockholders seceded. Morse's opposition to Trumbull's ideas of how to administer an academy continued for over a decade (see Morse 1933, with Trumbull's response). Flexner recounts the specific incident triggering the secession, with Trumbull, usually a hero in the story of the birth of American art, cast as villain. When some students complained that a studio was not open as scheduled, Trumbull, whose admiration of aristocratic models is clear in his request to Jefferson, replied that "beggars are not to be choosers" (Flexner 55).

7. Trumbull, Asher Durand, at the time an influential engraver, though not yet a landscapist, and another artist and patron, William H. Dunlap, later the author (in 1834) of the first history of American arts, came upon Cole's work in an art shop.

8. Avery (1986) and Howat (Metropolitan Museum, *American Paradise,* commentary on the painting, 246–50) describe the planning, publicizing, and reception of the exhibition, widely attended despite an unusually high admission charge. (Hereafter, the Metropolitan Museum's publication *American Paradise* will be cited only by its title.)

9. This fact of the painting's intellectual background has a correlative in the history of its production. It was painted, displayed, and praised during the nation's midcentury excursions into Latin American politics. Church had first visited the Andean region in 1853 in the company of Cyrus Vance, later the financial engineer of the transatlantic cable. Church undertook the journey in part to reproduce Humboldt's own visit to the region and to fulfill his vision of art as foreign exploration. Church was disappointed that Humboldt died just as *Heart of the Andes* left on its European tour, and so could not see what his writing had inspired. See Howat, *American Paradise* 248–50.

A transnationalist impulse is already evident in 1789, when Trumbull, in declining to serve as Jefferson's private secretary, instead solicited his support for a program of government art patronage that would "diffuse over the world" "monuments" to the nation's glorious history (McCoubrey 42).

10. Carrie Rebora contrasts Cropsey's painting with two earlier depictions emphasizing the viaduct's obtrusiveness in the landscape. She also demonstrates Cropsey's reliance on a description in a newspaper article about travel along the Erie line and on an illustration in a guidebook to the line (*American Paradise* 210–13). Novak describes the typical harmonization of trains with the landscape (166–76). See also Leo Marx (1985, 89–105).

11. As Rebora reports, the vantage of the viewers was not, as it may appear, attained by hiking. The outcropping on which they stand was near "an unscheduled train stop that afforded passengers the opportunity to get out and enjoy the scenery" (*American Paradise* 211).

12. Since it stimulated others to compete with its scale, *Heart of the Andes* almost single-handedly began a salon style of American landscape art (Roque 44–45). A year after Church's exhibit, Cropsey unveiled the comparably sized *Autumn—On the Hudson River* (figure 6), a painting bearing many similarities to *Progress.*

13. Two details of the painting's exhibition suggest Bierstadt's interest in the popular domestication of the exotic. On a "Great Picture" tour, it was displayed along with Bierstadt's collection of Indian artifacts (Gerald Carr, *American Paradise,* commentary on the painting, 286). When it was exhibited, facing *Heart of the Andes,* at the 1864 New York Metropolitan Fair to raise funds for the United States Sanitary Commission, Bierstadt organized a tableau vivant of an Indian village, with live Indian performances of ritual dances and sporting events (Carr 287; see also Baigell, *Bierstadt* 36).

14. The sense that a civilized aesthetic faculty entitles its proprietors to the landscape they appreciate, and establishes as well the value of landscape art, is expressed nicely in the leaflet Bierstadt had printed for exhibition customers: "Upon that very plain where an Indian village stands, a city, populated by our descendants, may rise and in its art galleries this picture may find a resting place" (quoted in Baigell, *Bierstadt* 11).

Bierstadt's depiction of the Indian tribe substantiates this point. The tribe is set off to the lower right; although nature nourishes the Indian community, the process of nourishment is invisible. The specific community being nourished is incidental to the scene depicted; or rather, the community actually nourished by the representation here is the leisure public attending it, represented not only by the figure for the recording instrument in the front left but by the couple sitting at the edge of the pool, which occupies the two lines of vision in the painting—one of sustenance and the other of aesthetic production. They sit at the point where the line from site of recording to the waterfall and the line from viewer to waterfall intersect. Finally, the setting belongs to *Lander* and the New York audience. In a public letter about this painting, Bierstadt made clear that he was in part commemorating the passing of a noble people. His formal realization of his purpose here commemorates the Indians' passing more profoundly than his letter intends.

15. The literature on the sublime is, of course, mammoth. My formulation of the issues does not pretend to be exhaustive but is limited in scope to the issues addressed in this book. My discussion is indebted to the work of Weiskel, who probes the sublime as a phenomenological and psychoanalytic problem (Bryan Jay Wolf's discussion of Thomas Cole [1982] follows his lead), and to the work of Frances Ferguson, who focuses on the sublime as a discourse of power, legitimation, taste, and domestication. Samuel Monk's *The Sublime* (1960) is a standard historical survey of the tradition.

16. Ferguson explores this point powerfully in "Legislating the Sublime" (1985). Chapters 3 and 9 will explore this issue as it figures in Twain and Cather.

17. I have slightly modified Bernard's translation; see *Kritik* 193. We should keep in mind the tangible quality of "unattainability," *Unerreichbarkeit,* un*reach*ability. Modification of future citations will be indicated by references to the *Kritik* following the page reference to Bernard's translation.

18. "Natural beauty," on the other hand, "brings with it a purposiveness in its form by which the object seems to be, as it were, preadapted to our judgment, and thus constitutes in itself an object of satisfaction" (83). Ferguson discusses this distinction in Burke as a distinction between what submits and what does not submit to our will (1981) (though in Burke, the submissiveness that we associate with the beautiful does not issue in genuine satisfaction).

19. Contesting the premises of associationist psychologists, William James puts this point succinctly: "self-feeling . . . is independent of the objective reasons we may have for satisfaction or discontent" (*Briefer Course* 164).

20. Allston's treatise is not widely available. Where possible, and when indicated, citations are to excerpts reprinted in McCoubrey. Citation of page numbers alone refers to Allston's treatise.

21. See *Capital* 1:81–96. Marx's understanding of value as a sublime relation will be discussed at length in the next chapter.

22. Similarly, James explains in the preface to *What Maisie Knew* that one reason he chose to write about a child is that "small children have many more perceptions than they have terms to translate them" (9). The "economy of process" art practices is this translation of the excess of perception into comprehensible form, specifically into the perception of the process of translation (11).

23. Kant's word choice emphasizes that sublimity is finally a capacity, or power, rather than a state, especially an external state. In general, and in the passages discussed here, Kant uses *Gemüt* rather than *Geist*. Although most conveniently translated as "mind," *Gemüt* is closer to temperament or, most aptly, disposition. That is, his word choice suggests that sublimity is dynamic, an inclination to movement.

Although Burke's more sensationalist account of the sublime identifies the sublime more with its sensible stimuli, he too is concerned with the effect of the sublime on human action, on the "labour" needed to overcome the threat that awe incites. For an extended argument on this point, see Ferguson (1985).

24. Ferguson has described the shell game of sublime threat, the way in which both Kant and Burke hold that we must already be physically safe to feel exaltation (1981; 1985).

25. Weiskel and Wolf (1982) aptly describe this clash as profoundly Oedipal.

26. Note Kant's legitimationist and contractarian language of command and consent. He seeks a judgment "welches zugleich *verlangen* darf, daß jedermann ihm *beipflichten* soll" (*Kritik* 206; emphasis mine).

27. This description of the sublime narrative applies as well to Burke's sensational exposition. With all their differences, we should remember, Kant was attempting to purify Burke's account, which he professed to admire.

28. Allston, who died in 1843, and whose *Lectures* was published in 1850, never formally belonged to the American School. But as the leading figure of the preceding generation—the 1839 exhibition of his work in Boston, attended and commented upon by virtually every important cultural figure, was the first single-artist exhibit devoted to an American (see Ellis)—and as friend to many of the younger artists, his ideas were both current and influential.

29. See Novak's introduction and her chapter "Changing Concepts of the Sublime" (34–44).

30. For a precise discussion of the picturesque and its relation to the sublime, see Fletcher's *Allegory* 252–61.

31. Commentators do not generally identify the cause of death. I call it pneumonia because Bryant attributed his death to "an inflammation of the lungs" (*Funeral Oration* 35).

32. Scholars have tended to identify Allston's romanticism with Coleridge's, but Allston's account of sublimity differs importantly from Coleridge's sense of the poetic imagination. For Coleridge, the poetic imagination "reveals itself in the balance or reconciliation of opposite or discordant qualities: of sameness, with difference" (269), finally achieving "the absolute identity of subject and object," "which it calls nature" (252). Describing a more Kantian political economy of judgment, Allston stresses the residual difference between self and nature, and the feeling of self-evacuation and the overwhelming of self by external might. While he advocates proportionality in composition, he defines sublime subjectification as an economy of imbalance. For an essay that, in a different context, distinguishes Allston's aesthetics from Coleridge's idealism on the one hand, and from Scottish common-sense realism on the other, see Doreen Hunter.

33. Though many modern commentators note an "ambivalence" in Cole and others toward the effects of progress on nature, Cole's moral and economic distinction between cultivation and utilitarian improvement has not been remarked. On one hand, thinking about the destruction of trees to make way for the railroad and towns, he could ask his patron Luman Reed to enlist Durand "to join with me in maledictions on all dollar-godded utilitarians" (Noble 160–61). On the other hand, he clearly conceived art as an appropriation and conquest of nature. When Durand was shifting vocations from engraving to painting in 1836 (a job retraining typical of the period), Cole expressed his pleasure "that you have attacked a landscape" (163). The next year, on a sketching expedition, finding a "prospect" blocked by "a mass of wood . . . I wished the axe had not been stayed" (179). The art of cultivation engages in analogous activities as improvement, a word Cole often used ironically, but is distinct for the different ends it has in view. "Essay on American Scenery" may be found reprinted in McCoubrey 98–110.

34. Though this technique has aesthetic significance, it also allowed Cole to travel unburdened by equipment and thus afforded him access to wilder scenes. Cole's emphasis on composition is evident in some of his titles, as in *Landscape, Composition, St. John in the Wilderness* (1827). It also suggests that Bryant was incorrect to praise the apparent immediacy of Cole's productions, as if they "owe nothing to revision" but rather were "transferred" to canvas "at first glance" (*Funeral Oration* 18). Cole's letters contain many discussions of what was for him the painful process of painting and repainting.

35. The title was bestowed by Cole's friend, the art critic Louis Legrand Noble, who in 1853 wove Cole's notebook entries and letters, along with commentary, into a narrative, *The Life and Works of Thomas Cole.* For Wolf's discussion of this entry and two other "sublime narratives" from Cole's journals, see 1982, 212–28.

36. Baigell, *Cole* 54; and *American Paradise* 126 (description of painting by Roque). I differ from Baigell and Roque, who believe that the Hebrew words mean only the divine promise to man.

37. My sense of the painting's conflicting statement about the capacity to possess nature is substantiated by its difference from preparatory sketches Cole made as early as 1829. These depict the scene more realistically from the vantage of the painter in the finished painting's foreground, viewing only the cultivated area. The composition's point of view ascends only in the last sketch, and the detailed thunderstorm is an innovation of the final painting. The development of the painting shows Cole qualifying the power of cultivation.

38. Noting contemporaneous political relations between Indians and whites intensifies this effect, though a Richard Ray would doubtless not have felt the connection. For a more detailed and somewhat different reading of *Progress* and its Native Americans, see Marx 1985, 104–7.

39. The panoramic style—usually views of city scenes—was an Enlightenment invention, credited to Robert Barker in Edinburgh in 1787 (Baigell 19).

40. Cole's plan for the series presumably originated in his notebook. He had included a similar description in a January 1832 letter to Gilmor ("Correspondence" 72–74). In his extensive correspondence with Reed, furnished by Noble, Cole regularly apologizes for delays in the series' completion caused by his continual replanning and revision. With all Cole's revisions, Reed died some months before his commission was delivered, an event greatly distressing to Cole and others in the New York art scene.

My understanding of Cole's series benefited greatly from a delightful morning studying them at the New York Historical Society with Richard Grusin.

41. I borrow the term *reclaim* from Bryant, who in his funeral oration for Cole, delivered before the National Academy of Design, described the drama of the final canvas as the landscape "reclaimed by nature" (24).

42. Roque tells us that this apocalyptic style was the trademark of the English painter John "Pandemonium" Martin (26).

43. We might compare this use of signature placement with Church's in *Heart of the Andes* (figure 1). Church signed this painting on an illuminated portion of a tree trunk in the left foreground, beside a sacredly symbolic patch of red vegetation. Significantly, Church's signature and the equally illuminated cross create a line to the sublime mountain peak.

44. In a striking discussion of the career of the sublime in Cole's series, Fletcher calls *Destruction* "perhaps the most perfectly sublime of the five paintings" (376). On one hand, in terms of the sublime tradition in general, this observation is certainly persuasive. In my account, though, the affect of the series, and Cole's ultimate aims, are conveyed by and culminate in *Desolation.*

45. We should note that Cole, too, even before the religious turn in his career in the 1840s, could produce the domestic sublime, in, for example, *View on the Catskill—Early Autumn* (1837).

46. James Fenimore Cooper, for example, called the series "not only . . . the work of the highest genius this country has ever produced, but . . . one of the noblest works of art that has ever been wrought" (cited in Noble 67).

47. Note one example of the popular domestication of Cole's cyclical composition. Merritt "wonders how Cole felt when his other pupil, Benjamin McConkey, wrote to him"

in 1847 of a series being planned by an artist named Sontag to complement Cole's series by depicting the "Progress of Civilization" (39).

Chapter 2

1. "Political Paradoxes" 13. Hereafter this essay will be cited as "Paradoxes."

2. Emerson himself was bemused by Greeley's declaration of allegiance, despite the younger man's avid promotion and publication of the likes of Margaret Fuller and Thoreau. Among his reasons for doubting the proselyte, Emerson does *not* include Greeley's protectionism but rather Greeley's Unitarianism and belief in the Doctrine of Miracles.

3. *Journals and Miscellaneous Notebooks* 8:314; McQuade 412. When possible, references to Emerson's writings will be taken from McQuade's widely available edition, cited only by page number. As is standard practice, references to the Harvard edition of *The Collected Works of Ralph Waldo Emerson* will be cited as *CW,* and references to *The Journals and Miscellaneous Notebooks of Ralph Waldo Emerson* will be cited as *JMN.*

4. This phrasing is used by, among others, Clay (1824, 295), Carey (*What Constitutes Currency?* 65), and Willard Phillips (203). In his influential 1820 speech on protecting home industry, Clay dramatically observes that without protection, "nothing is more uncertain than agriculture" (1820, 229).

5. Clay proclaimed that unlike free trade, protectionism was based in a true evaluation of "the nature of man" (1820, 223). Clay is actually more restrained than his followers in employing organic metaphors, but nevertheless compares statesmen to natural scientists (1824, 294) and physicians (255), and invokes natural law and "the immutable laws of God" (312) as the analytical basis of his policy.

6. Clay 1820, 221, 229. I am slighting one of the protectionist arguments, also one sounded today: since other nations were protectionist, American free-trade policies damaged, rather than promoted, home industry. This inequity is the reason free-trade policies, in the protectionists' view, denied labor its just reward.

7. Greeley, "The Grounds of Protection" 550–51, 544. To illustrate how people captivated by free-trade idolatry segregate their acts from others' acts and consumption from production, Greeley depicts the response of a typical consumer of foreign goods to the complaint by domestic farmers that his purchase lowers the price of their products: "'Why should I aid to keep up the price of Produce? I am only a *consumer* of it'" (544).

8. Colton, *The Rights of Labor* 19. Colton edited Clay's complete works.

9. Colton, *The Rights of Labor* 160. See also Carey, *Harmony* 154, 83, 192; and *Prospect* 8, 39.

10. *Prospect* 13–14; *Harmony* 211; "What Constitutes Real Freedom of Trade?" 128; *Letters to the President* 65. Clay mocked what he called the free traders' fallacy of independence (1820, 221). As we will see below, Carey attributed panics also to insufficient circulating currency. The (Jacksonian) free-trade vision partly entailed issuing only specie currency. Carey vigorously lobbied for bank notes and credit, the kinds of policies farmers and reformers would take up after the Civil War.

11. *Prospect* 6. Compare Brownson's remarks on Emerson's "Address": "his real object is . . . to induce men to think for themselves on all subjects, and to speak from their own full hearts and earnest convictions. His object is to make men scorn to be slaves to routine, to custom, to established creeds, to public opinion, to the great names of this age, of this country, or of any other" ("Emerson's Address," in Miller, *The Transcendentalists* 199).

It is worth noting that free traders, as we might predict, claimed to pursue freedom in commensurate terms. William Cullen Bryant, a leader of the Free-Trade League of New

York, said that he became a free trader by simply listening "to my own convictions." Free trade is "the simple discovery of truth" ("Freedom of Exchange," *Prose Writings* 2:242). "The law of God and nature" demands free trade rather than legislating "[our] own convictions" of divine truth (246, 242). It is significant, as well, to note the equally expansionist aims of free trade. Bryant argues: "When we make a new acquisition of territory, they [the protectionists] do not object that we are to have free-trade with the new region. On the contrary, they rejoice in a wider market" (243).

12. *Prospect* 4. Joseph Dorfman, in volume 2 of his monumental and enormously informative *The Economic Mind in American Civilization,* aptly titles his discussion of Carey "The 'Individualism' of Henry C. Carey" (2:789–805).

13. See, for example: Rusk 135; *JMN* 5:47. Before the election of 1832, Emerson muttered "that we shall all feel dirty if Jackson is reelected" (*JMN* 4:57).

14. For a recent discussion, however, of how Emerson's view of personaltiy is "racially circumscribed," see Cornell West (28–35). Even some of Emerson's arguments against slavery, West argues, are informed by a notion of racial hierarchies and evolution. For an extensive discussion of the problem, see Nicoloff.

15. For a discussion of what he calls the "antinomian" tendency of much "conservative thought," see Pocock 1973, 268–70.

16. More than Emerson's other sociological critics, West recognizes the intimacy between Emerson's goal of provocation and the stimulation, fluidity, and liquidity imbuing market activity—what I will later call its relentless transgression of forms. Yet, finally, West too (as he would readily admit) is an idealist critic, demanding that opposition must utterly transcend any specific arena of action. West thinks that the constitutive "convergence" between Emersonian provocation and market fluidity is an artificial limitation on the radical and communal possibilities of Emerson's thought. For example, West seems to think that Emerson's "powerful moral critiques of market culture" (26) must occupy an arena utterly distinct from the arena they would reform. Since they do not, and are appropriable by market culture, they are impotent. But if Emerson's (or anybody else's) criticisms did occupy a wholly distinct realm, and therefore could not be appropriated, they also could not make sense in the arena they criticize. That is, West radically opposes critique to participation; symptomatically, critique must originate outside the context it criticizes.

17. Lang cannily describes the affinity between Emerson's understanding of the individual and "free enterprise," a Smithian invisible-hand vision of collective harmony (128). Although I believe that this analogy somewhat overlooks the problematic quality of the harmony Emerson imagines, it is nevertheless generally apt. Yet for Lang, its very aptness marks Emerson's political failing, "confound[ing] economic and moral terms" (128). This phrase presumes precisely the kind of a priori moral distinction that Emerson, I think, resisted.

18. See Gilmore 18–34; and Porter 103–18. Elaborating the common, general thesis regarding the conservative drift of Emerson's career, Gilmore criticizes his increasing devotion to and hypostatization of the commodity and bourgeois wealth. Porter argues that "if one chooses to follow rigorously the course outlined in *Nature,* one has only two choices: to be either an Emersonian poet or a capitalist entrepreneur" (117). Emerson, I think, would not uphold such a radical, formal distinction between the two occupations.

19. Poirier's emphasis in *The Renewal of Literature* on the interventions of "genius" as inflections upon prior forms modifies the emphasis of his earlier book. See especially the middle sections of "Prologue: The Deed of Writing," and the chapters "The Question of Genius" and "Resistance in Itself."

20. Even in the more Thoreauvian (that is, critical) tones of "Man the Reformer," Emerson concedes that "the employments of commerce are not intrinsically unfit for a

man, or less genial to his faculties; but these are now in the general course . . . vitiated by derelictions and abuses at which all connive" (*CW* 1:147). Ian F. A. Bell has recently argued that Emerson's theory of correspondence expresses an absolute "antipathy toward trade and commerce" and a rejection of paper currency (735). Endorsement of specie would be the worst form of "squalid" fetishism and would certainly offer no adequate alternative to the problems Emerson engaged. (See the next section of this chapter.) Bell quotes a number of *Journal* passages criticizing trade practices and takes these as a criticism of all trade. Nearly every one of these passages accompanies a general endorsement of the necessity of trade as a spiritual activity, and criticizes not trade in principle but particular instances of commerce. In contrast, after a more extensive and rigorous analysis of *Journal* entries, B. L. Packer concludes that Emerson "came to understand that [capitalists'] aims were not so far from his own as they might appear . . . ; tycoons were poets who chose to write their epics in cash" (96).

21. This remark became in "The Individual": "All the *Trades* and *Callings* of men are the mode in which the concealed laws of the Mind are taught us through our hands" (*Early Lectures* 2:181).

22. Parker observes that both protectionists (manufacturers) and free traders (importers) are protectionists in the larger sense. Both pursue policies to protect their class interests, but neither attends to protecting employees and common laborers (453).

23. The interminability of relations is the main point of "Circles." In addition, in "Quotation and Originality" Emerson questions the absolute opposition between imitation and originality. "None escapes [debt]," he writes. "The originals are not original." Even "'Paradise Lost' had never existed but for [its] precursors" (172). "Original power" requires vast "assimilating power" (181).

24. Lang formulates a similar logic in her interesting and detailed discussion of Emerson's vexed relation to antinomianism. He explored "the logic that funnels the private into the public" (128).

25. *Capital* 1:82–85 passim. Hereafter this volume will be cited simply as *Capital.*

26. For Lukács, the reifying "basis" of the commodity-structure "is that a relation between people takes on the character of a thing and thus acquires a 'phantom objectivity,' an autonomy that seems so strictly rational and all-embracing as to conceal every trace of its fundamental nature: the relation between people" (83).

27. Since Lukács holds that reification is the "essence of the commodity-structure," he essentializes the alienating affect of the commodity more radically than Marx (83). Without, as Baudrillard does, organizing his argument around a distinction between the symbolic (differential relations) and semiological (code of relations) apprehension of sign systems, Thorstein Veblen anticipated a critique like Baudrillard's. See his essays of 1906 and 1907 on "The Socialist Economics of Karl Marx" (*The Place of Science* 409–30; 431–56; see especially 420–22).

28. This double transaction is the famous M-C-M and C-M-C delineated in *Capital.*

29. McQuade 125. See also the *Journal* entry of December 1836: "Do you not see that a man is a bundle of relations?" (*JMN* 5:266).

30. However brilliant his critique of the semiological fetishism of most analysis of commodification, Baudrillard finally exhibits a version of the same proclivity. He argues essentially two related points. First, commodity fetishism is a desire not for objects but for "the closed perfection of a system" that in hinting at complete transparency—at "the total commutability of all values" (93)—reminds us of our exclusion from "its internal logic or perfection" (96). The fetishist, in this view, has a perverse "autoerotic" desire, not for the gratification offered by commodities but for "the sanctuary of [one's] own alienation." Second, the fetishism of conventional analysis of commodities inverts essentially the same desire for

code. Here, the code is sought not to discover one's alienation but to close the gap between labor and enfranchisement that commodities signify. Thus, most ideological analysis would restore "the subject of private property" (96–97).

Baudrillard rejects this strategy for reproducing the ideological process itself, the reduction of the play or conflict of symbolic signification to comprehensible form (98). He would emphasize, instead, the revolutionary possibilities of insisting on the alienating "process of real labor," both productive and symbolic labor. He would retain the "ambivalence" structuring all exchanges (productive, symbolic, erotic) in order better to survey the process of disenfranchisement. But in thinking that retaining the ambivalent structure and affect of exchange may undo or resist ideological reduction, Baudrillard commits the mistake of formalism that is essentially what his charge against analytical fetishism amounts to. Baudrillard thinks the effects of his analysis follow from its form; he thinks that taking a different attitude toward the commodity form will in itself resist the political and psychic structures in which commodities circulate. Baudrillard's project inverts the fetishistic desire for closure he so cannily identifies. His analysis desires not the perfect closure that the sign system masquerades as having but the perfect dehiscence or ambivalence he thinks it really does have.

31. I discussed this aspect of sublime economy in the second section of the previous chapter. I am thinking here of something more specific than just Lacan's notion of the mirror stage. For example, Lacan invokes Russell's paradox: "The subject is the introduction of a loss in reality, yet nothing can introduce that, since by status reality is as full as possible. The notion of a loss is the *effect* afforded the instance of the trait . . ." (193; my emphasis). The recognition of a trait by the subject feels like a loss in the subject, because the particularity of the trait is known only by comparison with the incomprehensible whole.

32. I was disposed to note Willard's surprising characterization of his economic project by Pease's 1980 discussion of the catachrestic effect of Emerson's transparent eye-ball figure. (This part of his discussion seems absent from its adaption in Pease's later argument about visionary compacts.) Not only does the figure of catachresis seem logically apt, Phillips's formulation suggests that Pease's observation is historically appropriate as well.

33. Somewhat less pointedly than Emerson, Simmel calls money "the pure form of exchangeability," and observes that it "can never be enjoyed directly" (130, 128). Simmel intends this remark to apply to even obvious attempts to enjoy money directly, fetishism and miserliness. These "exceptions . . . negate [money's] specific character" (128). To the extent that Simmel accepts the intent of these "exceptions," he is mistaken. Even the fetishist or miser enjoys something else in enjoying money, taking pleasure in the *desire* to enjoy money directly.

Credit cards, we should note, seem to surpass money as an instance of metabolism; but in fact the operation of credit cards, if perhaps more conspicuously speculative, is structurally identical to that of money.

34. *Money* 4, 23, 29; see also *Letters to the President* 69, 76, 78.

35. Carey cites this phrase from Hume's "Of Money" often; see also *Money* 22, and *Letters to the President* 87. Carey viewed his protectionism as inimical to the specie policy of most economists, like Francis Bowen of Harvard, or, later, Henry Varnum Poor, both of whom had harsh words for Hume, Smith, Dugald Stewart, and Ricardo for denying that money could have intrinsic value. Hume's defense of credit and paper was especially controversial in the period. Carey's position, then, as I indicated earlier (see note 10), anticipates a central tenet of agrarian reform movements after the Civil War, soft money (nonspecie) and greater available currency; and Carey himself, albeit linked with Whigs, was a preeminent catalyst of the greenback movement. The precept of greater available currency is still respected today as having immediate benefit for laborers, but Carey's proposal was

vigorously criticized by liberal organs such as the *Democratic Review*. See "The Moral of the Crisis."

36. Thus Simmel writes: "the isolated individual who sacrifices something in order to produce certain products, acts in exactly the same way as the subject who exchanges, the only difference being that his partner is not another subject but the natural order and regularity of things, which, just like another human being, does not satisfy our desires without a sacrifice" (83).

37. See Michaels's discussion of the implications of Thoreau's attempt to discover Walden Pond's depth and source (1977).

38. Of all the members of the Transcendental Club, Hedge was perhaps the best acquainted with German idealism. Interestingly, Hedge compared fetishism favorably to "crass sensualism" and vulgar materialism, since the fetishist animates worshipped objects with spirit, worshipping "the demon's sake supposed to reside in them" (339).

39. I should note that the term *nature* means something different, the principle of limitation, throughout much of "Fate."

40. In a different context, Grusin has recently discussed Emerson's understanding of economy as a potlatch, and as related to Bataille's notion of economy as excess (1988).

41. Poirier quotes a June 1847 *Journal* entry that seems pertinent here: "Every thing teaches transition, transference, metamorphosis: therein is human power, in transference, not in creation . . ." (cited in 1987, 142).

42. Speaking of the basic desirousness of humans, Hobbes writes: "the Felicity of this life, consisteth not in the repose of a mind satisfied. . . . Felicity is a continual progresse of the desire, from one object to another; the attaining of the former, being still but the way to the later" (*Leviathan* 160). I discussed this overlap in the Introduction. Note that Emerson abstracts Hobbes's focus on desire to a question of sheer power.

43. Anticipating this passage from "Self-reliance," Emerson wrote in 1838: "it seems fittest to say *I Become* rather than *I am*. I am a *Becoming*. . . . I am nothing but a prophecy of that I shall be" (*JMN* 5:468).

44. The rest of Emerson's remark substantiates this point: "I say to you plainly there is no end to which your practical faculty can aim, so sacred or so large, that, if pursued for itself, will not at last become carrion and an offence to the nostril" (*CW* 1:133).

45. See Lang for a thorough, and differently inflected, analysis of Emerson's troubled adaption (and in some ways explicit excision) of antinomian principles, mainly due to their threat to social order (107–36). In *The Imperial Self,* Quentin Anderson places Emerson more firmly in the antinomian tradition, which he takes to be less subversive or even anarchic than Lang.

46. See Grusin's innovative chapters on the "Address" and on Emerson's resignation in *Transcendentalist Hermeneutics.* I am grateful to Grusin for discussions of this point.

47. For example, when preparing in the spring of 1838 for his July talks at the Harvard Divinity School and at Dartmouth College, he called the Transcendental Club "unnamed or misnamed," also lamenting that its "disciples" "speak in the phraseology of Swedenborg" and "do already dogmatize & rail at such as hold it not." That is, epigones adopt the language of spirit as merely a new doctrine (*JMN* 5:481).

48. See Burke, Bloom, and especially Poirier, Carpenter (164–78), and most recently Cornell West. There are strong historical reasons for this affiliation, Emerson's friendship with Henry James, Sr., and William's close scrutiny of Emerson's works. This connection has a conceptual source in *Nature*'s chapter "Idealism," for example, wherein Emerson argues that our "impotence" to test empirically "whether nature outwardly exists" makes no difference: "Whether [or not] nature enjoy a substantial existence without, . . . it is alike useful and alike venerable to me." In asking "what difference does it make?" (26–27) and

in contending that the ability or inability to validate beliefs in absolute terms makes no difference to the substance or efficaciousness of a belief, Emerson clearly anticipates the pragmatist conception of truth and validity.

49. While composing *Nature,* Emerson first formulated this idea as "Ethics again is to live Ideas" (*JMN* 5:146).

50. For example: "Ethics stand when wit fails." "[D]eal justly . . . & you do something and do invest the capital of your being in a bank that cannot break & that will surely yield ample rents" (*JMN* 5:344). Emerson is distinguishing this spiritual investment from the kind of banking that can fail, but the principle is the same.

51. In the same entry as the long passage cited above, Emerson writes that commerce cannot be founded "on this dangerous balloon of a credit" (412). Though not speaking of credit per se, in "The Transcendentalist" Emerson employs the image of a balloon sailing wildly out of control when discussing the naive empiricism of the "sturdy capitalist."

In "American Romanticism and the Depression of 1837," William Charvat argues that romantics, nervous about the possibility of revolt by labor, responded to the panic mainly as a patrician class unhappy with the irresponsible practice of the new business class supplanting them (49–67).

52. See Grusin's discussion of this entry as it pertains to the miracles controversy (1990, chapter 2).

53. In 1837, Emerson succinctly calls virtue "creation" in the arena of ethics, just as genius is creation in the intellectual arena; both demonstrate "the pure efflux of Deity" in the self (*JMN* 5:341).

Chapter 3

1. *Mark Twain's Letters* 1:101. Letter of 20 January 1866, San Francisco. Hereafter cited as *MTL.*

2. Henry Nash Smith, *The Development of a Writer* 78. Cox similarly points out that "the river is the element . . . giving perspective to all the values which emerge" (118).

3. Brooks 140; see also De Voto 1932, 106–7. Mark Twain himself did not formally divide the book into two parts.

4. Locke accepted the common, if not entirely literalistic, interpretation of Psalm 115's sentiment that the earth is "given . . . to the children of men."

5. Bascom taught at his alma mater, Williams College, and was a Unitarian and a free-trade advocate, though he mixed this policy with the American School optimism of the protectionist Carey. His book was used at Yale, among other institutions (Dorfman 2:752–54). Bascom and Locke differ from Aristotle on the source of value, and this difference defines the difference between Aristotelian, premercantile economic thought and classical economic thought. Aristotle believes that only nature creates value and that when he transforms nature's gifts into usable form, man merely is a caretaker or steward for them. Locke and Bascom hold that man *creates* value in the appropriation of nature.

6. *Mark Twain-Howells Letters* 1:33: "I have delayed thus long, hoping I might do something for the January number, & Mrs. Clemens has diligently persecuted me day by day with urgings to go to work & do something, but it's no use—I find I can't. We are in such a state of weary & endless confusion that my head won't 'go.' So I give it up." Hereafter this volume will be cited as *MTHL.*

7. He proposed that he could gather materials for it only in St. Louis. Albert Bigelow Paine's selection of Twain's letters does not include these details of the 1866 letter. The letter is held in the Mark Twain Papers at the Bancroft Library of the University of California at Berkeley and is cited in *MTHL* 1:35n; in Cardwell 284; and in Burde 878.

8. *The Love Letters of Mark Twain,* 27 November 1871, 166. Twain's book indeed became a standard work. One reviewer, Robert Brown, remarked that it contained to date "the best account" of the river's history (*Academy* 24 [28 July 1883]:58; reprinted in Anderson, ed. 117). Histories of trade or transportation on the river commonly cite Twain's book as an authoritative account (though sometimes to be disputed). I have even found an instance of mild plagiarism: Archer B. Hulbert writes in *The Paths of Inland Commerce* (1920) that the Mississippi "could shorten itself thirty miles at a single lunge" (177). Compare Twain: "More than once it has shortened itself thirty miles in a single jump" (14).

When the history of the composition of *Life on the Mississippi* is recounted, seldom is any discrepancy noted between the idea for "a standard work" and the idea for papers about piloting; when noted, this distinction is not considered significant. Henry Nash Smith and William Gibson, editors of the *Mark Twain-Howells Letters,* conclude that Twain's idea to write about his piloting experiences is merely a refinement and focusing of the earlier, more general Mississippi idea (35n). Cardwell follows Smith and Gibson (283). The *Atlantic* papers' concentration on piloting may be a refinement of focus, but the consistency with which Twain distinguished the two texts when referring to them indicates that he maintained in his mind two distinct ideas and works.

9. *MTHL* 1:26 (26 January 1875); 47 (3 December 1874); 88 (21 June 1875); 48 (4 December 1874); 52 (11 December 1874); 434 (10 July 1883). Canadian pirates shared Twain's and Howells's sense of "Old Times," titling the illegal Canadian edition *Piloting on the Mississippi* (*MTHL* 1:270–71, 17 September 1879).

10. The diary kept by one of De Soto's aides confirms Twain's sense of De Soto's attitude. Upon arriving at the banks of the Mississippi, De Soto "went to look at the river, and saw that near it there was much timber of which piraguas might be made, and a good site in which the camp might be placed." Here, the first view of the river is instrumentalist (quoted in Morris, ed., *The Mississippi River Reader* 9).

11. This is one aspect of the criticism that Walter Benn Michaels and Steven Knapp direct at attempts, as they phrase it, to "do theory" (see "Against Theory"). Knapp and Michaels do not use the term *romantic,* but essentially their charge is that theory, as they define it—an attempt to stand outside practice in order to govern practice—is romantic in the idealist sense. For an extended application of this argument to contemporary "oppositional" criticism, see Maslin.

12. In *Virgin Land,* Henry Nash Smith has demonstrated the appeal of primitivism to many popular authors and pamphleteers. See especially chapters 5–8.

13. Quoted in de Barbé-Marbois 215. The Americans were not overstating their contribution. Barbé-Marbois, French legate to the Louisiana Purchase negotiations, quotes Lord Hawkesbury: "To judge of the value of Louisiana in the hands of the French, let us recollect that they have heretofore possessed it for a long period, without being able to render it prosperous" (183). On the continent in general, Barbé-Marbois adds, Americans "had contributed more to all kinds of improvement than the European states had effected in the colonies subject to them during three centuries" (209).

14. For two subsequent invocations of the settlers' protest, see Gould 288, and Anderson 1890, 7.

15. Cumings 7. The 1822 edition does not contain this description.

16. The language here values the river because it seems constructed like any other artifact or tool, and Gould continues that it is "just as valuable as if artificially built" (333). During his 1882 visit to the river, Mark Twain employed the same conceptual vocabulary: "the river was as brand new to me as if it had been built yesterday and built while I was absent" (*Notebooks* 2:528).

17. "Report of the Committee on Commerce," 9 February 1843, 1–2.

18. "Report on the Improvement of the Navigation of the Ohio and the Mississippi

Rivers," 27 March 1828, 2; "Report of the Board of Engineers on the Ohio and Mississippi Rivers," 22 January 1823, 19–20. It is doubtless significant that Mark Twain named Huck's friend and mentor "Sawyer."

19. Morgan Neville, in *Western Souvenir* (1829), 106–7; quoted in Smith, *Virgin Land* 157.

20. Ellet 196; Frank, *The Development of the Federal Program of Flood Control on the Mississippi River,* chap. 1, especially 17–22. Frank's book was extremely helpful in describing issues behind debates and as a guide to research sources.

21. Frank 77. Any government program of flood control does, of course, constitute such a reclamation project, for distance from the river directly affected land values. Frank cites the fact that lands once nearly always flooded acquired great value after levees controlled water levels (78). He also points out, as did congressional debate, the Civil War's mutilation of the private levee system, and consequent damage to land values. Writing his excellent history in 1930, Frank is himself nervous about federal aid to individuals and attempts to downplay this effect of flood control; rather, he emphasizes, as did pro-aid politicians, the way flood-control precautions optimize the river's trade capacity, and thus its benefit to the nation. Like late-nineteenth-century legislators, Frank treats the inevitable private benefit as a nasty side effect.

22. *Senate Executive Documents,* No. 49, 32d Cong., 1st sess., 1852. Ellet's report appeared in expanded form in 1853 as *The Mississippi and Ohio Rivers,* to which I have already referred.

23. *Congressional Record,* 43d Cong., 1st sess., 21 April 1874, 3242; *Appendix to the Congressional Record,* 43d Cong., 1st sess., 4 June 1874, 411. Hereafter the *Congressional Record* will be cited as *CR.*

24. *CR* 43d Cong., 1st sess., 4 June 1874, 4569; *Appendix to the Congressional Record,* 43d Cong., 1st sess., 4 June 1874, 410.

25. *CR,* 43d Cong., 1st sess., 21 April 1874, 3242; ibid., 6 June 1874, 4658; *Appendix to the Congressional Record,* 43d Cong., 1st sess., 4 June 1874, 413.

26. "Report of the Mississippi River Commission" 4, 11; *Statutes at Large,* 28 June 1879, 38; "Preliminary Report of the Mississippi River Commission" 6.

Seeking "a system of observations" to underpin what the "Preliminary Report of the Mississippi River Commission" called a "comprehensive system of control" (*House Executive Documents,* 46th Cong., 2d sess., No. 58, 6), the nineteenth-century determination to control nature I am describing clearly evokes Foucault's thesis about surveillance and regulation in the carceral society in *Discipline and Punish,* from which some critics have evolved theses about literary production as projects in social control and containment. Critics such as D. A. Miller ("The Novel and the Police") and Mark Seltzer ("*The Princess Casamassima:* Realism and the Fantasy of Surveillance") have discussed the realist novel in these terms; and Frank Lentricchia has employed a generally Foucauldian model in thinking about Stevens's poetry. But the rhetoric of improvement legislation finally departs from most formulations of the carceral thesis; its goal is perfection, not simply control. That is, its management was enablement, "improvement," not merely constraint; or rather the constraint is itself a form of enablement. This logic is actually closer to that of Foucault's work after *Discipline and Punish,* like *The History of Sexuality,* vol. 1, or "The Subject and Power," which usefully revise the earlier carceral thesis.

27. Cox concludes that the main objective in Mark Twain's romance of piloting is to assuage skepticism, which had been the aesthetic of *Roughing It* (126). Forrest G. Robinson has more radically argued that "the river displaces depression" and a sense of sin "by emptying the mind of all content" (204). My argument concerns the positive content supplanting what these critics identify as evacuated.

28. See "What Paul Bourget Thinks of Us" 145–46, and "Down the Rhone" 143–45.

29. Hunter 242. This book was extremely helpful as a descriptive history and guide to research sources.

30. *Congressional Globe,* 32d Cong., 1st sess., 28 August 1852, 2427. Advocates of steamboat regulation recognize the interdependence of, rather than the absolute identity between, persons and property. They find it disgraceful that "human life would be recklessly and wantonly sacrificed to the cupidity of steamboat proprietors" (*Globe* 2426). No one is free if human concerns are aspects of commerce; such subordination would constitute a truer "interference with individual action" than government intervention (Gouge 425).

31. It is important to note that intervention is distinguishable from tariffs, a policy supported by some opponents of the Steamboat Act. As we saw in the previous chapter, proponents of tariffs felt they protected the self rather than interfered with self-maintenance.

32. On the points raised in this discussion, see Macpherson *The Political Theory of Possessive Individualism* 197–201, 229–38, 255–62. Macpherson writes, for example, that Locke's notion of individualism "asserts an individuality that can only fully be realized in accumulating property . . ." (255). For a related discussion of the Lockean self as an "identification" with its material properties and expressions, see Rapaczynski 195–207.

33. Hunter 307, 567. Eastern rivers also flooded, of course, but they had deeper channels, more resilient riparian soil, and more direct routes to the sea (reducing cutoffs). Thus, regular schedules were easier to follow, encouraging incorporation in the eastern industry.

34. Michaels's essay ("Romance and Real Estate") is a powerful interrogation of the problematic ideal of the Lockean fable of property, what Macpherson calls "full proprietorship of [one's] own person" (*Theory of Possessive Individualism* 231). In *The Antislavery Appeal,* Ronald G. Walters describes the ways in which abolitionists both recognized and effaced a free-labor market's apparent attenuation of the "self-ownership" they advocated and sought (122; see generally 111–23).

35. On the southern claim that, in Fitzhugh's words, "capital exercises a more perfect compulsion over free laborers than human masters over slaves" (32), and that slavery reduces the alienation of labor by functionally extending the family, see Genovese 151–235. For a full discussion of Stowe's complex attitude toward slavery as a form of free labor, and of the relation between Stowe's attitude toward slavery and her domesticity, see Gillian Brown, "Getting in the Kitchen with Dinah"; and Brown's chapter "Sentimental Possession" in *Domestic Individualism: Nineteenth-Century American Fictions of Self.* I am grateful to her for discussing with me the issues in this section of the essay.

36. On freedom as something that "comes," see also Washington 23, and Du Bois 47.

37. This argument that freedom in *Huck Finn* is a form of property applies as well to Huck's supposed escape to the Territory "ahead of the rest" (229). Unconditioned freedom is unavailable in the Territory not only because, as Holland has argued, by the time Huck arrives in 1884, the Territory is no "green continent" but "populated by Kings, Dukes, and Aunt Sallies" (75). The Territory was always an area designated by law for certain kinds of settlement. It is not unlike a contemporary Wilderness Area, exempted by government decree from development and excessive tourism. Untarnished nature and the freedom one finds there require government sanction and protection.

38. Even Uncle Mumford, despite his resentment, respects the scope of the Commission's intent: ". . . they are going to take this whole Mississippi, and twist it around and make it run several miles *upstream.* Well, You've got to admire men that deal in ideas of that size and can tote them around without crutches" (174).

39. Historical placards along the newly designed New Orleans riverfront area inform tourists of the river's continued erratic behavior, proudly citing Twain's and Uncle Mumford's remarks about the river's untamable lawlessness.

40. *Mark Twain in Eruption,* De Voto, ed., 18–19. I first encountered this quotation in Stanley Brodwin's "The Useful & the Useless River" 196.

41. See, for example, Rutherford B. Hayes's 1877 address to Congress (619); or "Letter from the Secretary of War," *House Executive Documents,* 28th Cong., 1st sess., No. 153, 26 February 1844, 2.

42. Ray Ginger recounts how quickly the transition from subsistence to merchant farming could occur. In 1879 in Bell County, Texas, farmers grew feed for their livestock, and "grew all their own breadstuffs," which was locally ground into flour. "They grew little cotton . . . for it cost too much to haul cotton by wagon to Gulf ports." The railroad arrived in 1880, bringing "flour of higher purity and reliable quality, ground by mass production methods in the Minnesota Mills of Charles A. Pillsbury or Cadwallader C. Washburn from grain grown in the burgeoning wheat belt on the Western Great Plains." By 1889, Bell County wheat production had declined 75 percent; cotton production was up 4000 percent (*Age of Excess* 22–23).

43. Hunter 488, 502, 519, 585; Haites and Mak 34; Fishlow 56; Taylor 1962, 102. For a differently nuanced account of the health of steamboating during the 1850s, see Erik Haites and James Mak, "The Decline of Steamboating in Antebellum Waters" 25–36.

44. See Rodgers for a discussion of the transformation to an ethos of consumption and an antipathy toward work. During an era of declining prices, surplus inventories, and unemployment, the ethos of consumption registers and responds to anxieties about overproduction and the inability to consume as much as industry could produce.

45. A corollary point of this episode is that it makes no sense to enjoin readers *not* to appeal to external evidence, for all evidence is equally external and internal; that is, it is evidence.

46. It is important to observe that this pragmatist account of persuasion—that persuasion is possible only in the context of the beliefs that we hold and that hold us—does not preclude change. But change occurs and makes sense, again, in the terms and beliefs one inhabits. William James is very specific on this point in the chapter "Habit" in the *Psychology.* He recommends developing the habit of resisting one's will and habits (see *Briefer Course* 137).

47. In his recent book, Richard Bridgman argues that one of the dominant psychological and thematic principles of organization (or lack of organization) in Twain's travel writings was Twain's anxiety about change.

Chapter 4

1. Howells, "Bibliographical," *A Hazard of New Fortunes* 4; *Letters of Henry James* 1:164, quoted in Carter, "Introduction" xxvii.

2. Kaplan argues that character is "the foundation of the realistic edifice" Howells sought to erect (35). My argument in this chapter is, obviously, consistent with Kaplan's view, though it explores different problems raised by this notion than she does.

3. Howells speaks of nervous distress as "the nervous woes of comfortable people" (430). While Howells suggests that the material distress of the poor may be "more" severe, because physically ineluctable, Edwin Cady and Kenneth Lynn point out in their biographies of Howells that the nervous woes of the comfortable were quite real to him. Much as Alice James did, Howells's daughter, Winny, suffered much of her brief life from neurasthenia, until she died during the composition of *A Hazard,* inducing in her father a mild collapse from exhaustion. Unlike in Alice James's case, Winny's diagnosed neurasthenia was not determined to be an organic malady until after her death. Of Howells's own peri-

odic nervous episodes, the most serious occurred during the composition of *The Rise of Silas Lapham.* As it was for Mark Twain, work was Howells's therapy and he continued to produce regular installments of the novel. See Lynn's discussion of Winny's illness and Howells's response.

4. On this term, see Taylor 1932. The other novels of the group are *Annie Kilburn* (1888), *The Quality of Mercy* (1892), and *The World of Chance* (1893).

5. The work of this period documenting the socialist and communitarian influences on Howells is extensive. See, for example, the work of Taylor, Getzels, Conrad Wright, and Arms. For related work continued in later decades, see Ekstrom, Budd, Cady, Lynn.

6. "Editor's Easy Chair," *Harper's Magazine* 124 (March 1912): 636. Subsequent references to this column will appear in the text abbreviated as EEC, followed by the date and page. References to Howells's earlier *Harper's* column, the "Editor's Study," will appear abbreviated as ES.

7. See also McMurray and Martin.

8. Perry 163. On the success of Perry's work, see Dorfman 3:4. As usual, Dorfman's study was an extraordinary research source. For a discussion of potato gospel, see Atkinson 1890, 1–7. On the stationary state, see Dobb 87–91.

9. For Carey's challenge to Ricardo's understanding of land and rent as inappropriate to American conditions, see his *Principles* 1:21–75, 129–42, 200–240.

10. Clark, *The Philosophy of Wealth* 32. All references to Clark's work in this chapter are to this 1887 work.

11. Atkinson sees human economy as slightly, though not severely, discordant with God's supreme harmony, but he envisions that soon (within two or three generations!) humanity will achieve divine concord. Other orthodox writers like Perry and Francis Amasa Walker (son of the free-trader Amasa Walker and another popular textbook author), and J. Lawrence Laughlin (a major contributor to bimetallism debates of the 1890s) find Americans only scarcely out of harmony.

12. In "The Past and Present of Political Economy," Richard Ely presents an exemplary account of the necessary dependence of citizens upon one another. "The phenomena of exchange . . . make . . . clear . . . [that] the first and foremost factor of modern economic life is dependence" (51; hereafter cited as "Past"). This principle is the starting point of Ely's *Social Aspects of Christianity* (1888). For Henry George's use of the solar and cellular metaphor, see *Progress and Poverty* 487–88.

13. This notion was inherited from Aristotle, who argued that the household was the first element of the polis (see his *Politics,* chap. 5). American economists shared Aristotle's historical assumption that society began with a few families banding together for mutual protection and support. These groups of families then developed into neighborhoods, tribes, and, later, towns and nations. On this model of primitive (and of course Christian) communism, Howells will write in 1895 "Who Are Our Brethren?"; moreover, Edward Bellamy's utopia in *Looking Backward* and Howells's own Altruria view all relations as familial.

14. *The Political Economy of Humanism* 5–7. This book reprints with two additional chapters Wood's 1894 *Natural Law in the Business World.*

15. In "Equality as the Basis of a Good Society" (1895), Howells suggests that justice develops only after individuals become the source of value: "Wherever men are remanded to a situation where personal worth has sway, social equality reappears among them" (64).

16. See Bellamy 170; and Ely "Recent American Socialism," especially 28, 39, 71. Ely levies some interestingly contradictory criticisms at radical groups. On one hand, he criticizes the Internationalist movement, notoriously (though not really) anarchist, as "individualism gone mad." On the other hand, he attacks the "terrible condition of a soul" that has, in the words of one revolutionary broadside, "'no personal interest, concerns or incli-

nations, no property, not even a name.'" Thus Ely condemns radicalism for being at once too individualistic and not individualistic enough.

17. Amy Kaplan discusses Josiah Strong's 1885 best-seller *Our Country: Its Possible Future and Its Present Crisis.* In this volume, Strong employs a similar analogy to opposite purpose: "the utmost depth of wretchedness exists not among savages, who have few wants, but in great cities, where, in the presence of plenty and of every luxury men starve" (quoted in Kaplan 45). Though Strong puts the vision of savages as desireless to more familiar use than Gladden—those without desire are innocent; we civilized are corrupted by desire— his use of the example of the savage with few wants suggests the typicality of Gladden's appeal to it.

18. Sklar 81–82; see, for example, Conant. This issue arises again with respect to the relation between the form of the trust and Debs's conception of industrial rather than craft unionism. See the fourth section of chapter 7. It may be evident from my brief discussion of this part of Sklar's argument that he views developments in corporate capitalism not as adjustments in the superstructure or practices that alienate persons from themselves, but as a transformation of "thought and feeling" (6).

19. For an analysis of how capitalism's need for proliferating desire always to outstrip the means of satisfaction informs Theodore Dreiser's *Sister Carrie,* and of what implications this concept might have for realism, see Walter Benn Michaels, "*Sister Carrie*'s Popular Economy," in *The Gold Standard* 31–58.

20. The line of argument that Deleuze and Guattari have importantly influenced is directly attacked by Foucault in *The History of Sexuality,* vol. 1. Desire, Foucault reiterates frequently, is not extrinsic to a system but constituted within it. In fact, Deleuze and Guattari's "schizophrenia" is a far more elegant and complex relation than the one Foucault attacks. Desire for them is a necessary element of capitalism, though a continual threat to it. Yet I do think their two-stage account retains a residual idealist component, of desire as independent of its culture, even though necessary to it. It "wells up," then is contained. The revolution they associate with desire can come only from outside the matrix of relations. If so, however, on what basis could it have any effect? Foucault is correct, I think: even if desire, or any other feature of an economy or psychology, seems to threaten an institutional order, its origin, content, and trajectory are not independent of it. Certainly, the economists and moralists I am discussing held this position.

21. "Why Is Economics Not an Evolutionary Science?" (1898), in *The Place of Science in Modern Civilization* 73–74. Hereafter cited as "Why."

22. "Some Neglected Points in the Theory of Socialism" (1892), in *The Place of Science* 408.

23. While there is much overlap between Veblen's and Marx's concepts of history, and while their recommendations for economic reform are similarly founded on the idea that technology itself should provide a basis for economic organization, nevertheless Veblen levels the same criticism against Marx. His concepts of history and character are teleological, with the classical convention of the hedonistic self ultimately and naturally achieving perfection in an economy, albeit socialist rather than capitalist. See "The Socialist Economics of Karl Marx, I" (1906), in *The Place of Science* 409–30, especially the first third of the essay.

24. "The Evolution of the Scientific Point of View" (1908), in *The Place of Science* 44. Hereafter cited as "Evolution."

25. *The Theory of the Leisure Class* 132. Hereafter cited as *LC.*

26. Veblen would add an important (nonpositivist) caveat: the induction gained by the study of adaptation is always retrospective, conditioned by old habits of thought and belated at least a split second after the newly evolving habits of thought. This position on cognition

resembles that of Veblen's contemporaries James and Peirce, but it also resembles the contemporary poststructuralist insight about the mediated nature of perception and consciousness.

27. "Literary curiosity" can be compared to Veblen's concept of "idle curiosity," which he defines as "irrelevant" attention rather than "pragmatic attention" to objects. The latter is a "tropismatic reaction," immediately categorizing phenomena according to one's purpose. Idle curiosity's "interpretation of the facts" is less appropriative. It "formulates its response to stimulus not in terms of an expedient line of conduct, . . . but in terms of the sequence of activities going on in the observed phenomena." Though still animistic in that it dramatizes phenomena according to cultural expectations, the idle curiosity is Veblen's attempt to formulate a relatively neutral mode of perception ("The Place of Science in Modern Civilization," in *The Place of Science* 6–7).

I should add that Veblen desires to achieve the immediacy of idle curiosity, and this desire is analogous to his somewhat amorphous social ideal that engineers and machinists should manage institutional organizations since they are most familiar with technology and therefore can best fit institutions to current stages of technological development. Veblen thinks the reason for social inequity is, basically, the analogous temporal discrepancy between exigency and habits of thought. This cognitive discrepancy enables the social discrepancy of exploitation. Idle curiosity and management by engineers might cure this endemic discrepancy. It is precisely in this hope that Veblen departs from the pragmatism of James and, most especially, Peirce, who, for all his fascination with scientific models, always insisted on the mediated nature of knowledge and action. If Veblen's ideal is epistemologically idealistic, it is also historically fanciful; he knew very well, as evident in *The Engineers and the Price System* (1921), that engineers do not conduct themselves and pursue their professions as social innocents.

28. Berthoff criticizes the "accommodation in the tidying up at the end" (55), and Tanner calls Howells's narrative strategy "a contrived and melodramatic chance beyond the bounds of pure realism" (xvii). Vanderbilt calls the ending not just "inconclusive" but "uncommitted," suggesting "a basic disharmony in Howells' esthetic" (186).

29. Amy Kaplan powerfully argues that the entire project of realism was a way of "settling" the relation between middle class and immigrant laboring class.

30. *"Alles Interesse verdirbt das Geschmacksurteil und nimmt ihm seine Unparteilichkeit. . . ."* Bernard translates *"Alles"* as "all" (58).

31. Gronlund 1885, 75. For Gronlund, feudalism's explicit enslavement creates "an intense feeling of *Unity*" among a society's members. Because he longs for the divine unity from which we have fallen into capitalism, and toward which we are headed in socialism, Gronlund, as Marx occasionally did, contrasts the harmony of the eras of slavery and feudalism with the present discord.

32. "Complicity" here is not, as some have argued, simply egalitarian sympathy (Boardman; Vanderbilt; Bennett), but rather, as Henry Nash Smith has observed, "the inescapable involvement of everyone in society as a whole" (1964, 85).

33. It is this trust that Beaton lacks. Beaton does not know love but mere infatuation; he feels subject to "the mood, the fancy of a girl" (Christine Dryfoos), much as the common man of Basil's complaint is prey to the mood and indigestion of an employer. In what we might call his purely sordid state, Beaton has "nothing to trust to" (473).

34. In the chapter "Of Power" in *Essay Concerning Human Understanding*, Locke explains that the will is not an inherent faculty of mind but "a power to begin or forbear actions . . . of our minds, and motions of our bodies." Freedom, too, is a power belonging to an agent, the "power to think or not to think, to move or not to move, according to the direction of [one's] mind." Since willing and liberty are both exercises of power, already

signaling the subjection of an agent within physical and cultural conditions, Locke dismisses as "unreasonable, because unintelligible" "that long agitated" question, "Whether man's will be free or no?" Freedom is not anything that belongs to the will. Liberty "is but a power, belongs only to *agents,* and cannot be an attribute or modification of the will, which is also but a power" (1:313, 315, 319–20). For an extended discussion of this and related points, see Rapaczynski 126–50.

35. James cites Mill to say that character is "a completely fashioned will." Freedom from habits means, for James, the ability to exercise "attention and effort" in the face of habit, which is a "material law" of the brain. Nor is this power metaphysical, but also a habit, and James recommends practice at doing "something for no other reason than that you would rather not do it." Then, in an hour of need, you may be better able to respond to unexpected circumstances (*Principles* 1:125–26).

36. In the chapter "Self" James provides a similar account of stability amid fluctuation. He probes how we can feel like the same person even though identity is something we recognize as a shifting "assemblage" of material properties, feelings, memories. These compose the Me, which the I, the ego, cognizes. If the Me is always shifting, how can the I, which cannot be experienced (unless it has become a fluctuating Me) have any consistent identity?

Eschewing the classic dualism of the rationalist tradition, James answers that we experience a perpetual "feeling of our vitality." His example is waking up: "Each of us when he awakens says, Here's the same old Me again." Since he wants to dismiss metaphysical questions, James is careful to specify that this principle of identity is not a "transcendent principle." But while we have no "*substantial* identity," we do have "a functional identity," an "attribute of continuity" that forms "an adequate vehicle for all the experience of personal unity and sameness." He goes further. Classic absolute identity is an incoherent concept, but since no other stability of identity or principle of judgment exists other than the functional (pragmatist) criterion, may we not call the ways in which we experience continuity or identity "essential"? (see *Briefer Course* 180–84).

Chapter 5

1. Pizer 1966, 175–76. My first sentence paraphrases the beginning of Pizer's conclusion.

2. Walker 291–92. Marchand emphasizes the "law" of supply and demand (170), whereas Walker stresses the wheat itself as the force.

3. What became known as the Beard-Hacker thesis was first proposed by Mary and Charles Beard in *The Rise of American Civilization* (1927), and was promoted by Louis Hacker in *The Triumph of American Capitalism* (1940). For a discussion of its subsequent qualification by econometricians, see Lee and Passell (226–38).

4. See Veblen 1904, 88, 131–32, 133–36, 150, 151–54, 275.

5. *Congressional Record,* 52d Cong., 1st sess., 1892, 6883. Subsequent references to this volume will be cited as *CR* 1892.

6. *Responsibilities of the Novelist* 194, 204. Subsequent references to this volume will be cited as *R.*

7. Emery writes that "the greatest speculation in produce which the world has ever seen has grown up in Chicago" (7).

8. White 530–31. White's article presents the findings of the committee Governor Hughes of New York established after the panic of 1907 to examine the effects of speculation and recommend procedures to stabilize conditions.

9. See Emery 119; Hamlin 413–14; Vrooman 424; *CR* 1892, 6442, 6881, 6883; Cowing

4, 34. Cowing's book, *Populists, Plungers, and Progressives,* was an extremely helpful guide to research sources.

10. William Hatch, "Dealing in Fictitious Farm Products." *House Reports,* 52d Cong., 1st sess., 1892, No. 969, 3. Hereafter cited as Hatch.

11. Hamlin 413–14. See 413: "The risks inseparable from real work, from moving the crops and the manufacture of raw material, are right, but a risk which is all risk and no work is gambling. . . . Necessary hazard is as right as needless hazard is wicked."

12. Vrooman 416. See also Vrooman 426; Hamlin 414; White 534.

13. Hatch (7) and Emery (217) employ this phrase. See also White 534–35: "It is known that 75 per cent of the trades on the stock [produce and commodity exchanges] (some say 90 per cent) are of the gambling type."

14. See Emery 98–101; White 534; Hamlin 414; *CR* 1892, 6439.

15. To understand Walker's discomfort better, we might note his book's 1932 publication date.

16. See *The Politics of Aristotle,* 18–29 passim.

17. Taylor, *History of the Board of Trade* 932, 953, 963. Subsequent references to this volume will be cited as Taylor. Charles Kaplan examines Norris's adaptation of Leiter's corner, and argues that he did not change the main pattern of the manipulation.

18. To see this point about Hough's novel, one need only note the advertised subtitle: "How The Star of Good Fortune Rose and Set and Rose Again, by A Woman's Grace, for One John Law of Lauriston."

19. Irwin Unger's *The Greenback Era* is an excellent discussion of these debates. Unger quotes a correspondent to the *Ohio Farmer* in 1868: "The Greenback of today is not a real value" (200). Harvey's lectures on the need for specie and silver culminated in his *Coin's Financial School* (1894), along with a sequel. Among orthodox economists, Henry Carey was the most prominent agitator for greenbacks. William Cullen Bryant, a free-trade activist, supported hard-money policies.

20. Veblen 1904, 146. Cowing points out that during "the antebellum days . . . economic control had been more local and therefore was more personal" (4).

21. See Pizer 1966, 166, and especially Walker 286, 296, and Marchand 86. By 1932, the date of Walker's study, *The Pit* had outsold *McTeague* and *The Octopus* combined, and Marchand reports in 1942 that these proportions had not significantly altered. The Edenic impulse to which I am attributing *The Pit*'s popularity is criticized by Henry Nash Smith in *Virgin Land:* "Agrarian theory," bound up with "the myth of the garden," "encouraged men to ignore the industrial revolution altogether, or to regard it as an unfortunate and anomalous violation of the natural order of things" (259).

22. I am deliberately citing two conservatives here, in order to stay the charge of fetishism commonly made against classical economists since Marx. These authors may have thought social relations were, as it were, natural; many of them, certainly, hoped to justify and preserve current social hierarchies. But in fact few denied that value was a social relation, or that money—even, frightfully enough, gold—was a commodity. To paraphrase Atkinson, their account of money was as metaphysical as Marx's. I raise this point to suggest that their conservative politics are not always, and certainly not necessarily, traceable to their definition of value and money.

Chapter 6

1. 1969, 123, 110. Unless specified, all references to Tarbell are to the more widely available, single-volume, abbreviated Norton edition, edited by David Chalmers.

2. Sklar writes that "concentration by itself may be big business, but it is not yet corporate capitalism." He distinguishes the mere existence of "the corporate form of enterprise," long predominant in the railroad industry and conspicuously growing in influence since the 1830s, from a widespread "system of property relations and administered markets" coming to infuse the culture (44).

3. Since I am about to argue that the trust form continued the Transcendentalist tradition that included antebellum protectionist thought, it is worth noting that Jenks wrote a brief book assessing the contributions of Henry Carey to American economic thought.

4. Many authors noted that trusts were not always the scandal of large-scale practice. In 1903 Charles F. Beach, noting that "nothing within the range of" economic discussion is accorded such public attention and concern as the trust, but recalling that restraint of trade had been a subject of adjudication since the early fifteenth century, observes that were he speaking a century earlier, his topic would not be trusts, but "'corporations, a menace to the commonwealth.' Then the people believed that the creation and growth of corporations threatened their undoing," as trusts are now thought to do (2).

5. Cowperwood's is clearly an androcentric use of "man," and the trust problem was primarily debated by men. As we will see in this and the next chapter, people with disparate political and aesthetic concerns worried over the "manhood" of agents. In chapter 8, however, I will suggest that Cather, for example, shared the transcendental logic of selfhood I am about to delineate. This logic informed projects traversing a broad range of partisan affiliations, and was an instrument for opposing as well as justifying patriarchal power.

6. McQuade 87–90. Throughout this chapter, as throughout this book, when convenient and unless otherwise specified, citations from Emerson will be taken from this widely available edition.

7. A canny remark by Cavell is appropriate to Emerson's attempt, as I read it, to answer skepticism by denying its premises: "Romanticism is understandable in part as an effort to overcome both skepticism and philosophy's responses to skepticism" (1984, 34).

8. In contrast, see Poirier's discussion of change as an "inflection" or modification of conditions and convention, in *The Renewal of Literature,* discussed in chapter 2 above. Grusin (1985) makes a compatible point (against arguments such as Bloom's in *Agon*) about Emerson's notion of "apocalypse."

9. My sense that Emerson's sense of self, however romantic, inherits much from the Enlightenment is borne out by Michael Warner's argument about Benjamin Franklin and republicanism. Warner cites a contemporaneous pamphlet praising Franklin "as a perfect republican citizen because he is 'void of all partial, or all private ends'" (112). Later Warner describes "the paradoxical logic of literal intellection": "the 'I,' although it must be entirely occulted as the designing agent . . . , is also seen as perfectly transparent." "Not to be any particular man . . . is to possess a character of integrity" (121).

See also Quentin Anderson's account of "the drama of the imperial self" (44). This drama involves a contradiction: "the attempt to incarnate the universal in the particular," which leaves Emerson's self "reduced and attenuated" (21). My view differs in diminishing the narcissism of transcendence. Incarnation, per se, is not the goal; the particular and universal, as Anderson senses but does not finally accept, do not, for Emerson, "intersect." A disparity remains, and the self remains, unlike in Anderson's view, "transitive" (47).

10. Many have taken the resilience of the ego to undermine the aim of the passage. Poirier best poses the objection that the speaker never does fully "relinquish his particular identity" as he comes to "assume an ever more inclusively general one" (*World* 66).

11. In contrast to Bloom, Packer contends that in the moment of transcendent self-reliance, "the petty dialectic of self and society, self and past acts of the self, fades away into insignificance" (145). My position has affinities as well with Pease's argument (1980 and

1987) that the eye-ball passage, an analogy for "the very activity of making metaphor, the transition of one term into another," dramatizes the means by "which something is itself by becoming something else." Yet the deconstructive theoretical implications Pease finds in the passage in his 1980 article (less so in its revised form in *Visionary Compacts* 215–34), share the dualism of Bloom's dialecticism. The "death scene" enacted in the passage is not, as Pease argues it is, "a loss of self-present identity and the recovery of universal relations" (62); rather, a merely local presence transmogrifies into a universal presence that transcends its boundaries and incorporates all particulars. All experience becomes literal and intelligible because everything both signifies and is the universal signifier. In the sublime moment, the world is perfected in its dissolution. For Pease, the sublime—the "transition between" terms—"cannot be localized" (1980, 58); my point, Emerson's, is that to be experienced it *must* be localized. It is not that "no one exists to experience it [transcendence]" (61). Rather, even if no analogies are exactly fungible with this experience, a self emerges that is material yet purified of its mean materiality. The residual deconstructive distinction in Pease's 1980 argument between identity or property and discursiveness is finally more conventionally idealist than Emerson's thought.

In *Visionary Compacts* Pease converts his earlier deconstructive critique of the (im)possibility of property and self into a more positive, communal account of the adequate (transitive, rather than static or positivist) conditions of property and self. This later version of the argument retains the same structure but revises the conditions in which communication and community may justly transpire. Before, the "transformative power" of language subverted the possibility of propriety and community; now true community is defined as accommodating transformation. But both Pease's initial and revised accounts posit the same moral hierarchy for what constitutes true community.

12. "Channel," Packer's word and a word Emerson sometimes uses (especially as he ages) to describe the state "through which absolute power is flowing" (145), perhaps depersonalizes transcendence too radically.

Jehlen's understanding of what she calls Emerson's "Augustinian solution" eliminates the force of individual agency and equates individual with the divine far more absolutely. In Jehlen's view Emerson's "man [is] at one with nature and the world," "a complete realization of the one" (*Incarnation* 1986, 80). No willful intervention is necessary or, for that matter, permitted, for any deed distorts (85). In Jehlen's view, Emerson's vision of transcendence is intrinsically conservative, in the substantive political sense of the word. It should be clear that I believe this interpretation misreads the residual tension between self and the divine. This tension is notoriously retained in the syntactical oxymoron "I am nothing; I see all," and critics as different as Porter, Poirier, Bloom, and Packer note this problem. In short, Catholicism is the wrong religious tradition in which to place Emerson's vision. It is Protestant, even Puritan. Augustine regards all self or will not just as the mark of one's distance from God but as sin, and virtue as the utter eradication of will. Like other Protestants, Emerson is seeking the conditions in which the will may be virtuous, redemptive though irreducibly individual.

13. It is worth noting that James excises this appeal to the Kantian pure ego from the chapter "Self" in the later *Briefer Course.*

14. In his *Random Reminiscences* Rockefeller corroborates Tarbell's assessment. Upon entering the industry, he recalls, "I thought that I saw great opportunities in refining oil, and did not realize at that time that the whole oil industry would soon be swamped by so many men rushing into it" (56). With the glut of producers, "the price went down and down until the trade was threatened by ruin" (59). Many legislative and judicial witnesses and the many pamphlets about the trust problem—even those critical of the Standard—are deeply preoccupied with the instability overproduction created in the oil industry.

15. "The Standard Oil Trust Agreement" 1222. Reprinted in the *Preliminary Report on Trusts* by the U.S. Industrial Commission. Hereafter cited as "Standard Oil Trust Agreement."

16. *A Brief History of the Standard Oil Trust* 20. Hereafter cited as *Brief.* See also "Standard Oil Trust Agreement" 1226–27.

17. Jones describes the arrangement this way: "It should be noted . . . that these stocks were held by the trustees for the joint account rather than for the individual account of the certificate holders; a stockholder in any one company lost by the trust agreement his title to the stock of the particular company, and secured instead a proportionate interest in all the stocks and property held by the trustees" (19–20). In effect, William Cook wrote, anticipating the much better known Jenks by over a decade: "The new combination could succeed only by depriving the parties of their power to withdraw consent" (4). Similarly, in his speech at the 1899 Chicago Conference on Trusts, the communitarian socialist Lawrence Gronlund observed that "in every trust the owners virtually abdicate all their powers in favor of the managers" (571).

18. Beach 10. Also placing discussions of corporate development in an idealist tradition, Walter Benn Michaels has recently observed legal theorists' difficulty determining whether corporations represented or embodied stockholders (1987, 198–206). The trust was meant to obviate these conceptual problems by permitting the trustees themselves, from a legal point of view, to transcend ordinary, "natural rights" notions of personhood and representation.

19. Nevins observes that "in effect, though not in law, one great company" had been created (1:394).

20. Because trust certificates represented proportionate interest in the entire combination rather than tangible stock in individual firms, reduction in any single firm's production could actually increase the dividends of former shareholders in that firm by adjusting market conditions to increase the trust's overall profits.

21. See also, for example, Sklar 47–53; and Chandler 419: as the size of corporations grew, "the separation of management and ownership widened." As we saw in the previous chapter, Veblen based his argument in *The Theory of Business Enterprise* on the shift in definitions of ownership and capital brought on by increased economies of scale, when even major stockholders, finally, owned the rights to dividends but not the actual tangible or putative property of a company.

22. We will see later (in note 31) that Fleming was an arch-conservative critic of the trusts who sought to restrain the concentration of capital as a violation of pure capitalism.

23. In *Random Reminiscences* Rockefeller recalls that before the Standard "centraliz[ed] the administration of" diverse firms, "our plans were constantly changed by changed conditions" (59).

24. Ralph and Muriel Hidy have written that the trust pioneered "a new type of central administrative organization" (56). Nevins observes that the trust "marked a new departure in the history not only of the Standard, but of industrial organizations in the United States" (1:393); the trust's "committee system," so typical now, "was something new in American corporate management" (2:22).

25. Dodd employed the language of unification in an 1881 letter held in the files of the Socony-Vacuum Company (Nevins 1:392).

26. Nevins reports Rockefeller's remark to W. O. Ingles (1:402). Trachtenberg discusses "Rockefeller's tearless farewell to the classic American individualism and its doctrine of free and virtuous labor" in *The Incorporation of America* (86).

27. We may contrast Dodd's Transcendentalist defense with two classically individualist defenses of the trust. Testifying before the 1879 Hepburn Committee, the one to which

Rogers denied that individual agency determined the Standard family's prices, Commodore Vanderbilt spoke of the "combination of men" composing the Standard as a combination of very smart individuals: "I don't believe by any legislative enactment . . . you can keep such men down" (quoted in Moody 116). And George Gunton, publisher of the conservative *Gunton's Merchant's Magazine,* qualified Dodd's emphasis in saying that in the trust individuals become "a fractional part of a large productive concern." Neither Vanderbilt nor Gunton thinks, with Dodd, that the trust involves a drowning of individuality (232–33).

28. After the Standard's existence was publicized, Nevins among others points out, the word *trust* became, as it currently is, synonymous with *monopoly,* an enterprise traditionally deemed undemocratic (Nevins 1:394).

29. Dodd was criticized for this autotelic definition of the trust, especially by John Moody in 1904. Typical of those who popularly equated trusts with monopolies, Moody insisted that a trust "is an association formed with the intent to monopolize trade and fix prices" (xiii).

30. Gompers had not yet completed what radicals in the labor movement, the allies of Eugene Debs, regarded as his conservative turn. Speaking on "legitimate, sound trusts" (569–74), the communitarian Gronlund affirmed that trusts "are economic necessities, due to our complex civilization" (569). The 1899 conference was more evenhanded, and more critical of the trusts, than the 1908 conference of the same title.

31. This conceptual association of trusts with the union movement is evident as well in the conservative critic of trusts Robert Fleming, who equated trusts with socialism, and wanted both eliminated (235).

32. This party would evolve a few years later into the Socialist Party, headed by Eugene Debs.

33. In "*Santa Clara* Revisited," Morton J. Horwitz carefully and precisely traces the uneven shifts in the acceptance and understanding of corporate personality between, roughly, the *Santa Clara* decision (1886) and *Lochner* (1905). In fact, Horwitz argues, at the beginning of this period, American jurists scarcely possessed the assumptions and vocabulary to comprehend this notion, which meant something quite different, and accomplished a very different purpose, by *Lochner.*

34. Rockefeller himself testified that "the power conferred by combination may be abused" ("Answers to Interrogations," *Preliminary Report on Trusts* 797). William Jennings Bryan, although more than most "inclined to doubt the possibility of a good trust" (Bancroft 8), concluded his speech at the 1899 Chicago Conference on Trusts by anticipating the conclusion of the 1899 U.S. Industrial Commission investigating trusts: legislation should not eradicate the trusts but "preserve the benefits of the corporation and take from it its possibilities for harm" (593). The commission's chief attorney wrote the next year: "Their power for evil should be destroyed and their means for good preserved" (*Preliminary Report* 5).

35. Message at the Opening of the Second Session of the Fifty-Eighth Congress, 7 December 1903.

36. The phrase "uniform rules" is Ely's. See also Beach: the Sherman Act is effective mainly "in compelling them [trusts] slightly to change *the form* of their organization" (28). See Benjamin (1910 and 1912) for a rare (nearly unique) rigorous attempt to devise per se rules for determining monopolistic practice, and thus for distinguishing "normal" from "abnormal" corporate persons, corporations from trusts.

37. In the 1897 forum in *The Independent,* the economist John Bates Clark observed that even "partnerships between two or more master workmen were once dreaded and forbidden; they were combinations in restraint of trade" (266).

38. This argument, central to the questions of what constitutes restraint of trade and, a corollary point, whether federal restraint on freedom of contract is legitimate, was rehearsed by numerous critical as well as conservative voices. For a summary of these two issues, respectively, see Sklar 133–39, 103–17. In the opening section of his well-known *Monopolies and Trusts,* Ely rehearses the historical difficulty of addressing monopolies; they have been defined since the fifth century like all other property rights, as possession "to the exclusion of other possessors" (1902, 20).

Chapter 7

1. Subsequent references to *The Financier* will be cited as *F*; references to *The Titan* will be cited as *T.*

2. Walcutt (202) and Asselineau (107) discuss Cowperwood as exemplary individualist.

3. Some critics have noted the affinity between Dreiser's naturalism and Emerson's thought. See Walcutt, Lehan, Asselineau, and Matthiessen.

4. In *Crumbling Idols,* in which he regularly pays obeisance to both Howells and Whitman, Garland proposes his "veritism" to combat "feudalistic" literature, whose effect is the "benumbing of the faculties," "the enslavement of our readers and writers to various . . . imitative forms" (10–11). In phrases he dedicates to Whitman, he announces the goal of veritism as "the abolition of all privilege, the peaceful walking together of brethren, equals before nature and before the law" (99).

5. "Editor's Study," *Harper's Magazine* 76 (December 1887): 153–55. Subsequent citations to "Editor's Study" and "Editor's Easy Chair" columns will use the abbreviations ES and EEC, respectively, followed by the date and page. Though my citations are to *Harper's,* these columns have been helpfully collected by James S. Simpson in *Editor's Study.*

6. Walter Benn Michaels discusses what he calls the genteel aesthetics of Sewell's and Howells's economy of pain in "Sister Carrie's Popular Economy" (*The Gold Standard* 36–41).

7. Here I would modify an interesting remark by Amy Kaplan, that "fictionality itself—rather than the particular form of the romance—seems to be the underlying enemy of realism," since it makes "modern life . . . indistinguishable from fiction" (19). Howells would partially agree; nevertheless, he does not propose an alternative to imitation, rather only an alternative model for readers to imitate, as they inevitably do.

8. Notably, in *Silas Lapham,* the word "shrinkage" is interchangeable with "vicissitude." This semantic conflation makes sense, of course, because shrinkage is what we really fear about market fluctuation.

9. Howells added this material to the December 1887 "Editor's Study," which he made the opening chapters of *Criticism and Fiction.*

10. Note, in this context, Howells's praise of James's *The Princess Casamassima:* "Mr. James's knowledge of London is one of the things that strike the reader most vividly, but the management of his knowledge is vastly more important" (ES, April 1887, 829).

11. Norris did not, nor do modern critics generally, turn the charge of mere objectivity or empiricism into a moral condemnation; but others, competing with Howells, Flaubert, et al. for the public ear, did so. What Hugh Wright Mabie called the "analytic" method was often censured as "sensualism." According to Mabie, realism's "disinterested ideal" of the "reproduction of . . . the real," because lacking "a wise husbanding of resources," was incapable of being more than "superficial," and in danger of displaying, because it treats without comment, "moral pathology" (297, 300, 302, 301, 304). Typical of the inflammatory (and well-documented) attacks on realism were W. S. Lilly's denunciatory "The New Natural-

ism," and the successful efforts of the National Vigilance Committee to bring to trial Henry Vizetelly, London publisher of translations of Flaubert and Zola. Henry Wood, too, Christian economist and sometime novelist, attacked sensualism not only at the start of his novel *Victor Serenus* (v) but also in an economic treatise (1901, 6). George J. Becker, of course, has edited the excellent collection *Documents of Modern Literary Realism*. It should be noted that Howells was not always fully censured. Mabie, for example, modifies his criticism of Howells, and O. B. Frothingham excludes Howells (and James as well) from the list of those who produce "the morally objectionable in literature," as he titled his 1882 essay in the *North American Review*.

12. Tolstoi is Howells's model of aesthetic fidelity, "who would feel the slightest unfaithfulness to his subject a sin" (ES, February 1887, 480). For an interesting account of the nontranscriptiveness of Howells's notion of realism, see Max Westbrook's and also Everett Carter's discussions of the subject. More recently, Amy Kaplan and Michael Davitt Bell (1984) have also considered Howells's notion of realism as a social project.

13. *Hey Rub-A-Dub-Dub* 74. Subsequent references to this essay will be cited as AF.

14. Carl Smith makes a similar point (71). See also Gerber's chapter on Dreiser's use of Yerkes's career (87–110).

15. Without speaking of the technicalities of Cowperwood's financiering, Carl Smith makes the related observation that for Cowperwood "mastering" commercial culture "paradoxically seems to be the only way to escape it" (76).

16. The appreciative William Marion Reedy, acute enough to see that Cowperwood "is not immoral, only unmoral," is typical in objecting to the pointlessness (others call it the weariness) of what he calls *The Financier*'s "maze" of financial detail (*Reedy's Mirror*, 2 January 1913, in Salzman 124). See also, in Salzman, the reviews from the *New York World* (98), the *New York Evening Mail* (101), *Bookman* (115–17), the *Chicago Record-Herald* (129–30), the *Dial* (130–31), the *Independent* (132–34), and *McClure's Magazine* (134–35). Even Mencken, while appreciating the effect of the novel, is struck by the "slow plodding through jungles of detail" (103). See also Pizer (1976), and even Matthiessen. Even Michaels has not investigated the aesthetic and historical resonances of Cowperwood's specific practice, hypothecation. See "Dreiser's *Financier*: The Man of Business as a Man of Letters," in *The Gold Standard* 59–83.

17. Pizer has argued that the three main affairs, with Rita, Stephanie, and Berenice, ascend a ladder of aesthetic sophistication and idealism toward "an almost Platonic union with the spirit of absolute beauty" (1976, 193). Frank surely pursues absolute beauty, but thinking of his pursuits as an ascent, either philosophical or ethical, seems to me to exaggerate Dreiser's ambitions. One of Frank's last affairs is with a prostitute, its purpose to place Frank near to her adolescent daughter Berenice (a parody of Beatrice?). Pizer's sense that Frank's "perverse logic" is philosophically cleansed seems a difficult argument to sustain. See, alternatively, Michaels's discussion of the relation between mistresses and natural overproduction in the essay "Dreiser's *Financier*" in *The Gold Standard*.

18. Dreiser's project typifies the transformation to an ethos of personality, rather than character, that Susman and Amy Kaplan discuss. In addition, Dreiser defends personality, unusually for the period, though predictably for him. He offers the ethos of personality as an alternative to the moralism he so deplored.

19. Although he does not pursue the relevance of this point to Cowperwood's notion of self, morality, and art, Carl Smith aptly observes that Dreiser emphatically "links the financier's growing transportation empire with the idea of the city as a center of power" (119).

20. "The relationship between power and freedom's refusal to submit cannot . . . be separated," Foucault writes in "The Subject and Power." "Power is exercised only over free

subjects, and only insofar as they are free . . . freedom must exist for power to be exerted." In contrast, "slavery is not a power relationship," but an absolute "physical determination" (790).

21. Debs has been the beneficiary of two thorough and admiring biographies, by Ray Ginger (originally published as *The Bending Cross* in 1949) and more recently by Nick Salvatore. I have generally relied on these interesting volumes. James P. Cannon's introduction to *Eugene V. Debs Speaks* and Stephen Marion Reynolds's biography in *Debs: His Life, Writings and Speeches* are also useful.

22. Gompers, of course, defended himself against these charges, at least by charging that industrial unionism amounted to coercion of individual unions and of individuals. Sympathetic striking was in any case, he argued, the less effective tactic (*Gompers* 29–30).

23. See "How I Became a Socialist," in *Eugene V. Debs Speaks,* 43–49. This volume contains many of the writings and speeches collected, as well as some not collected, in *Debs: His Life, Writings and Speeches,* issued in 1908 to counter popular and journalistic misrepresentation of his positions. Since it is more widely available, I will cite *Debs Speaks* when possible, without naming the volume, by simply listing the page and sometimes (when convenient and significant) the name of the essay or speech in the text. References to other volumes will be indicated. *Life, Writings and Speeches* will be cited as *Speeches.*

For a discussion of Debs's initial remarks on "Fraternization" in 1890, see Salvatore 88–90. Ginger explains that at least at first Debs found Marx dull, and hence relied more on Gronlund and Kautsky (189). Debs himself suggests that Gronlund was more prominent in his mind.

24. "Industrial Unionism" 126. This important 1905 speech at New York's Grand Central Palace will be cited in the text as IU.

25. "What the workingmen of the country are profoundly interested in is the private ownership of the means of production and distribution . . ." ("The Outlook for Socialism" 64).

26. Mother Jones identified a similar pattern. With so many "leaders wining and dining with the aristocracy" (241), labor politics, like ordinary politics, is "only the servant of industry." Therefore, labor must be organized "along industrial lines" (204).

27. "Unionism and Socialism," *Speeches* 128. In his "Speech at the Founding Convention of the Industrial Workers of the World," Debs called the AFL leaders "lieutenants of capitalism" (*Debs Speaks* 113).

Ginger discusses the transformation in 1900 of the seven-year-old Chicago Civic Federation, whose report on the Pullman strike had been favorable to the strikers, into the National Civic Federation (232–34). Sklar examines the different objectives of the two organizations, particularly differences between their respective conferences on trusts (204–21).

28. "Sound Socialist Tactics" 191–94. This article was published in the *International Socialist Review* in February 1912 in anticipation of the conflicts that might arise in formulating the party platform for that year's presidential campaign.

29. Salvatore titles his chapter covering Debs's conversion to socialism "Transcending the Brotherhoods." He means by this term little more than "going beyond." But, as I am here beginning to argue, I think this term is appropriate in a specific, technical sense to Debs's rhetoric and logic of value and action.

30. "Labor's Struggle for Supremacy," *International Socialist Review* (1911), 141. Subsequent references to essays from this journal will be named, followed by *Review* and the date.

31. "The Socialist Party's Appeal (1908)" 168; IU 132; "Working Class Politics" 174; "Arouse, Ye slaves" 147.

32. See also "Canton Speech," where Debs yokes "manhood and womanhood" together with doing one's "duty" (278).

33. Debs's "The American Movement," a long pamphlet, was collected in *Speeches* 95–117. Debs's language asks us to take this title seriously, to think about labor organization as dynamic, as movement and agitation. The following quotations are from 117.

34. Ginger reports an incident that further substantiates this point. Interviewed by Lincoln Steffens during the 1908 presidential campaign, Debs allowed that he would simply "take" the trusts if elected (289). That is, he would confiscate their property and organization rather than return to an individualistic basis of enterprise.

35. "Plea" 207; "Labor's Struggle," *Review* 1911, 142.

36. After the massacre in 1914 of striking miners at Ludlow, Colorado, by troops called in by a Rockefeller corporation, Debs advocated "arm[ing union] members against the gunmen of the corporations" ("The Gunmen and the Miners" 230). Historians generally, and rightly, take this remark to indicate Debs's horrified and desperate recognition that sometimes violence, which he generally abhorred, was necessary to counter corporate tactics. It suggests as well, I think, his understanding that class conflict takes place between two all-encompassing, similarly organized armies.

37. Sklar has recently argued that American political economists and scientists anticipated, and perhaps were more sophisticated than, their Continental counterparts in conceptualizing the connection between expansive capital concentrations and imperialism. Sklar argues that before the turn of the century, imperialism became not just the expansion of markets but a complete cultural transformation. See the second section of chapter 4, above, for a discussion of Sklar's point that American writers and policymakers first understood that corporate capitalism needed to import modern civilization to noncapitalist societies "by transforming them into capitalist societies" (82).

38. "Tactics" 196; "Working Class Politics" 175; "Appeal" 107.

39. Quoted in the biography of Debs by Stephen Marion Reynolds in *Speeches* 71.

40. "Debs' Speech of Acceptance," *Review* 1904, 692. Debs had been put up for president in 1900, but by the Social Democratic Party. The Socialist Party of America was not formed until the next year.

41. As Mother Jones writes of a Debs speech: "The churches were empty that night, and that night the crowd heard a real sermon by a preacher whose message was one of human brotherhood" (117).

Chapter 8

1. *Report of the Public Lands Commission* 445–46, 535, 584–87.

2. "I felt . . . as if we had come to the end of everything—it was a kind of erasure of personality," she told a Philadelphia newspaper in 1913 (Slote 448). In her recent, monumental biography of Cather's early life and career, Sharon O'Brien discusses this remark extensively (63) and in general charts Cather's career as a development from this anxiety over self-erasure to a confidence in authentic voice. My argument differs in seeing in Alexandra and Cather a desire for self-erasure as a form of empowerment.

3. The phrase "interior structure . . . of property" is adapted from Elaine Scarry, part of whose undertaking in the magisterial *The Body in Pain* is to make us "reacquainted with the interior structure of material objects" (243). I will be stressing that to speak of property is to speak of generative relations rather than of things. Similarly, Scarry notes that "material objects" must be understood not as static entities but as reflective relations, "the locus of a reciprocal action" between producer and artifact (257).

4. Ely, *Property and Contract* 1:96. Subsequent references to this treatise will be cited as Ely, followed by volume and page number.

5. Moreover, in a scene that partakes of both aristocratic and natural-rights traditions,

and indeed suggests their intermixture, her father had on his deathbed bequeathed Alexandra management of the farm. The scene legitimates the redistribution of patrimony by revising the Biblical episode it recalls, Jacob's usurpation of Esau's birthright. Alexandra is the oldest child, and her managerial skills merit responsibilities usually assumed by sons. Brook Thomas suggested to me the need to consider the importance of this scene.

6. That gender anxiety underlies the brothers' position is supported by O'Brien's discussion of Carl, whom Cather describes as having features "too sensitive for a boy's" (10). O'Brien notes the way ordinarily polarized gender attributes are mixed in the introductions of Carl and Alexandra (435).

7. Contemporary debate about the basis of rights often recasts the vocabulary of the debate I am rehearsing between legal/conventional and natural-rights accounts of property, but it retains essentially the same terms and oppositions. Leo Strauss contrasts natural-rights justifications to historical, institutional, or conventional accounts of rights. D. M. Armstrong contrasts natural rights with what he calls "regularity theory," a notion of rights based on Hume's notion of inferential causality. Most writers on the problem employ the familiar term associated with Bentham, *utilitarianism.* Margaret MacDonald and H.L.A. Hart (1984), for example, contrast (or in Hart's case, question the contrast of) utilitarian to natural justifications of rights. For an overview of the specific issues in contemporary debate, see Jeremy Waldron, "Introduction." In the period I am studying, it seems, *utilitarianism* was not common in American debates, as it was in English, as in D. G. Ritchie, *Natural Rights* (1894).

8. C. B. Macpherson writes that common usage notwithstanding, a property right is "a claim that will be enforced by society or the state, by custom or convention or law" (1978, 3). In a 1913 essay, reprinted posthumously as the title chapter of his often cited book, Wesley Hohfeld laments the ambiguity even in legal uses of the term *property,* which typifies the general "looseness of our legal terminology." "Sometimes it is employed to indicate [a] physical object," rather than "to denote the legal interest (or aggregate of legal relations) appertaining to such physical object" (28). Hohfeld cites even some judicial holdings defining property as things (27–30).

9. Most recently, Roberto Mangabeira Unger has criticized, and sought to refine, the way both classical liberals and radicals have identified rights in general "with a particular style of entitlement, . . . consolidated property right" (1987, 509).

10. Lindsay 68–69. Lindsay complains that such essentialism reduces society to "a system of mechanical and external alliances."

11. Hohfeld's attempt to clarify the "correlative" nature of rights and duties—against the imprecision of even the influential Pollock and Maitland treatise, *History of English Law* (1905)—is frequently cited in contemporary debates (35–64).

12. Numerous authors noted the anachronistic quality of the myth of property as an absolute extension of the ego. As the past few chapters have emphasized, what they called "the regime of individual property" was giving way to corporate and collective capital and property (Ely 1:269). The personality theory of property evolved during the transition to a mercantile economy, wrote Henry Scott Holland. After custom settled, "the corporate [that is, popular] imagination" retained "the vague impression of a law of nature . . . within and behind all particular laws" arising from the essential nature of personhood (172; see also Macpherson 1978, 2, 10). This idealism, Holland writes, passed into "the structure of human thought," propelling policy and rhetoric though no longer appropriate to conditions. No wonder, as Ely writes, "people at the present time are puzzled concerning property" (1:237). Veblen offers a compatible account in *The Theory of Business Enterprise* (chaps. 6, 8, 9). Richard Schlatter describes the enduring popularity of natural-rights assumptions in America even as they were coming under attack by English utilitarians and

the German historical school (chap 9). More specifically, Hart describes Americans' rejection of Bentham's offers to help write legal codes (between 1817 and 1830), despite acknowledging the force and influence of his arguments (1982, 73–78).

13. Attempts to extricate from positive law some generative principle of law, whether or not it is called natural law, continue today. See note 7, above. See also Ernst Bloch, *Natural Law and Human Dignity;* and Ronald Dworkin, "Rights as Trumps." Every author alludes to the empirical and experiential commingling of founding principles of law with positive law. Dworkin's massive and growing oeuvre represents probably the contemporary scene's most substantial attempt to establish logically a foundation of rights antecedent to determinate forms of law. See *Taking Rights Seriously,* especially chaps. 1, 6, 7, 12, 13.

14. Ely comments caustically on the circularity of justifications of property rights in various state constitutions: "The extracts from these Constitutions are in themselves a commentary on the [personality] theory [of property]," which is simply "dogmatism in disguise." "No reason is given, but the statement is set up as its own reason. We cannot discover any natural rights existing prior to Constitutions among men" (2:534).

15. The sense that rights are inferences suggests the importance of Hume's notion of inferential causality to the conventionalist insight. Schlatter argues that Bentham added little to Hume's contribution to the definition of property (245). See in Book III, Part II of the *Treatise of Human Nature,* Hume's chapters on whether justice is "a natural or artificial virtue"; and "Of the Origin of Justice and Property." Like Hume, Bentham, Holmes, and Hohfeld, Roberto Unger defines rights as a set of stable "expectations," which are the "indispensable expression" of the constitutional and institutional forms constituting the social matrix (1987, 508–9).

16. In an essay entitled (and appropriating the title of Kafka's story) "Before the Law," Derrida has analyzed the relation of *différance* (difference and belatedness) that positive law has to universal law. Positive law seeks "its provenance" in universal law (1987, 134). But universal law, which "exceeds all boundaries" (137), remains in itself illegible and inaccessible. Indeed, the very readability of "singular" law attests to the remoteness, and finally the authority, of universal law.

17. Adapting C. B. Macpherson's work to Hawthorne's understanding of romance, Walter Benn Michaels has concisely stated for literary critics the problem of private property: "Property, to be property, must be alienable" (1987, 112). I take this insight as a starting point appropriate to Cather's late-realist, early-modernist aesthetics, to the problem of homesteading, and to the debate over property that subtends these two subjects. Emphasizing that property is not possessions but enforceable claims to the benefits of resources, my argument extends Michaels's point to consider the definitive alienability of any right. Rights seem more inalienable—more intrinsic to a private or presocial self—than the possessions we happen to own; but in fact rights arise in the same representational or artifactual structure as property.

18. In the *Principles of Psychology,* James contemplates what feels most intimate to us. About even "our bodies themselves" he wonders, "are they simply ours, or are they *us?*" (1:291).

19. *Congressional Globe,* 29th Cong., 2d sess., Appendix, Senator McClernand of Illinois, 10 July 1846, 36. Cited hereafter as *Globe.*

Tilling the soil was a "necessity" innate in the "the law of [human] nature" (*Globe,* 31st Cong., 1st sess., Congressman Johnson of Tenn., 25 July 1850, 1449), without which the soil, God's gift to man, would have no value (*Globe,* 33d Cong., 1st sess., Appendix, Congressman Dawson of Pa., 16 February 1854, 180). Throughout this section specific references to debates or publications of the period are cited. I am generally indebted to studies of the history of public-land policy by Paul W. Gates, Benjamin Hibbard, and Roy

M. Robbins, whose 1942 full-length work remains the standard work on the subject. Henry Norris Copp, Benjamin Terry, and Seymour D. Thompson wrote contemporary treatises on homesteading. For brief accounts of major events and issues, see Lawrence Friedman, who capsulizes the development and problems of public-land law, and Eric Foner, who provides a typically excellent and concise account of political developments leading to the Homestead Act.

20. Gates 1968, 396–99; "Incongruous" 323–24. Gates writes: "the American Emigrant Company, which had bought the swamplands of a number of Iowa counties, summed up under the caption 'Better than a Free Homestead' all the disadvantages of free land: 'Under the homestead law the settler must, in order to get a good location, go far out into the wild and unsettled districts, and for many years be deprived of school privileges, churches, mills, bridges, and in fact of all the advantages of society.' Landlookers were told by the Burlington and Missouri Railroad in 1878: 'You can judge for your self whether it is not better to purchase land at four or five dollars per acre, on ten years' credit, and six percent interest, that is good land and near a railroad, and will quickly advance to 25 dollars per acre, rather than go upon the far western or southern plains and homestead land upon which you are never certain of half a crop, and which will never advance in value'" (1968, 397).

21. One registry agent said he doubted "if the trees standing on any timber-culture entry west of the hundredth meridian would retard a zephyr" (Robbins 248–49).

22. It is necessary to note that Copp's next listing is "Intentions are not the equivalent of actual residence and improvement." The conflict between this ruling and the one preceding it are intended by Copp, I think, not to be cautionary but to serve as models of interpretive possibility. He's suggesting the flexibility with which the law might be read. In another guide, William E. Preston advises settlers and soldiers how to use preempting and homesteading to bolster each other, and that only alternate six months' residency was necessary to prove up (19–21).

23. One contemporaneous treatise writer expressed "disgust" that the "confused and almost inexplicable system, indicative of differing intentions, theories, and designs, on the part of the lawmakers," made a "conflict of judicial construction and interpretation" inevitable and "a systematic treatise" nearly impossible to compile (Thompson vi).

24. Friedman himself has written in his massive *History of American Law* that the "complex and contradictory" quality of the homestead policy testifies to "the multiplicity of interests that had a voice in enactment and administration" (419).

25. Richard Slotkin's argument is a recent instance of the popular mythology of homesteading. In his view, "the original intent" of the homestead program simply embodied "the vision of a 'fee-simple empire,'" but the actual operation of the program transgressed this vision (285). At least with respect to the spectrum of debates about homesteading, my argument, like Henry Nash Smith's in *Virgin Land,* suggests that it is a mistake to imagine the fee-simple vision as simple in the first place.

26. *Globe,* 33d Cong., 1st sess., Appendix, Congressman Dawson of Pa., 14 February 1854, 184; *Globe,* 29th Cong., 2d sess., Appendix, Senator McClernand of Ill., 10 July 1846, 39; *Gales & Seaton's Register of Debates,* 19th Cong., 1st sess., Senator Benton of Mo., 16 May 1826, 732. We can find imperial rhetoric even in the speeches of the most radical reformer in Congress, Galusha Grow. Grow, who denied that the government owned or even was trustee of the land, began a long discussion of Roman land policy this way: "The first step in the decline of empires is the neglect of their agricultural interest, and with its decay crumbles national power" (*Globe,* 33d Cong., 1st sess., Appendix, 21 February 1854, 242). It is important to keep in mind that land grant principles were inherited from English crown policies to colonize the New World (Terry 10).

27. *Globe,* 33d Cong., 1st sess., Appendix, Senator Cass of Mich., 18 July 1854, 1089; *Globe,* 33d Cong., 1st sess., Appendix, Congressman Grow of Pa., 21 February 1854, 241. Congressman Sapp of Ohio remarked: "The laboring man, without land or property, does not feel much interest in the welfare and prosperity" of the nation (*Globe,* 33d Cong., 1st sess., Appendix, 16 February 1854, 178–79). The tendency of cultivators to identify with their property is explicit, as were the three general premises of the movement, in speeches of Senator Thomas Hart Benton of Missouri, land reform's first congressional voice (see *Register,* 19th Cong., 1st sess., 16 May 1826, 727–33).

28. See speeches by Andrew Johnson in the *Congressional Globe:* 31st Cong., 1st sess., 25 July 1850, 1450; 31st Cong., 1st sess., Appendix, 25 July 1850, 250; 32d Cong., 1st sess., Appendix, 29 April 1852, 529; 36th Cong., 1st sess., 11 April 1860, 1654. See also 32d Cong., 1st sess., Appendix, Congressman Hall of Mo., 20 April 1852, 437; 33d Cong., 1st sess., Senator Dawson of Pa., 14 February 1854, 180–81, 185 (Dawson brought this session's Homestead Bill to the floor from committee); 29th Cong., 1st sess., Congressman Bowlin of Mo., 9 July 1846, 1060.

29. *Globe,* 32d Cong., 1st sess., Appendix, Congressman Cleveland of Conn., 1 April 1852, 574; *Globe,* 29th Cong., 1st sess., Congressman Bowlin of Mo., 9 July 1846, 1059.

30. See, for example, *Globe,* 33d Cong., 1st sess., Appendix, Senator Dawson of Pa., 20 July 1854, 1106. Not until conditions improved in 1846 could many take advantage of the opportunities afforded by the law.

31. *Globe,* 29th Cong., 1st sess., Congressman McConnell of Ala., 9 March 1846, 473; U.S. Department of Interior 14.

32. Galusha Grow, *Globe,* 33d Cong., 1st sess., Appendix, 21 February 1854, 242.

33. Johnson, *Globe,* 31st Cong., 1st sess., Appendix, 25 July 1850, 951; *Globe,* 33d Cong., 1st sess., Appendix, Congressman Dawson of Pa., 14 February 1854, 182. By similar reasoning Grow answered the charge that after five years settlers would be free to mortgage their land in speculation. A possibility, Grow admitted, but "the man who has kept his quarter section five years, and has surrounded it with the comforts of the fireside, and has connected with it all the associations of home, is not likely to leave it unless it be for the purpose of bettering his condition in some far distant location.... But even if the settler does sell, it would be to some one who wanted to cultivate the land, and that would secure its continued settlement and cultivation...." (*Globe,* 33d Cong., 1st sess., Appendix, 21 February 1854, 243). Grow's response went unchallenged, but I find it unpersuasive.

34. Note the similar conclusions to the two novels. Holmes: "And so with the moonlight and starlight falling upon the old homestead, and the sunlight of love falling upon the hearts of its inmates, we bid them adieu" (114). In Micheaux, the protagonist asks his wife if she had loved her former fiancé, and she indicates its impossibility by exclaiming the man's name. "Something dark passed before him—terrible years when he had suffered much. She was speaking again. 'You know I never loved any one in the world but you'" (533).

35. Homesteading opportunities were not extended to Indians until the 1880s. Ironically, Indians' eligibility to homestead, included as a mechanism for citizenship in the Dawes Act of 1887 (stipulating that Indian homesteads were to be inalienable for twenty-five years), accelerated the dissolution of the communal organization of the Indian tribes by opening Indian lands to sale and settlement (Robbins 283).

36. *Globe,* 33d Cong., 1st sess., Senator Weller of Calif., 18 July 1854, 1090. See Shannon's "The Homestead Act and the Labor Surplus."

37. See, for example, *Globe,* 33d Cong., 1st sess., Appendix, 181–82, 1088, 1090, 1106; 32d Cong., 1st sess., Appendix, 434, 436; 29th Cong., 1st sess., Appendix, 778.

38. Most proponents of homesteading either denied the unfairness of eleemosynary

policy, which would stimulate national productivity, or extolled the duty of the state toward its poor. This latter argument was challenged, not surprisingly, as a faulty interpretation of natural law and the Declaration of Independence. See, for example, Senate floor debate in *Globe,* 36th Cong., 1st sess., 10 April 1860, 1629–31.

39. See, for example, *Globe,* 36th Cong., 1st sess., Johnson, 5 April 1860, 1555; *Globe,* 36th Cong., 1st sess., Ibid., Johnson, 11 April 1860, 1653; *Globe,* 33d Cong., 1st sess., Appendix, Congressman Dawson of Pa., 14 February 1854, 182; *Globe,* 33d Cong., 1st sess., Senator Cass of Mi., 18 July 1854, 1050; *Globe,* 31st Cong., 1st sess., Appendix, Johnson, 25 July 1850, 951; Evans 5, 59. In his polemical defense of the Grange, *History of the Grange Movement,* James McCabe (under the pseudonym of Edward Winslow Martin), associated agrarianism and communism in the same breath (537).

40. See editions of women pioneers' letters and diaries, by Elizabeth Hampsten, Lillian Schlissel, and Joanna Stratton.

Emphasizing women's "receptivity" rather than the nineteenth century's characterization of women as passive, Sharon O'Brien locates Cather, especially in her professional and personal relationship with Jewett, in "a less competitive paradigm" than male appropriativeness toward nature and Bloomian anxiety about antecedents and rivals (364–65); the gift giving of female and folk traditions (347) encourages a "communal ideal" (383). Drawing a connection between Alexandra's love of the land and Cather's lesbianism, Judith Fryer "would suggest that loving the land, as Willa Cather who loved women did, is for women a different experience from the dominance and mastery suggested by critics like Richard Slotkin" (376 n. 56).

41. This advertisement appeared in a national magazine during the Dakotas rush between 1909 and 1913 (Marie MacDonald 33). For a discussion of the New Woman in professional and academic arenas, see Rosenberg, chap. 3.

42. I will refer to Stewart as Pruitt because that is how she signs her letters, even though she marries Mr. Stewart about one-third through the book.

43. Stewart 134; *Svendsen,* Farseth and Blegen 44, 42. Special holdings by the Department of the Interior and the attorney general were necessary to determine that widows and single women who spent up to six months of the year working as housekeepers retained rights to their claims, if procedural conditions were met. For the case of an unmarried Bohemian girl who periodically "worked out for others," see Copp 1875, 234.

44. Various commentators have discussed the middle-class origins of antebellum women's reform, a heterogeneous alliance of promoters of Enlightenment ideals and advocates of moral reform (see Buhle and Buhle 6–10; and Rossi 244–50, 265–74). The postbellum women's movement saw heterogeneity often dissolve into discord (Buhle and Buhle 18–33), but the logic of natural rights continued to inform much of its language, especially that of its best-remembered advocates.

45. See Marx's discussion of rights in "On the Jewish Question," *Early Writings* 21–28. In "Alienated Labour," Marx identifies property as both the means and result of "alienated labour, alienated life, estranged man" (131). The structure of property is "incorporate[d] . . . in the very essence of man," who is known and constituted as "a phenomenon of *being external to oneself*" (148). "The supersession of private property" (160) would also be "the supersession of self-estrangement" (152); that is, the supersession of property would constitute man's "return to himself" (155). Scarry calls Marx's goal that of "restoring the referent" (273), in other words, returning disembodied labor to its source. See notes 49 and 50, below.

46. Evans consistently held "that the land should not be a matter of traffic, gift, or will . . . that the land is not property, and, therefore, should not be transferable like the products of man's labor." Skidmore generally agreed, but a rift developed between the two men in

1829 after Skidmore installed in the platform of the New York Working Men a resolution that upon entering society "man gives up to others his original right of soil" (Evans 2:7–9). See Skidmore, *The Right of Man to Property!*

47. See Derrida, "Law of Genre" 63–66; Cohen 204–6. In another context, turn-of-the-century debates about money and art (and finally the nature of minds), Walter Benn Michaels has importantly delineated a related though more elaborate structure of representation, which he calls the logic of naturalism. The structure of mimesis "epitomiz[es] the distinction between what we are made of and what we are" (1987, 171). To summarize briefly a very complex argument: phenomena (nature, objects, minds, persons) appear to be themselves not themselves, more specifically by appearing to be imitations of themselves (154–61; 169–74).

48. In exemplifying the structural alienation of identity, property evokes Lacan's sense that the subject is known and indeed emerges as loss. He speaks, of course, of "the gap that constitutes the subject. The subject is the introduction of a loss in reality" ("Of Structure as an Inmixing" 193). In both Lacanian subjectivity and the structure of property, expenditure (in labor or cognition) is never redeemed in identical terms.

49. I adapt this point from Ely's analysis of natural-rights notions of the individual. See Ely 1:66, 106, 142. This point fully applies to Marx, who, Veblen saw, "does not take a critical attitude toward the underlying principles of Natural Rights." It is precisely in terms of the ontological and moral precepts of natural rights, Veblen writes, that Marx was appalled that "actual exchange value of goods systematically diverges from their real (labor-cost) value" (Veblen, "The Economics of Karl Marx: I," in *The Place of Science* 411, 422). This definition of real value, and the outrage that exchange-value can never be absolutely fungible with it, hold as an ideal the natural-rights identification of property and self.

50. This conflation and circularity infects the arguments of even authors criticizing the personality theory of property and urging in its place what was called the "social theory of property." Authors like the Reverend Washington Gladden, leader of the Social Gospel School of Economics, his ally Richard Ely, and the contributors to *Property, Its Duties and Rights* define private property as a stewardship "established and maintained for social purposes" (Ely 1:165). This definition has the virtue, for its proponents, of recognizing that the "right over one's person" and property is "limited," not "absolute" (175–76), "a relative and dependent" stewardship of resources divinely supplied (Gore xi). A similar notion had been urged at an 1829 meeting of Mechanics in New York, a group that would evolve into the Working Men, led by George Evans and Thomas Skidmore: "all men hold their property by the consent of the great mass of the community, and by no other title" (Evans 3). Like the Working Men, later authors promoted the social theory of property as reducing if not resolving the conflict between individual and state potential in a social-contract view of society. But it has this effect necessarily only if the identification between self and property is extended to an identification between the individual and the state; the social theory of property elevates rather than eliminates the elisions of natural-rights logic.

Marx's goal to supersede private property also imagines the "resolution" of the "contradiction" in property, which he views, typically, as a metaphor for the structure of cognition and self. Marx thus foresees an elision between the poles of self and self-representation. The return of man "to himself" that would effectively end politics, withering away the state and *Recht,* is a fantasy of the end of representation (*Early Writings* 132, 155). Scarry calls Marx's goal of recuperating alienated labor that of "restoring the referent" (see note 45, above). Perhaps more radically, and romantically, Marx seeks the end of referentiality, of the difference between self and its identifying productions. This condition, however, would not constitute the liberation of subjects into consciousness, as Marx claims, but the end of the conditions of consciousness. In the Lacanian sense (see note 48, above)—which is a conventionalist rather than purely natural-rights logic—the structural difference of self from itself and of cognition from things cognized is the condition of knowledge and possibility as well as their limitation.

51. Thomas Gill, "Landlordism in America" *North American Review* 142 (January 1886): 62; see also 52–67; Desmond, "America's Land Question," 142 (February 1886): 153–58; George, "More About American Landlordism," 142 (April 1886): 387–401. Desmond submits that soon "the words of the song: 'Uncle Sam has land enough to give us all a farm,' will suggest nothing but a sorrowful reminiscence" (158). These authors were disputing the claims of Thomas Donaldson, "The Public Lands of the United States" 133 (August 1881): 204–13; Horace Strong, "American Landlordism, I" 142 (March 1886): 246–53; David Bennett King, "American Landlordism, II" 142 (March 1886): 254–57; and John A. Martin, "The Progress of Kansas" 142 (April 1886): 348–55. As remedy, these conservatives urged primarily that failing farmers practice "thrift" (King). Donaldson, author of the most influential treatises on the public domain, charged that the "worst monopolists" were settlers who managed to acquire extra quarter sections (205). Attacking donation policy as socialist and disruptive of "law and order," Strong urged in rhetoric enjoying a resurgence today that "the people, left to themselves will obey economic law and settle economic problems in the most efficient way. All the government has to do is keep the peace" (253).

52. This insight, however, leaves the particular distribution of property and enfranchisement subject, of course, to political contingencies.

53. I note the overlap between Cather and James here not just because I believe there is a textual coincidence. Cather criticism has often noted, as Cather often noted, her early apprenticeship to James's style, from which Cather felt she had to graduate to establish her own identity as a writer. Recently, most powerfully in the work of O'Brien, Cather's graduation is seen to have feminist implications. I do not deny the feminist point, nor that, as Cather believed, ceasing to imitate James strengthened her writing. Nevertheless, I am suggesting that, although she did not employ the same style, Cather's account of the structure of individual authority was similar to James's.

54. In thinking that Cather means literally that no new stories can be written, I agree with Fryer (259–60). In contrast, O'Brien argues that Cather "was telling a new story in American women's writing" (428). O'Brien's characterization of the newness of the narrative may be true, but we should distinguish it from Cather's conviction that no truly new stories are possible.

55. Letter to Mary Austin, 9 November 1927, *Willa Cather on Writing* 12. Cited hereafter as *OW*. Fryer aptly calls her narratives "discontinuous" (226). In her letter to Mary Austin, Cather suggested that the novel form was impossible. I draw my concluding account of the logic of Cather's aesthetics from her remarks about writing. I do not mean to suggest that all Cather's novels follow this sensibility. Critics have often remarked the preindustrial impulse of her work, a trait that could be characterized as a nostalgia for an Edenic moment before exchange. Sometimes this is the case. *My Ántonia, Song of the Lark,* or *A Lost Lady,* for example, either imagine a harmony between human production and the land, or they repudiate a culture that impedes such harmony. Like *O Pioneers!,* I would argue, *The Professor's House, Death Comes for the Archbishop,* or *Shadows on the Rock,* for example, exhibit Cather's fascination with dispossession. At least portions of each do concern preindustrial experience (in the Southwest, for example). But the effect of these episodes is to examine not how securely we might own limited quantities of possessions, but how delicate is the chore of surrendering them. I am grateful to Amy Kaplan for suggesting the need to attend to the question of Cather's nostalgia.

56. Cather's destruction of her letters and the prohibitions of her will have raised various questions. Was she trying to protect her friends, her privacy, her lesbianism from public consumption? By my account of Cather's concerns, the emphasis on privacy, especially as a key to her corpus, is misplaced. In the spirit of Stewart's suggestion, one might compare Cather's gesture to Henry Adams's conception of his *Education* as "my last Will and Testament" (letter to Charles Gaskell, 508). He advised Henry James "to take your own life in the same way in order to prevent biographers from taking it in theirs" (513)—in other words, commemorate your cessation in order to stake out the field first. In destroying her letters, Cather bounded the field of public inquiry.

Most theatrically, Cather's actions obstruct inquiry into her sexual practices and personal relationships. Critics have noted her refusal to identify herself as a lesbian writer, even as a lesbian. I think we should take her determination seriously: to emphasize lesbianism would emphasize gender, specificity of identity, and finally the identification of self with physiological features or personal proclivities or attributes. Albeit ineluctably proprietary, her notion of self departed from this liberal notion of self-proprietorship, the possession of intrinsic and private characteristics anterior to the conditions of their assignability.

Afterword

1. For a related understanding of James's notion of self, which he examines to elucidate that of Charlotte Gilman Perkins, see Michaels 1987, 8–13.

2. It is worth noting, however, that the fact that property and cognition signify what we are not and what is sacrificed does not mean that the self or cognition is necessarily inadequate. Rather, Lacan writes, "by status reality is as full as possible" (193). The pragmatist James illuminates what this sentiment means. He asks: "That something which at every moment goes out and knowingly appropriates the *Me* of the past, and discards the non-me as foreign, is it not a permanent abiding principle of spiritual activity identical with itself wherever found?" Our "I" consists of a continuity in the feeling of warmth and intimacy. In James's view, the Lacanian cognition of self as different from itself is the very condition of a "functional" identity, even, since it may readily suffice, an "essential" identity (*Briefer Course* 180).

3. See Michael Hudson's introduction to McCabe's *History* (6).

Works Cited

Adams, Henry. *The Education of Henry Adams.* Edited by Ernest Samuels. Boston: Houghton Mifflin, Riverside, 1973.

———. "The New York Gold Conspiracy." In *Chapters of Erie,* 101–36. Ithaca: Cornell
University Press, 1956.

Adorno, Theodor W. *Prisms.* Translated by Samuel and Sherry Weber. Cambridge: MIT
Press, 1982.

Allston, Washington. *Lectures on Art, and Poems.* Edited by Richard Henry Dana, Jr. New
York: Baker & Scribner, 1850.

Althusser, Louis. "Ideology and Ideological State Apparatuses (Notes Towards an Investigation)." In *Lenin and Philosophy, and Other Essays,* 127–86. New York: Monthly
Review Press, 1971.

Althusser, Louis, and Balibar, Etienne. *Reading Capital.* London: Verso, 1970, 1978.

Anderson, Alex D. *The Mississippi and Its Forty-Four Navigable Tributaries.* Washington,
D.C.: Government Printing Office, 1890.

Anderson, Frederick, ed. *Mark Twain: The Critical Heritage.* New York: Barnes & Noble,
1971.

Anderson, Quentin. *The Imperial Self: An Essay in American Literary and Cultural History.* New York: Random House, 1971.

Arac, Jonathan. *Critical Genealogies: Historical Situations for Postmodern Literary Studies.* New York: Columbia University Press, 1987.

Aristotle. *The Nicomachean Ethics.* Translated and introduced by David Ross. New York:
Oxford University Press, 1980.

———. *The Politics of Aristotle.* Translated and edited by Ernest Barker. New York:
Oxford University Press, 1958.

Arms, George. "Further Inquiry into Howells's Socialism." *Science and Society* 3 (Spring
1939): 245–48.

———. Introduction to *A Hazard of New Fortunes,* by William Dean Howells, vii–xviii.
New York: Dutton, Everyman's, 1952.

———. "The Literary Background of Howells's Social Criticism." *American Literature*
14 (November 1942): 261–76.

Armstrong, D. M. *What Is a Law of Nature?* New York: Cambridge University Press, 1983.

Asselineau, Roger. *The Transcendentalist Constant in American Literature.* New York:
New York University Press, Gotham, 1971.

Atkinson, Edward. *The Distribution of Products, or the Mechanism and the Metaphysics of
Exchange.* New York: Putnam's, 1885.

———. *The Industrial Progress of the Nation: Consumption Limited, Production Unlimited.* New York: Putnam's, 1890.

Mary Austin Collection. Huntington Library.

Avery, Kevin A. "*The Heart of the Andes* Exhibited: Frederic E. Church's Window on the Equatorial World." *American Art Journal* 18 (1986): 52–72.

———. "A Historiography of the Hudson River School." In *American Paradise: The World of the Hudson River School,* by the Metropolitan Museum of Art, 3–20. New York: Harry N. Abrams, 1987.

Baigell, Matthew. *Albert Bierstadt.* New York: Watson-Guptill, 1981.

———. *Thomas Cole.* New York: Watson-Guptill, 1981.

Bancroft, Edgar A. "Destruction or Regulation." Address before the Trust Conference of the National Civic Federation, Chicago, 23 October 1907. Chicago, 1907.

Bascom, Rev. John. *Political Economy: Designed as a Text-Book for Colleges.* Andover, Mass.: W. F. Draper, 1859.

Baudrillard, Jean. *For a Critique of the Political Economy of the Sign.* Translated and introduced by Charles Levin. St. Louis, Mo.: Telos Press, 1981.

Beach, Charles F. "Recent and Pending 'Trust' Legislation and Litigation in the United States." Address at the University of Minnesota, 13 February 1903. St. Paul, 1903.

Beard, Charles A. *An Economic Interpretation of the Constitution of the United States.* New York: Macmillan, 1913.

Beard, Charles A. and Beard, Mary R. *The Rise of American Civilization.* New York: Macmillan, 1927.

Becker, George, J., ed. *Documents of Modern Literary Realism.* Princeton: Princeton University Press, 1963.

Bell, Ian F. A. "The Hard Currency of Words: Emerson's Fiscal Metaphor in *Nature.*" *ELH* 52 (Fall 1985): 733–53.

Bell, Michael Davitt. *The Development of American Romance: The Sacrifice of Relation.* Chicago: University of Chicago Press, 1980.

———. "The Sin of Art and the Problem of American Realism: William Dean Howells." In *Prospects: An Annual of American Culture Studies,* vol. 9, edited by Jack Salzman, 115–42. New York: Cambridge University Press, 1984.

Bellamy, Edward, *Looking Backward.* New York: New American Library, Signet, 1960.

Benjamin, Reuben Moore. *Normal Corporations vs. Continental Monopolies.* Bloomington, Ill., n. p., 1910.

———. *The Sherman Anti-Trust Act: Its Efficiency and Inefficiency.* Bloomington, Ill.: Pantagraph Printing, 1912.

Bennett, George N. *The Realism of William Dean Howells, 1889–1920.* Nashville: Vanderbilt University Press, 1973.

———. *William Dean Howells: The Development of a Novelist.* Norman: Oklahoma University Press, 1959.

Bentham, Jeremy. *Theory of Legislation.* Vol. 1. Translated from the French of Etienne Dumont by Richard Hildreth. Boston: Weeks, Jordan, 1840.

Bercovitch, Sacvan. "Afterword." In *Ideology and Classic American Literature,* edited by Bercovitch and Myra Jehlen, 418–42. New York: Cambridge University Press, 1986.

———. *The American Jeremiad.* Madison: University of Wisconsin Press, 1978.

———. Preface to *Reconstructing American Literary History.* Edited by Sacvan Bercovitch. Cambridge: Harvard University Press, 1986.

———. "The Problem of Ideology in American Literary History." *Critical Inquiry* 12 (Summer 1986): 631–53.

Bercovitch, Sacvan, and Jehlen, Myra, eds. *Ideology and Classic American Literature.* New York: Cambridge University Press, 1986.

Bergmann, Hans. "Panoramas of New York, 1845–1860." In *Prospects: An Annual of American Culture Studies,* vol. 10, edited by Jack Salzman, 119–37. New York: Cambridge University Press, 1985.

Berthoff, Warner. *The Ferment of Realism: American Literature, 1884–1919.* New York: Free Press, 1965.

Bliss, W.D.P. "What to Do Now?" *Dawn* (July 1890). Reprinted in *Socialism in America: A Documentary History,* compiled by Albert Fried. New York: Doubleday, 1970.

Bloch, Ernst. *Natural Law and Human Dignity.* Translated by Dennis J. Schmidt. Cambridge: MIT Press, 1986.

Bloom, Harold. "Mr. Emerson." *New York Review of Books* 31 (22 November 1984): 20–24.

Boardman, Arthur. "Social Point of View in the Novels of William Dean Howells." *American Literature* 39 (March 1967): 42–59.

Bourdieu, Pierre. *Outline of a Theory of Practice.* New York: Cambridge University Press, 1977.

Bowen, Francis. *The Principles of Political Economy, Applied to the Condition, the Resources, and the Institutions of the American People.* Boston: Little, Brown, 1859.

Bridgman, Richard. *Traveling in Mark Twain.* Berkeley: University of California Press, 1987.

A Brief History of the Standard Oil Trust: Its Method and Its Influence, n. p., 1888.

Brodwin, Stanley. "The Useful and Useless River: *Life on the Mississippi* Revisited." *Studies in American Humor* 2 (January 1976): 196–208.

Brooks, Van Wyck. *The Ordeal of Mark Twain* (1920). New York: Meridian, 1955.

Brown, Gillian. *Domestic Individualism: Nineteeth-Century American Fictions of Self.* Berkeley: University of California Press, 1990.

———. "Getting in the Kitchen with Dinah: Domestic Politics in *Uncle Tom's Cabin.*" *American Quarterly* 36 (Fall 1984): 503–23.

Brownson, Orestes. "The Laboring Classes." *Boston Quarterly Review* 3 (July 1840): 358–95.

———. "Our Future Policy." *Boston Quarterly Review* 4 (January 1841): 68–112. Reprinted in *The Works of Orestes A. Brownson,* collected and arranged by Henry F. Brownson, 15:113–49. Detroit: Thorndike Nourse, 1884.

Bryant, William Cullen. *A Funeral Oration Occasioned by the Death of Thomas Cole.* Delivered before the National Academy of Design, 4 May 1848. New York: D. Appleton, 1848.

———. *Prose Writings of William Cullen Bryant.* Edited by Parke Godwin. 2 vols. New York: D. Appleton, 1884.

———, ed. *Picturesque America; or, the Land We Live In: A Delineation by Pen and Pencil.* Vol. 1. New York: D. Appleton, 1872.

Budd, Louis J. "William Dean Howells' Debt to Tolstoy." *American Slavic and East European Review* 9 (December 1950): 292–301.

Buhle, Mari Jo, and Buhle, Paul, eds. *The Concise History of Woman Suffrage: Selections from the Classic Work of Stanton, Anthony, Gage, and Harper.* Urbana: University of Illinois Press, 1978. "Introduction: Woman Suffrage and American Reform," 1–45.

Burde, Edgar J. "Mark Twain: The Writer as Pilot." *PMLA* 93 (October 1978): 878–92.

Burke, Edmund. *A Philosophical Enquiry into the Origin of Our Ideas of the Sublime and Beautiful.* Edited by James T. Boulton. Notre Dame: University of Notre Dame Press, 1968.

Burke, Kenneth. *A Grammar of Motives.* Berkeley: University of California Press, 1945.

———. "I, Eye, Ay—Concerning Emerson's Early Essay on 'Nature' and the Machinery of Transcendence." In *Language as Symbolic Action: Essays on Life, Literature, and Method,* 186–200. Berkeley: University of California Press, 1968.

Cady, Edwin. *The Realist at War.* Syracuse: Syracuse University Press, 1958.

———, ed. *W. D. Howells as Critic.* Boston: Routledge & Kegan Paul, 1973.

Cannan, Edwin, ed. "Editor's Introduction." In *An Inquiry into the Nature and Causes of the Wealth of Nations,* by Adam Smith. New York: Random House, Modern Library, 1937.

Cardwell, Guy A. "*Life on the Mississippi:* Vulgar Facts and Learned Errors." *ESQ* 14 (1973): 283–93.

Carey, Henry C. *The Harmony of Interests: Agricultural, Manufacturing, and Commercial.* In *Miscellaneous Works of Henry C. Carey, LL.D.* Vol. 1. New York: Burt Franklin, 1966.

————. *Letters to the President, on the Foreign and Domestic Policy of the Union, and Its Effects, as Exhibited in the Condition of the People and the State.* Philadelphia: M. Polock, 1858.

————. *Money.* Lecture delivered before the New York Geographical and Statistical Society, February 1857. Philadelphia: Henry Carey Baird, 1860.

————. *The Past, the Present, and the Future.* 1847. Reprint. New York: Augustus M. Kelley, 1967.

————. *Principles of Political Economy.* 3 vols. Philadelphia: Carey, Lea, & Branch, 1937–1940.

————. *The Prospect: Agricultural, Manufacturing, Commercial, and Financial at the Opening of the Year, 1851.* Philadelphia: J. S. Skinner, 1851.

————. "Two Letters to a Cotton Planter of Tennessee." In *Plough, Loom, and the Anvil.* New York: Myron Finch, 1852.

————. *What Constitutes Currency?* Philadelphia: Lea & Blanchard, 1840.

————. "What Constitutes Real Freedom of Trade?" *American Whig Review* 12 (August 1850): 127–40; (September 1850): 228–40; (October 1850): 353–66; (November 1850): 456–67.

Carnegie, Andrew. *The Empire of Business.* New York: Doubleday, 1904.

————. *The Gospel of Wealth, and Other Timely Essays.* Edited by Edward C. Kirkland. Cambridge: Harvard University Press, 1962.

Carpenter, Frederic Ives. *Emerson Handbook.* New York: Hendricks House, 1953.

Carr, Gerald L. Commentary on Albert Bierstadt, *The Rocky Mountains, Lander's Peak.* In *American Paradise: The World of the Hudson River School,* by the Metropolitan Museum of Art, 284–88. New York: Harry N. Abrams, 1987.

Carstensen, Vernon, ed. *The Public Lands: Studies in the History of the Public Domain.* Madison: University of Wisconsin Press, 1963.

Carter, Everett. *Howells and the Age of Realism.* Philadelphia: Lippincott, 1954.

————. Introduction to *A Hazard of New Fortunes,* by William Dean Howells. Edited by David J. Nordloh et al., xi-xxix. Bloomington: Indiana University Press, 1976.

Cather, Willa. *My Ántonia.* Boston: Houghton Mifflin, 1918.

———— *O Pioneers!* Boston: Houghton Mifflin, 1913.

————. Preface to *Alexander's Bridge.* New ed. Boston: Houghton Mifflin, 1922.

————. *Willa Cather on Writing: Critical Studies on Writing as an Art.* Foreword by Stephen Tennant. New York: Knopf, 1962.

Cavell, Stanley. "Genteel Responses to Kant? In Emerson's 'Fate' and in Coleridge's *Biographia Literaria.*" *Raritan* 3 (Spring 1984): 34–61.

————. *The Senses of Walden.* Expanded ed. San Francisco: North Point Press, 1981.

Chandler, Alfred D., Jr. *The Visible Hand: The Managerial Revolution in American Business.* Cambridge: Harvard University Press, Belknap Press, 1977.

Charvat, William. *The Profession of Authorship in America, 1800–1870: The Papers of William Charvat.* Edited by Matthew J. Bruccoli. Columbus: Ohio State University Press, 1968.

Clark, John Bates. *The Philosophy of Wealth.* Boston: Ginn, 1887.

————. "Trusts and the Law." *Independent* 49 (4 March 1897): 265–66.

Clay, Henry. "On Protection of Home Industry." House of Representatives, 26 April 1820. In *The Works of Henry Clay, Comprising His Life, Correspondence and Speeches.* Edited by Calvin Colton. Vol. 6. New York: Putnam's, 1904.

————. "Speech on American Industry." House of Representatives, 30 and 31 March 1824. In *State Papers and Speeches on the Tariff,* edited by F. W. Taussig, 252–316. Cambridge: Harvard University Press, 1892.

Cohen, Ralph. "History and Genre." *New Literary History* 17 (Winter 1986): 203–18.

Cole, Thomas. "Correspondence between Thomas Cole and Robert Gilmor, Jr." In *Studies in Thomas Cole, an American Romanticist. Annual* 2. Appendix I, 41–82. Baltimore: Baltimore Museum of Art, 1967.

————. "Essay on American Scenery." *American Monthly Magazine* 1 (January 1836): 1–12.

Coleridge, Samuel Taylor. *Selected Poetry and Prose of Coleridge.* Edited by Donald A. Stauffer. New York: Random House, Modern Library, 1951.

Colton, Calvin. *The Junius Tracts and the Rights of Labor* (1842, 1847). Edited by Michael Hudson. New York: Garland, 1974.

Commons, John R. *The Distribution of Wealth.* New York: Macmillan, 1893.

Conant, Charles A. "The Economic Basis of 'Imperialism.'" *North American Review* 167 (September 1898): 326–40.

Cook, William. *"Trusts": The Recent Combinations in Trade.* New York: L. K. Strouse, 1888.

Cooper, James Fenimore. *The American Democrat.* New York: Vintage, 1956.

Copp, Henry Norris. *The American Settler's Guide: A Popular Exposition of the Public Land System of the United States of America.* 9th ed. Washington, D.C.: Published by the editor, 1885.

————. *Public Land Laws.* Washington, D.C.: General Land Office, 1875. Reprint. New York: Arno Press, 1979.

Cowing, Cedric B. *Populists, Plungers, and Progressives: A Social History of Stock and Commodity Speculation, 1890–1936.* Princeton: Princeton University Press, 1965.

Cox, James M. *Mark Twain: The Fate of Humor.* Princeton: Princeton University Press, 1966.

Cumings, Samuel. *The Western Pilot.* Cincinnati: George Conklin, 1847.

de Barbé-Marbois, François. *The History of Louisiana.* Translator unnamed. Philadelphia: Carey & Lea, 1830.

Debs, Eugene V. *Debs: His Life, Writings and Speeches.* Chicago: Charles H. Kerr & Co. Co-operative, 1908.

————. "Debs' Speech of Acceptance." *International Socialist Review* 4 (May 1904): 692–94.

————. *Eugene V. Debs Speaks.* Edited by Jean Y. Tussey and introduced by James P. Cannon. New York: Pathfinder Press, 1970.

————. "Labor's Struggle for Supremacy." *International Socialist Review* 12 (September 1911): 141–43.

————. *Walls and Bars.* 1927, Reprint. Chicago: Charles H. Kerr, 1973.

De Leon, Daniel. "The Socialist View of Trusts." *Independent* 49 (4 March 1897): 293–94.

Deleuze, Gilles, and Guattari, Félix. *Anti-Oedipus: Capitalism and Schizophrenia.* Minneapolis: University of Minnesota Press, 1983.

Derrida, Jacques. *"Devant La Loi."* Translated by Avital Ronell. In *Kafka and the Contem-*

porary Critical Performance, edited by A. Udof, 129–49. Bloomington: Indiana University Press, 1987.

———. "The Law of Genre." *Critical Inquiry* 7 (Autumn 1980): 55–81.

———. "Signature Event Context." *Glyph* 1. Baltimore: The Johns Hopkins University Press, 1977: 172–97.

De Voto, Bernard. *Mark Twain's America.* Boston: Little, Brown, 1932.

———, ed. *Mark Twain in Eruption.* New York: Harper, 1940.

Dimock, Wai-chee. *Empire for Liberty: Melville and the Poetics of Individualism.* Princeton: Princeton University Press, 1989.

Dobb, Maurice. *Theories of Value and Distribution since Adam Smith: Ideology and Economic Theory.* New York: Cambridge University Press, 1973.

Dodd, Samuel C. T. "Aggregated Capital" (1893). In *Trusts,* 1–34. New York, 1900.

———. *Memoirs of S.C.T. Dodd: Written for His Children and Friends.* New York: Robert Grier Cooke, 1907.

———. "The Present Legal Status of Trusts." *Harvard Law Review* 7 (October 1893): 157–69.

———. "Uses and Abuses of Combinations" (1888). In *Trusts,* 35–55. New York, 1900.

Dorfman, Joseph. *The Economic Mind in American Civilization, 1606–1865.* Vol. 2. New York: Viking, 1946. Vol. 3 (1865–1918). New York: Viking, 1949.

Douglas, Ann. *The Feminization of American Culture.* New York: Avon, 1977.

Dreiser, Theodore. *An Amateur Laborer.* Edited and introduced by Richard W. Dowell. Philadelphia: University of Pennsylvania Press, 1983.

———. *The Financier.* New York: Thomas Y. Crowell, Apollo, 1974.

———. *Hey Rub-A-Dub-Dub: A Book of the Mystery and Wonder and Terror of Life.* New York: Boni & Liveright, 1920.

———. *A Hoosier Holiday.* New York: John Lane, 1916.

———. *Notes on Life.* Edited by Marguerite Tjader and John J. McAleer. University: University of Alabama Press, 1974.

———. *Theodore Dreiser: A Selection of Uncollected Prose.* Edited by Donald Pizer. Detroit: Wayne State University Press, 1977.

———. *The Titan.* New York: Thomas Y. Crowell, Apollo, 1974.

Du Bois, W.E.B. *The Souls of Black Folk.* New York: New American Library, Signet, 1969.

Dworkin, Ronald. *Taking Rights Seriously.* Cambridge: Harvard University Press, 1978.

———. "Rights as Trumps." In *Theories of Rights,* edited by Jeremy Waldron, 153–67. New York: Oxford University Press, 1984.

Eagleton, Terry. *Literary Theory: An Introduction.* Minneapolis: University of Minnesota Press, 1983.

Ekstrom, William F. "The Equalitarian Principle in the Fiction of William Dean Howells." *American Literature* 24 (March 1952): 40–50.

Eliot, T. L. *The Ethics of Gambling.* San Francisco: C. A. Murdock, 1886.

Ellet, Charles, Jr. *The Mississippi and Ohio Rivers.* Philadelphia: Lippincott, 1853.

Ellis, Elizabeth Garrity. "The 'Intellectual and Moral Made Visible': The 1839 Washington Allston Exhibition and Unitarian Taste in Boston." In *Prospects: An Annual of American Culture Studies,* vol. 10, edited by Jack Salzman, 39–75. New York: Cambridge University Press, 1985.

Ely, Richard T. *Monopolies and Trusts.* New York: Macmillan, 1902, Reprint. New York: Arno Press, 1973.

———. "The Past and Present of Political Economy." *Johns Hopkins University Studies in Historical and Political Science* 2 (March 1884): 1–64.

———. *Property and Contract in Their Relations to the Distribution of Wealth.* 2 vols. New York: Macmillan, 1914.

————. "Recent American Socialism." *Johns Hopkins University Studies in Historical and Political Science* 3 (April 1885): 1–74.

————. "The Situation and the Remedy." *Independent* 49 (4 March 1897): 268–70.

————. *Social Aspects of Christianity.* Boston: W. L. Greene, 1888.

Emerson, Ralph Waldo. *The Collected Works of Ralph Waldo Emerson.* Vol. 1, *Nature, Addresses, and Lectures.* Edited by Alfred R. Ferguson. Cambridge: Harvard University Press, 1971.

————. "The Fortune of the Republic." In *Miscellanies,* vol. 11 of *Emerson's Complete Works,* 395–425. Cambridge: Riverside Press, 1883.

————. "The Individual." In *The Early Lectures of Ralph Waldo Emerson.* Vol. 2. Edited by Robert E. Spiller, Stephen E. Whicher, and Wallace E. Williams, 173–88. Cambridge: Harvard University Press, 1964.

————. *The Journals and Miscellaneous Notebooks of Ralph Waldo Emerson.* Vol. 4. Edited by Alfred R. Ferguson. Cambridge: Harvard University Press, Belknap Press, 1965; Vol. 5. Edited by Merton M. Sealts, Jr., Cambridge: Harvard University Press, Belknap Press, 1965; Vol 6. Edited by Ralph H. Orth. Cambridge: Harvard University Press, Belknap Press, 1966; Vol. 8. Edited by William H. Gilman and J. E. Parsons. Cambridge: Harvard University Press, Belknap Press, 1970; Vol. 12. Edited by Linda Allardt. Cambridge: Harvard University Press, Belknap Press, 1976.

————. *The Letters of Ralph Waldo Emerson.* Vol. 3. Edited by Ralph L. Rusk. New York: Columbia University Press, 1939.

————. "The Lord's Supper." In *Miscellanies,* vol. 11 of *Emerson's Complete Works,* 9–29. Cambridge: Riverside Press, 1883.

————. "Natural History of Intellect." In *Natural History of Intellect, and Other Papers,* vol. 12 of *Emerson's Complete Works,* 1–59. Cambridge: Riverside Press, 1883.

————. "Quotation and Originality." In *Letters and Social Aims,* vol. 8 of *Emerson's Complete Works,* 167–94. Cambridge: Riverside Press, 1883.

————. *Selected Writings of Emerson.* Edited by Donald McQuade. New York: Random House, Modern Library, 1981.

————. "Trades and Professions." In *The Early Lectures of Ralph Waldo Emerson.* Vol. 2. Edited by Robert E. Spiller, Stephen E. Whicher, and Wallace E. Williams, 113–29. Cambridge: Harvard University Press, 1964.

Emery, Henry Crosby. *Speculation on the Stock and Produce Exchanges of the United States,* vol. 7 of *Columbia Studies in History, Economics, and Public Law.* New York: Columbia University Press, 1896.

Evans, George Henry. *The Radical, in Continuation of the Working Man's Advocate, Devoted to the Abolition of the Land Monopoly, and Other Democratic Reforms.* 1841–1843.

Everett, Alexander. *Journal of the Proceedings of the Friends of Domestic Industry, and British Opinions on the Protection System.* Edited by Michael Hudson. New York: Garland, 1975.

Everett, Edward. "Accumulation, Property, Capital, and Credit." *Hunt's Merchant's Magazine* 1 (July 1839): 20–29.

Farseth, Pauline, and Blegen, Theodore C., trans. and eds. *Frontier Mother: The Letters of Gro Svendsen.* Northfield, Minn.: Norwegian-American Historical Association, 1950.

Ferguson, Frances. "Legislating the Sublime." In *Studies in Eighteenth-Century British Art and Aesthetics,* edited by Ralph Cohen, 128–47. Berkeley: University of California Press, 1985.

————. "The Nuclear Sublime." *Diacritics* 14 (Summer 1984): 4–11.

————. "The Sublime of Edmund Burke, or the Bathos of Experience." *Glyph* 8. Baltimore: The Johns Hopkins University Press, 1981: 62–78.

Fish, Stanley. "Consequences." In *Against Theory: Literary Studies and the New Pragmatism*, edited by W.J.T. Mitchell, 106–31. Chicago: University of Chicago Press, 1985.

———. "Critical Legal Studies (II): Roberto Unger's Transformative Politics." *Raritan* 7 (Winter 1988): 1–24.

Fishlow, Albert. *American Railroads and the Transformation of the Ante-Bellum Economy.* Cambridge: Harvard University Press, 1965.

Fitzhugh, George. *Cannibals All!, or, Slaves without Masters.* Edited by C. Vann Woodward. Cambridge: Harvard University Press, Belknap Press, 1960.

Flagg, Jared B. *The Life and Letters of Washington Allston.* New York: Scribner's, 1892.

Fleming, Robert. *Depraved Finance: Remedy for Trusts.* New York: Robert Fleming, 1904.

Fletcher, Angus. *Allegory: The Theory of a Symbolic Mode.* Ithaca: Cornell University Press, 1964.

Flexner, James Thomas. *Nineteenth-Century American Painting.* New York: Putnam's 1970.

Flint, Timothy. *The History and Geography of the Mississippi Valley.* Cincinnati: Flint & Lincoln, 1832.

———. *Recollections of the Last Ten Years in the Valley of the Mississippi* (1826). Edited by C. Hartley Grattan. New York: Knopf, 1932.

Foner, Eric. *Free Soil, Free Labor, Free Men: The Ideology of the Republican Party before the Civil War.* New York: Oxford University Press, 1970.

Foucault, Michel. *Discipline and Punish: The Birth of the Prison.* Translated by Alan Sheridan. New York: Random House, Vintage, 1979.

———. *The History of Sexuality.* Vol. 1, *An Introduction.* Translated by Robert Hurley. New York: Random House, Vintage, 1980.

———. *Power/Knowledge: Selected Interviews and Other Writings, 1972–77.* Edited by Colin Gordon. New York: Pantheon, 1980.

———. "The Subject and Power." *Critical Inquiry* 8 (Summer 1982): 777–95.

Frank, Arthur De Witt. *The Development of the Federal Program of Flood Control on the Mississippi River.* New York: Columbia University Press, 1930.

Frederic, Harold. *The Damnation of Theron Ware* (1896). Edited by Everett Carter. Cambridge: Harvard University Press, Belknap Press, 1960.

Frederiksen, Ditlew M. "The Common Law and the New Trusts." *Michigan Law Review* 3 (December 1904): 1–21.

Fried, Albert, comp. *Socialism in America: A Documentary History.* New York: Doubleday, 1970.

Friedman, Lawrence M. *A History of American Law.* 2d ed. New York: Simon & Schuster, Touchstone, 1985.

Frothingham, O. B. "The Morally Objectionable in Literature." *North American Review* 135 (September 1882): 323–38.

Fryer, Judith. *Felicitous Space: The Imaginative Structures of Edith Warton and Willa Cather.* Chapel Hill: University of North Carolina Press, 1986.

Fuller, Henry Blake. *With the Procession* (1895). Chicago: University of Chicago Press, 1965.

Garland, Hamlin. *Crumbling Idols* (1894). Edited by Jane Johnson. Cambridge: Harvard University Press, Belknap Press, 1960.

Gates, Paul W. "The Homestead Law in an Incongruous Land System." *American Historical Review* 41 (July 1936): 652–81. Reprinted in *The Public Lands: Studies in the History of the Public Domain,* edited by Vernon Carstensen, 315–48. Madison: University of Wisconsin Press, 1963.

————. *History of Public Land Law Development.* Washington, D.C.: Government Printing Office, 1968. Written for the Public Land Law Review Commission.

Geertz, Clifford. *The Interpretation of Cultures.* New York: Basic Books, 1973.

Gelfant, Blanche H. Introduction to *O Pioneers!,* by Willa Cather. New York: Penguin, 1989.

Genovese, Eugene D. *The World the Slaveholders Made.* New York: Random House, Vintage, 1971.

George, Henry. "More about American Landlordism." *North American Review* 142 (April 1886): 387–401.

————. *Progress and Poverty* (1879). New York: Robert Schalkenbach Foundation, 1966.

Gerber, Phillip L. *Theodore Dreiser.* New York: Twayne, 1964.

Getzels, Jacob Warren. "William Dean Howells and Socialism." *Science and Society* 2 (Summer 1938): 376–86.

Gill, Thomas. "Landlordism in America." *North American Review* 142 (January 1886): 52–67.

Gilman, Charlotte Perkins. *Women and Economics.* Edited by Carl Degler. New York: Harper, 1966.

Gilmore, Grant. *The Ages of American Law.* New Haven: Yale University Press, 1977.

Gilmore, Michael T. *American Romanticism and the Marketplace.* Chicago: University of Chicago Press, 1985.

Ginger, Ray. *Age of Excess: The United States from 1877 to 1914.* New York: Macmillan, 1965.

————. *Eugene V. Debs: A Biography—The Making of an American Radical.* New York: Macmillan, Collier, 1962. Originally published as *The Bending Cross: A Biography of Eugene V. Debs.* New Brunswick, N.J.: Rutgers University Press, 1949.

Gladden, Washington. *Applied Christianity: Moral Aspects of Social Questions.* Boston: Houghton Mifflin, 1886.

Gompers, Samuel. "The Control of Trusts." In *Chicago Conference on Trusts,* 329–30. Chicago: Civic Federation of Chicago, 1900.

————. *Gompers.* Edited by Gerald Emanuel Stearn. Englewood Cliffs, N.J.: Prentice-Hall, 1971.

Gore, Charles, ed. *Property, Its Duties and Rights: Historically, Philosophically, and Religiously Regarded.* New ed. London: Macmillan, 1915.

Gouge, William M. "Report on the Steamboat Act." *House Executive Documents.* No. 10. 34th Cong., 1st sess., 6 November 1855.

Gould, Emerson. *Fifty Years on the Mississippi: Gould's History of River Navigation.* St. Louis: Nixon-James, 1889.

Greeley, Horace. *The American Laborer, Devoted to the Cause of Protection to Home Industry* (1843). Edited by Michael Hudson. New York: Garland, 1974.

————. "The Grounds of Protection." Speech at the Tabernacle, New York, 10 February 1843. In *Recollections of a Busy Life,* 528–53. New York: J. B. Ford, 1968.

————. *Recollections of a Busy Life.* New York: J. B. Ford, 1968.

Greenblatt, Stephen. "Capitalist Culture and the Circulatory System." In *The Aims of Representation: Subject/Text/History,* edited and introduced by Murray Krieger, 257–73. New York: Columbia University Press, 1987.

————. *Shakespearean Negotiations: The Circulation of Social Energy in Renaissance England.* Berkeley: University of California Press, 1988.

Griswold, Rufus W. "Henry C. Carey: The Apostle of the American School of Political Economy." *American Whig Review* 7 (January 1851): 79–86.

Gronlund, Lawrence. *The Co-operative Commonwealth.* London: Modern Press, 1885.

————. "Legitimate, Sound Trusts." In *Chicago Conference on Trusts,* 569–74. Chicago: Civic Federation of Chicago, 1900.

Grusin, Richard A. "'Monadnoc': Emerson's Quotidian Apocalypse." *ESQ* 31 (1985): 149–63.

————. "'Put God in Your Debt': Emerson's Economy of Expenditure." *PMLA* 103 (January 1988): 35–44.

————. *Transcendentalist Hermeneutics: Institutional Authority and the Higher Criticism of the Bible.* Durham, N.C.: Duke University Press, 1990.

Gunton, George. *Trusts and the Public.* New York: D. Appleton, 1899.

Hacker, Louis. *the Triumph of American Capitalism: The Development of Forces in American History to the End of The Nineteenth Century.* New York: Simon & Schuster, 1940.

Haites, Erik, and Mak, James. "The Decline of Steamboating in Antebellum Waters." *Explorations in Economic History* 11 (Fall 1973): 25–36.

Hamlin, Rev. C. H. "Gambling, or Theft by Indirection." In "Gambling and Speculation: A Symposium." *Arena* 11 (1895): 413–16.

Hampsten, Elizabeth. *Read This Only to Yourself: The Private Writings of Midwestern Women, 1880–1910.* Bloomington: Indiana University Press, 1982.

Harper, Frances E. W. *Iola Leroy; or, Shadows Uplifted.* Boston: Beacon Press, 1987.

Hart, H.L.A. "Are There Any Natural Rights?" In *Theories of Rights.* Edited by Jeremy Waldron, 77–90. New York: Oxford University Press, 1984.

————. *Essays on Bentham: Studies in Jurisprudence and Political Theory.* New York: Oxford University Press, 1982.

Hartz, Louis. *The Liberal Tradition in America: An Interpretation of American Political Thought since the Revolution.* New York: Harcourt, Harvest, 1955.

Hatch, William. "Dealing in Fictitious Farm Products." *House Reports.* No. 969. 52d Cong., 1st sess., 1892.

Harvey, William Hope. *Coin's Financial School* (1894). Edited by Richard Hofstadter. Cambridge: Harvard University Press, Belknap Press, 1963.

Hawthorne, Nathaniel. *The House of Seven Gables.* New York: Norton, 1967.

Hayes, Rutherford B. "Annual Report to Congress" (1877). In *A Compilation of the Messages and Papers of the Presidents, 1789–1897,* edited by James D. Richardson. Vol. 7. Washington, D.C.: Published by the Authority of Congress, 1897.

Hedge, Frederic Henry. "Coleridge." In *The Transcendentalists: An Anthology,* compiled by Perry Miller, 66–72. Cambridge: Harvard University Press, 1950.

————. "The Philosophy of Fetichism." In *Atheism in Philosophy, and Other Essays,* 337–53. Boston: Roberts Bros., 1884.

Hegel, G.W.F. *Natural Law: The Scientific Ways of Treating Natural Law, Its Place in Moral Philosophy, and Its Relation to the Positive Sciences of Law.* Translated by T. M. Knox. Philadelphia: University of Pennsylvania Press, 1975.

Herron, George. From *The New Redemption.* Reprinted in *Socialism in America: A Documentary History,* compiled by Alfred Fried. New York: Doubleday, 1970.

Hibbard, Benjamin Horace. *A History of the Public Land Policies.* New York: Macmillan, 1924.

Hidy, Muriel E., and Hidy, Ralph W. *Pioneering in Big Business, 1882–1911.* New York: Harper, 1955.

Hirsch, E. D., Jr. *Validity in Interpretation.* New Haven: Yale University Press, 1967.

Hobbes, Thomas. *Leviathan.* Edited and introduced by C. B. Macpherson. New York: Penguin, 1968.

Hohfeld, Wesley Newcombe. *Fundamental Legal Concepts as Applied in Judicial Reasoning, and Other Legal Essays.* Edited and introduced by Walter Wheeler Cook. New Haven: Yale University Press, 1919.

Holland, Rev. Henry Scott. "Property and Personality." In *Property, Its Duties and Rights: Historically, Philosophically, and Religiously Regarded,* edited by Charles Gore, 169–92. New ed. London: Macmillan, 1915.

Holland, Laurence B. "'A Raft of Trouble': Word and Deed in *Huckleberry Finn.*" *Glyph* 5. Baltimore: The Johns Hopkins University Press, 1979: 69–87. Reprinted in *American Realism: New Essays,* edited by Eric J. Sundquist, 66–81. Baltimore: The Johns Hopkins University Press, 1982.

Holmes, Mrs. Mary J. *The Homestead on the Hillside, and Other Tales.* New York: Carleton, 1867.

Holmes, Oliver Wendell. *The Common Law.* Edited by Mark De Wolfe. Boston: Little, Brown, 1963.

———. *The Formative Essays of Justice Holmes; The Making of an American Legal Philosopher.* Edited and introduced by Frederic Rogers Kellogg. Westport, Conn.: Greenwood Press, 1984. Essays cited: "Codes, and the Arrangement of the Law," 77–99, from *American Law Review* 5 (October 1870); "Possession," 167–99, from *American Law Review* 12 (July 1878).

Horwitz, Howard. "'I Can't Remember': Skepticism, Synthetic Histories, Critical Action." *South Atlantic Quarterly* 87 (Fall 1988): 787–820.

Horwitz, Morton J. "*Santa Clara* Revisited: The Development of Corporate Theory." *West Virginia Law Review* 88 (1985): 173–224.

Hough, Emerson. *The Mississippi Bubble.* Indianapolis: Bowen-Merrill, 1902.

———. *The Passing of the Frontier: A Chronicle of the Old West.* New Haven: Yale University Press, 1921.

Howard, Jean. "The New Historicism in Renaissance Studies." *English Literary Renaissance* 16 (Winter 1986): 13–43.

Howat, John K. "A Climate for Landscape Painters." In *American Paradise: The World of the Hudson River School,* by the Metropolitan Museum of Art, 49–70. New York: Harry N. Abrams, 1987.

———. Commentary on Frederic Church, *Heart of the Andes.* In *American Paradise: The World of the Hudson River School,* by the Metropolitan Museum of Art, 246–50. New York: Harry N. Abrams, 1987.

———. *The Hudson River and Its Painters.* New York: American Legacy Press, 1972.

Howells, William Dean. *Criticism and Fiction.* New York: Hill & Wang, 1962.

———. "Editor's Easy Chair." *Harper's Magazine,* 1900–1920.

———. "Editor's Study." *Harper's Magazine,* 1886–1892.

———. *Editor's Study.* Edited by James S. Simpson. Troy, N.Y.: Whitson, 1983.

———. "Equality as the Basis of a Good Society." *Century Magazine* 51 (November 1895): 63–67.

———. *A Hazard of New Fortunes.* Edited by David J. Nordloh et al. Bloomington: Indiana University Press, 1976.

———. "Henrik Ibsen." *North American Review* (July 1906). Reprinted in *W. D. Howells as Critic,* edited by Edwin H. Cady. Boston: Routledge & Kegan Paul, 1973.

———. Introduction to *Through the Eye of the Needle.* In *The Altrurian Romances,* edited by Clara and Rudolph Kirk. Bloomington: Indiana University Press, 1968.

———. *My Mark Twain: Reminiscences and Criticisms.* Edited and introduced by Marilyn Austin Baldwin. Baton Rouge: Louisiana State University Press, 1967.

————. "The Nature of Liberty." *Forum* 20 (November 1895): 401–9.

————. "An Opportunity for American Fiction." *Literature* 1, nos. 16, 17 (28 April, 5 May 1899): 361–62, 385–86.

————. *The Rise of Silas Lapham.* New York: Norton, 1982.

————. *A Traveller from Altruria.* In *The Altrurian Romances,* edited by Clara and Rudolph Kirk. Bloomington: Indiana University Press, 1968.

————. "Who Are Our Brethren?" *Century Magazine* 51 (April 1896): 932–36.

Hulbert, Archer B. *The Paths of Inland Commerce.* New Haven: Yale University Press, 1920.

Hume, David. *Essays, Moral, Political, and Literary.* Edited by Eugene F. Miller. Indianapolis: Liberty Classics, 1985.

————. *A Treatise of Human Nature.* Edited by L. A. Selby-Bigge. 2d ed. Oxford: Clarendon Press, 1978.

Humphreys, A. A., and Abbott, H. L. *Report upon the Physics and Hydraulics of the Mississippi.* Philadelphia: Lippincott, 1861.

Hunter, Doreen. "America's First Romantics: Richard Henry Dana, Sr. and Washington Allston." *New England Quarterly* 45 (March 1972): 3–30.

Hunter, Louis C. *Steamboats on the Western Rivers: An Economic and Technological History.* Cambridge: Harvard University Press, 1949.

"Imagination and Fact." *Graham's Magazine* 40 (January 1852): 39–43.

Irwin, John. *American Hieroglyphics: The Symbol of the Egyptian Hieroglyphics in the American Renaissance.* Baltimore: Johns Hopkins University Press, 1980.

James, Henry, Jr. *Letters of Henry James.* Edited by Percy Lubbock. London, 1941.

————. *The Spoils of Poynton.* New York: Oxford University Press, 1982.

————. *What Maisie Knew.* New York: Penguin, 1984.

James, William. *Pragmatism, and the Meaning of Truth.* Introduced by A. J. Ayer. Cambridge: Harvard University Press, 1975.

————. *The Principles of Psychology.* 2 vols. 1890, Reprint. New York: Dover, 1950.

————. *Psychology: Briefer Course* (1892). Cambridge: Harvard University Press, 1984.

Jameson, Fredric. "Interview." *Diacritics* 12 (Winter 1982): 72–91.

————. *The Political Unconscious: Narrative as a Socially Symbolic Act.* Ithaca: Cornell University Press, 1981.

Jay, Martin. *Marxism and Totality: The Adventures of a Concept from Lukács to Habermas.* Berkeley: University of California Press, 1984.

Jehlen, Myra. *American Incarnation: The Individual, the Nation, and the Continent.* Cambridge: Harvard University Press, 1986.

————. "Introduction: Beyond Transcendence." In *Ideology and Classic American Literature,* edited by Sacvan Bercovitch and Jehlen, 1–18. New York: Cambridge University Press, 1986.

Jenks, Jeremiah W. *The Trust Problem.* New York: McClure, 1903.

Jones, Eliot. *The Trust Problem in the United States.* New York: Macmillan, 1929.

Jones, Mary Harris. *The Autobiography of Mother Jones.* Chicago: Charles H. Kerr, 1925, 1976.

Kant, Immanuel. *Critique of Judgment.* Translated and edited by J. H. Bernard. New York: Hafner, 1951.

————. *Kritik der Urteilskraft.* Edited by Wilhelm Weischedel. Frankfurt: Suhrkamp, 1957.

Kaplan, Amy. *The Social Construction of American Realism.* Chicago: University of Chicago Press. 1989.

Kaplan, Charles. "Norris's Use of Sources in *The Pit.*" *American Literature* 25 (March 1953): 75–84.

Kay, Carol. *Political Constructions: Defoe, Richardson, and Sterne in Relation to Hobbes, Hume, and Burke.* Ithaca: Cornell University Press, 1988.

Kellogg, Frederic Rogers, ed. and intro. *The Formative Essays of Justice Holmes; The Making of an American Legal Philosopher.* Westport, Conn.: Greenwood Press, 1984.

Knapp, Steven, and Michaels, Walter Benn. "Against Theory." In *Against Theory: Literary Studies and the New Pragmatism,* edited by W.J.T. Mitchell, 11–31. Chicago: University of Chicago Press, 1985.

Kolodny, Annette. *The Land before Her: Fantasy and Experience of the American Frontiers, 1630–1860.* Chapel Hill: University of North Carolina Press, 1984.

Lacan, Jacques. "Of Structure as an Inmixing of an Otherness Prerequisite to Any Subject Whatsoever." In *The Structuralist Controversy: The Languages of Criticism and the Sciences of Man,* edited by Richard Macksey and Eugenio Donato, 186–200. Baltimore: Johns Hopkins University Press, 1972.

Lang, Amy Schrager. *Prophetic Woman: Anne Hutchinson and the Problem of Dissent in the Literature of New England.* Berkeley: University of California Press, 1987.

Lee, Susan Previant, and Passell, Peter. *A New Economic View of American History.* New York: Norton, 1979.

Lehan, Richard. *Theodore Dreiser: His World and His Novels.* Carbondale: Southern Illinois University Press, 1969.

Lentricchia, Frank. *After the New Criticism.* Chicago: University of Chicago Press, 1980.

———. *Ariel and the Police: Michel Foucault, William James, Wallace Stevens.* Madison: University of Wisconsin Press, 1988.

Letwin, William. *Law and Economic Policy in America.* New York: Random House, 1965.

Lewis, Edith. *Willa Cather Living: A Personal Record.* New York: Knopf, 1953.

Lewis, R.W.B. *The American Adam: Innocence, Tragedy, and Tradition in the Nineteenth Century.* Chicago: University of Chicago Press, 1955.

Lilly, W. S. "The New Nationalism." *Fortnightly Review* 38 (1 August 1885). Reprinted in *Documents of Modern Literary Realism,* edited by George J. Becker, 274–95. Princeton: Princeton University Press, 1963.

———. "The Shibboleth of Liberty." *Forum* 10 (January 1891): 508–16.

Lindsay, A. D. "The Principle of Private Property." In *Property, Its Duties and Rights: Historically, Philosophically, and Religiously Regarded,* edited by Charles Gore, 65–81. New ed. London: Macmillan, 1915.

Lloyd, Henry Demarest. "The Oil Combination." *Independent* 49 (4 March 1897): 266–67.

Locke, John. *An Essay Concerning Human Understanding.* 2 vols. New York: Dover, 1959.

———. *The Second Treatise of Civil Government.* In *Two Treatises of Government,* edited by Thomas I. Cook. New York: Hafner, 1973.

Lukács, Georg. *History and Class Consciousness: Studies in Marxist Dialectics.* Cambridge: MIT Press, 1971.

Lynn, Kenneth S. *William Dean Howells: An American Life.* New York: Harcourt, 1970.

Mabie, Hugh Wright. "A Typical Novel." *Antioch Review* 4 (November 1885). Reprinted in *Documents of Modern Literary Realism,* edited by George J. Becker, 296–309. Princeton: Princeton University Press, 1963.

McCabe, James Dabney [Edward Winslow Martin]. *History of the Grange Movement, or the Farmer's War against Monopolies.* Philadelphia: National Publishing Co., 1873. Reprint. Introduced by Michael Hudson. New York: Augustus M. Kelley, 1969.

McCoubrey, John W. *American Art, 1700–1960: Sources and Documents.* Englewood Cliffs, N.J.: Prentice-Hall, 1965.

MacDonald, Margaret. "Natural Rights" (1949). Reprinted in *Theories of Rights.* edited by Jeremy Waldron, 21–40. New York: Oxford University Press, 1984.

MacDonald, Marie. *After Barbed Wire: A Pictorial History of the Homestead Rush in the Northern Great Plains, 1900–1919.* Glendive, Mont.: Frontier Gateway Museum, 1963.

McLellan, David. *Ideology.* Minneapolis: University of Minnesota Press, 1986.

McMurray, William. *The Literary Realism of William Dean Howells.* Carbondale: Southern Illinois Press, 1967.

Macpherson, C. B. *The Political Theory of Possessive Individualism: Hobbes to Locke.* London: Oxford University Press, 1962.

————. "Introduction: The Meaning of Property." *Property: Mainstream and Critical Positions.* Toronto: University of Toronto Press, 1978.

McQuade, Donald, ed. *Selected Writings of Emerson.* New York: Random House, Modern Library, 1981.

Marchand, Ernest. *Frank Norris: A Study.* Stanford: Stanford University Press, 1942.

Marsh, George Perkins. *Man and Nature; or, Physical Geography as Modified by Human Action* (1864). Edited by David Lowenthal. Cambridge: Harvard University Press, Belknap Press, 1965.

Martin, Jay. *Harvests of Change: American Literature, 1865–1914.* Englewood Cliffs, N.J.: Prentice-Hall, 1967.

Marx, Karl. *Capital.* Vol. 1. Translated by Samuel Moore and Edward Aveling, and edited by Frederick Engels. New York: Random House, Modern Library, 1906.

————. *Capital.* Vol. 3. Edited by Frederick Engels. New York: International Publishers, 1967.

————. *A Contribution to the Critique of Political Economy.* Translated by S. W. Ryazanskaya and edited by Maurice Dobb. New York: International Publishers, 1970.

————. *Early Writings.* Translated and edited by T. B. Bottomore. New York: McGraw-Hill, 1963.

———— and Engels, Frederick. *The German Ideology.* Edited and introduced by C. S. Arthur. New York: International Publishers, 1978.

Marx, Leo. *The Machine in the Garden: Technology and the Pastoral Ideal in America.* New York: Oxford University Press, 1964.

————. "The Pilot and Passenger: Landscape Conventions and the Style of *Huckleberry Finn.*" *American Literature* 28 (May 1956). Reprinted in *Mark Twain: A Collection of Critical Essays,* edited by Henry Nash Smith, 47–63. Englewood Cliffs, N.J.: Prentice-Hall, 1963.

————. *The Pilot and the Passenger: Essays on Literature, Technology, and Culture in the United States.* New York: Oxford University Press, 1988.

————. "The Railroad-in-the-Landscape: An Iconological Reading of a Theme in American Art." In *Prospects: An Annual of American Culture Studies,* vol. 10, edited by Jack Salzman, 77–117. New York: Cambridge University Press, 1985.

Maslan, Mark. "Foucault and Pragmatism." *Raritan* 7 (Winter 1988): 94–114.

Matthiessen, F. O. *American Renaissance: Art and Expression in the Age of Emerson and Whitman.* New York: Oxford University Press, 1941.

————. *Theodore Dreiser.* New York: Dell, 1951.

Merrick, George Byron. *Old Times on the Upper Mississippi: The Recollections of a Steamboat Pilot from 1854 to 1863.* Cleveland: Arthur H. Clarke, 1909.

Merritt, Howard S. "'A Wild Scene': Genesis of a Painting." In *Studies in Thomas Cole, an American Romanticist. Annual* 2, 7–40. Baltimore: Baltimore Museum of Art, 1967.

Metropolitan Museum of Art. *American Paradise: The World of the Hudson River School.* Introduced by John K. Howat. New York: Harry N. Abrams, 1987.

Michaels, Walter Benn. *The Gold Standard and the Logic of Naturalism: American Literature at the Turn of the Century.* Berkeley: University of California Press, 1987.

————. "Is There a Politics of Interpretation?" In *The Politics of Interpretation,* edited by W.J.T. Mitchell, 335–45. Chicago: University of Chicago Press, 1983.

————. "*Walden*'s False Bottoms." *Glyph* 1. Baltimore: Johns Hopkins University Press, 1977: 132–49.

Michaels, Walter Benn, and Knapp, Steven. "Against Theory." In *Against Theory: Literary Studies and the New Pragmatism,* edited by W.J.T. Mitchell, 11–31. Chicago: University of Chicago Press, 1985.

Micheaux, Oscar. *The Homesteader, a Novel.* College Park, Md.: McGrath, 1969.

Miller, D. A. "The Novel and the Police." *Glyph* 8. Baltimore: Johns Hopkins University Press, 1981: 127–47.

Miller, Perry. *Errand into the Wilderness.* Cambridge: Harvard University Press, Belknap Press, 1956.

————. *The Life of the Mind in America, from the Revolution to the Civil War: Books One Through Three.* New York: Harcourt, 1965.

————. *Nature's Nation.* Cambridge: Harvard University Press, Belknap Press, 1967.

————. *The Transcendentalists: An Anthology.* Cambridge: Harvard University Press, 1950.

Mississippi River Improvement Convention. *A Memorial to Secure the Adequate Appropriation for a Prompt and Thorough Improvement of the Mississippi River.* St. Louis: John J. Day, 1877.

————. *Official Report of the Proceedings.* St. Louis: Great Western, 1881.

Mitchell, Lee Clark. *Witnesses to a Vanishing America: The Nineteenth-Century Response.* Princeton: Princeton University Press, 1981.

Mitchell, W.J.T., ed. *Against Theory: Literary Studies and the New Pragmatism.* Chicago: University of Chicago Press, 1985.

————. ed. *The Politics of Interpretation.* Chicago: University of Chicago Press, 1983.

Monk, Samuel Holt. *The Sublime: A Study of Critical Theories in Eighteenth-Century England.* Ann Arbor: University of Michigan Press, 1960.

Montgomery, Marion. "The New Romantic vs. the Old: Mark Twain's Dilemma in 'Life on the Mississippi.'" *Mississippi Quarterly* 11 (Spring 1958): 79–82.

Montrose, Louis. "Renaissance Literary Studies and the Subject of History." *English Literary Renaissance* 16 (Winter 1986): 5–12.

Moody, John. *The Truth about Trusts: A Description and Analysis of the Trust Movement.* 1904. Reprint. New York: Greenwood Press, 1968.

"The Moral of the Crisis." *The United States Magazine and Democratic Review* 1 (October 1837): 108–22.

Moretti, Franco. *Signs Taken for Wonders: Essays in the Sociology of Literary Forms.* Translated by Susan Fischer, David Forgacs, and David Miller. Rev. ed. London: Verso, 1988.

Morris, Wright, ed. *The Mississippi Reader.* Garden City, N.Y.: Doubleday, Anchor, 1962.

Morse, Samuel F. B. *Examination of Colonel Trumbull's Address, in Opposition of the Projected Union of the American Academy of Fine Arts and the National Academy of Design.* New York: Clayton & Van Norden, 1833.

Nash, Roderick. *Wilderness and the American Mind.* Rev. ed. New Haven: Yale University Press, 1973.

Nation 44 (5 May 1887).

Nevins, Allan. *Study in Power: John D. Rockefeller, Industrialist and Philanthropist.* 2 vols. New York: Scribner's, 1953.

New York *Times.*

Nicoloff, Philip. *Emerson on Race and History.* New York: Columbia University Press, 1961.

Noble, Louis Legrand. *The Life and Works of Thomas Cole* (1853). Edited by Elliot S. Vessell. Cambridge: Harvard University Press, 1964.

Norris, Frank. *The Pit.* 1902. Reprint. Columbus, Ohio: Merrill, 1970.

———. *The Responsibilities of the Novelist.* New York: Hill & Wang, 1962.

North, Douglass C. *The Economic Growth of the United States, 1790–1860.* New York: Norton, 1966.

North American Review. Various volumes.

Novak, Barbara. *Nature and Culture: American Landscape and Painting, 1825–1875.* New York: Oxford University Press, 1980.

Noyes, Alexander Dana. *Thirty Years of American Finance.* New York: Putnam's, 1898.

O'Brien, Sharon. *Willa Cather: The Emerging Voice.* New York: Oxford University Press, 1987.

Ogg, Frederic Austin. *The Opening of the Mississippi: A Struggle for Supremacy.* New York: Macmillan, 1904.

Packer, B. L. *Emerson's Fall: A New Interpretation of the Major Essays.* New York: Continuum, 1982.

Parker, Theodore. "A Sermon of Merchants." In *The Transcendentalists: An Anthology,* edited by Perry Miller, 449–57. Cambridge: Harvard University Press, 1950.

Patterson, C. Stuart. *The Problem of the Trusts.* New York: American Philosophical Society, 2 April 1903.

Pease, Donald. "Emerson, Nature, and the Sovereignty of Influence." *boundary 2,* no. 8 (Spring 1980): 43–71.

———. *Visionary Compacts: American Renaissance Writings in Cultural Context.* Madison: University of Wisconsin Press, 1987.

Perry, Arthur Lapham. *Introduction to Political Economy.* New York: Scribner's, 1882.

Phillips, Willard. *Propositions Concerning Protection and Free Trade.* 1850. Reprint. New York: Augustus M. Kelley, 1968.

Pizer, Donald. *The Novels of Frank Norris.* Bloomington: Indiana University Press, 1966.

———. *The Novels of Theodore Dreiser.* Minneapolis: University of Minnesota Press, 1976.

———, ed. *Theodore Dreiser: A Selection of Uncollected Prose.* Detroit: Wayne State University Press, 1977.

Plato. *The Symposium.* Translated by Walter Hamilton. New York: Penguin, 1951.

Pocock, J.G.A. *Politics, Language, and Time: Essays on Political Thought and History.* New York: Atheneum, 1973.

———. *Virtue, Commerce, and History: Essays on Political Thought and History, Chiefly in the Eighteenth Century.* New York: Cambridge University Press, 1985.

Poirier, Richard. *The Performing Self: Compositions and Decompositions in the Languages of Contemporary Life.* New York: Oxford University Press, 1971.

———. *The Renewal of Literature: Emersonian Reflections.* New York: Random House, 1987.

———. *A World Elsewhere: The Place of Style in American Literature.* New York: Oxford University Press, 1966.

———. "Writing Off the Self." *Raritan* 1 (Summer 1981): 106–33.

"Political Paradoxes." *American Whig Review* 6 (July 1850): 2–16.

Poor, Henry Varnum. *Money and Its Laws.* 1877. Reprint. New York: Greenwood Press, 1969.

Porter, Carolyn. *Seeing and Being: The Plight of the Participant Observer in Emerson, James, Adams, and Faulkner.* Middletown, Conn.: Wesleyan University Press, 1981.

Potter, Alonzo. *Political Economy: Its Objects, Uses, and Principles.* New York: Harper Bros., 1862.

Preston, William E. *The Soldier's, Settler's and Inventor's Guide, Containing the Pre-emption and Homestead Laws of the United States, in Force 1872.* Cleveland: Published by the author, 1872.

Rae, John. "Genius and Its Application" (1839). In *John Rae, Political Economist, an Account of His Life and a Compilation of His Main Writings.* Vol. 1, *Life and Miscellaneous Writings.* Edited by Robert Warren James, 320–26. Toronto: Toronto University Press, 1965.

————. *Statement of Some New Principles on the Subject of Political Economy* (1834). Vol. 2 of *John Rae, Political Economist, an Account of His Life and a Compilation of His Main Writings.* Edited by Robert Warren James. Toronto: Toronto University Press, 1965.

Rapaczynski, Andrzej. *Nature and Politics: Liberalism in the Philosophies of Hobbes, Locke, and Rousseau.* Ithaca: Cornell University Press, 1987.

Rashdall, The Rev. Hastings. "The Philosophical Theory of Property." In *Property, Its Duties and Rights: Historically, Philosophically, and Religiously Regarded.* New ed. Ed. Charles Gore, 33–64. London: Macmillan, 1915.

Ray, Richard. *An Address Delivered before the American Academy of the Fine Arts* (17 November 1825). New York: G. & C. Carvilli, 1825.

Raymond, Daniel. *The Elements of Political Economy.* 1823. Reprint. New York: Augustus M. Kelley, 1964.

Rebora, Carrie. Commentary on Jasper Cropsey, *Starrucca Viaduct, Pennsylvania.* In *American Paradise: The World of the Hudson River School,* by the Metropolitan Museum of Art, 210–13. New York: Harry N. Abrams, 1987.

Republican Association of Washington. "Lands for the Landless." Washington, D.C.: Congressional Republican Executive Committee, 1859.

Reynolds, Stephen Marion. "Life of Eugene V. Debs." In *Debs: His Life, Writings and Speeches,* by Eugene V. Debs, 1–76. Chicago: Charles H. Kerr & Co. Co-operative, 1908.

Richardson, James D. *A Compilation of the Messages and Papers of the Presidents, 1789–1897.* Vols. 5, 7. Washington, D.C.: Government Printing Office, 1897.

Riis, Jacob. *The Making of an American.* New York: Macmillan, 1901.

Ripley, George. "Martineau's Rationale." In *The Transcendentalists: An Anthology.* compiled by Perry Miller, 129–32. Cambridge: Harvard University Press, 1950.

Ritchie, David George. *Natural Rights: A Criticism of Some Political and Ethical Conceptions.* London: G. Allen and Urwin, 1894.

Robbins, Roy M. *Our Landed Heritage: The Public Domain, 1776–1936.* Princeton: Princeton University Press, 1942.

Robinson, Forrest G. *In Bad Faith: The Dynamics of Deception in Mark Twain's America.* Cambridge: Harvard University Press, 1986.

Rockefeller, John D. *Random Reminiscences of Men and Events* (1908–1909). Reprint. Tarrytown, N.Y.: Sleepy Hollow Press and Rockefeller Archive Center, 1984.

Rodgers, Daniel T. *The Work Ethic in Industrial America, 1850–1920.* Chicago: University of Chicago Press, 1976.

Rogin, Michael Paul. *Fathers and Children: Andrew Jackson and the Subjugation of the American Indian.* New York: Random House, Vintage, 1975.

————. Review of *American Incarnation,* by Myra Jehlen. *American Literature* 61 (May 1989): 409–13.

————. *Subversive Genealogy: The Politics and Art of Herman Melville.* New York: Knopf, 1983.

Roosevelt, Theodore. *The Roosevelt Policy: Speeches, Letters, and State Papers Relating to Corporate Wealth and Closely Allied Topics.* Vol. 1. New York: Current Literature, 1908.

Roque, Oswaldo Rodriguez. "The Exaltation of American Landscape Painting." In *American Paradise: The World of the Hudson River School,* by the Metropolitan Museum of Art, 21–48. New York: Harry N. Abrams, 1987.

Rose, Anne C. *Transcendentalism as a Social Movement, 1830–1850.* New Haven: Yale University Press, 1981.

Rosenberg, Rosalind. *Beyond Separate Spheres: Intellectual Roots of Modern Feminism.* New Haven: Yale University Press, 1982.

Rossi, Alice S., ed. *The Feminist Papers, From Adams to de Beauvoir.* With introductory essays. New York: Bantam Books, 1973. "Social Roots of the Woman's Movement in America," 241–81.

Royce, Sarah. *A Frontier Lady: Recollections of the Gold Rush and Early California.* Edited by Ralph Henry Gabriel. New Haven: Yale University Press, 1932.

Rusk, Ralph L. *The Life of Ralph Waldo Emerson.* New York: Columbia University Press, 1949.

Said, Edward. *The World, the Text, and the Critic.* Cambridge: Harvard University Press, 1983.

Salvatore, Nick. *Eugene V. Debs: Citizen and Socialist.* Urbana: University of Illinois Press, 1982.

Salzman, Jack, ed. and intro. *Theodore Dreiser: The Critical Reception.* New York: David Lewis, 1972.

Sanford, Mollie Dorsey. *Mollie: The Journal of Mollie Dorsey Sanford in Nebraska and Colorado Territories, 1857–1866.* Edited by Donald F. Danker. Lincoln: University of Nebraska Press, 1959.

Scarry, Elaine. *The Body in Pain: The Making and Unmaking of the World.* New York: Oxford University Press, 1985.

Scheir, Miriam, ed. *Feminism: The Essential Historical Writings.* New York: Random House, Vintage, 1972.

Schlatter, Richard. *Private Property: The History of an Idea.* New Brunswick, N.J.: Rutgers University Press, 1951.

Schlissel, Lillian. *Women's Diaries of the Westward Journey.* Preface by Carl N. Degler. New York: Schocken, 1982.

Schmidt, Paul. "River vs. Town: Mark Twain's *Old Times on the Mississippi.*" *Nineteenth-Century Fiction* 15 (September 1960): 95–111.

See, Fred. *Desire and the Sign: Nineteenth-Century American Fiction.* Baton Rouge: Louisiana State University Press, 1987.

Seltzer, Mark. "*The Princess Casamassima:* Realism and the Fantasy of Surveillance." *Nineteenth-Century Fiction* 35 (1981). Reprinted in *American Realism: New Essays,* edited by Eric J. Sundquist, 95–119. Baltimore: Johns Hopkins University Press, 1982.

Shannon, Fred A. "The Homestead Act and the Labor Surplus." *American Historical Review* 41 (July 1936): 637–51. Reprinted in *The Public Lands: Studies in the History of the Public Domain,* edited by Vernon Carstensen, 297–313. Madison: University of Wisconsin Press, 1963.

Shaw, Luella. *True History of Some of the Pioneers of Colorado.* Hotchkiss, Colo.: W. S. Coburn, John Patterson, and A. K. Shaw, 1909.

Shell, Marc. *Money, Language, and Thought: Literary and Philosophic Economies from the Medieval to the Modern Era.* Berkeley: University of California Press, 1982.

Simmel, Georg. *The Philosophy of Money.* Translated by Tom Bottomore and David Frisby. Boston: Routledge & Kegan Paul, 1978.

Simpson, James S, ed. and intro. *Editor's Study,* by William Dean Howells. Troy, N.Y.: Whitson, 1983.

Skidmore, Thomas. *The Right of Man to Property!* 1829. Reprint. New York: Burt Franklin, 1964.

Sklar, Martin J. *The Corporate Reconstruction of American Capitalism, 1890–1916: The Market, The Law, and Politics.* New York: Cambridge University Press, 1988.

Slote, Bernice, ed. *The Kingdom of Art: Willa Cather's First Principles and Critical Statements, 1893–96.* Lincoln: University of Nebraska Press, 1966.

Slotkin, Richard. *The Fatal Environment: The Myth of the Frontier in the Age of Industrialism.* New York: Atheneum, 1985.

Smith, Adam. *An Inquiry into the Nature and Causes of the Wealth of Nations.* Edited by Edwin Cannan. New York: Random House, Modern Library, 1937.

Smith, Carl S. *Chicago and the American Literary Imagination, 1880–1920.* Chicago: University of Chicago Press, 1984.

Smith, Henry Nash. *Mark Twain: The Development of a Writer.* 1962. Reprint. New York: Atheneum, 1972.

————. "The Search for a Capitalist Hero." In *The Business Establishment,* edited by Earl F. Cheit, 77–112. New York: Wiley, 1964.

————. *Virgin Land: The American West as Symbol and Myth.* Cambridge: Harvard University Press, 1950.

Stanton, Elizabeth Cady. "Solitude of Self." In *The Concise History of Woman Suffrage: Selections from the Classic Work of Stanton, Anthony, Gage, and Harper,* edited by Mari Jo Buhle and Paul Buhle, 325–28. Urbana: University of Illinois Press.

Stewart, Elinore Pruitt. *Letters of a Woman Homesteader.* 1913. Reprint. Boston: Houghton Mifflin, 1982.

Stewart, Susan. *On Longing: Narratives of the Miniature, the Gigantic, the Souvenir, the Collection.* Baltimore: Johns Hopkins University Press, 1984.

Stowe, Harriet Beecher. "A Family Talk on Reconstruction." In *The Chimney Corner,* by Christopher Crowfield (pseud.). Boston: Ticknor and Fields, 1868. Reprinted in *Household Papers and Stories,* 274–99. Vol. 8 of *The Writings of Harriet Beecher Stowe.* Riverside ed. Boston: Houghton Mifflin, 1896.

Stratton, Joanna L. *Pioneer Women: Voices from the Kansas Frontier.* Introduced by Arthur M. Schlesinger, Jr. New York: Simon & Schuster, Touchstone, 1981.

Strauss, Leo. *Natural Right and History.* Chicago: University of Chicago Press, 1953.

Sundquist, Eric J., ed. *American Realism: New Essays.* Baltimore: Johns Hopkins University Press, 1982.

————. *Home as Found: Authority and Genealogy in Nineteenth-Century Literature.* Baltimore: Johns Hopkins University Press, 1979.

Susman, Warren I. "Personality and the Making of Twentieth-Century Culture." In *Culture as History: The Transformation of American Society in the Twentieth Century,* 271–85. New York: Pantheon, 1984.

Tanner, Tony. Introduction to *A Hazard of New Fortunes,* by William Dean Howells, vii–xxxv. London: Oxford University Press, 1965.

Tarbell, Ida M. *The History of the Standard Oil Company.* Edited by David M. Chalmers. New York: Norton, 1969.

————. *The History of the Standard Oil Company* (1904). Vol. 2. Appendix 51. Gloucester, Mass.: Peter Smith, 1963.

Taussig, F. W. *The Tariff History of the United States.* 1894. 8th rev. ed., 1931. Introduced by David M. Chalmers. New York: Capricorn, 1964.

Taylor, Charles H. *History of the Board of Trade of the City of Chicago.* Chicago: Robert O. Law, 1917.

Taylor, George Rogers. *The Transportation Revolution, 1815–1860.* New York: Holt, Rinehart, 1962.

Taylor, Walter Fuller. *The Economic Novel in America.* Chapel Hill: University of North Carolina Press, 1942.

———. "On the Origin of Howells' Interest in Economic Reform." *American Literature* 2 (March 1930): 3–14.

———. "William Dean Howells and the Economic Novel." *American Literature* 4 (May 1932): 103–13. Reprinted in *Howells: A Century of Criticism,* edited by Kenneth E. Eble, 183–95. Dallas: Southern Methodist University Press, 1962.

Terry, Benjamin Stites. *The Homestead Law Agitation.* Freiburg, Germany, 1892.

Thompson, Seymour D. *A Treatise on Homestead and Exemption Laws.* San Francisco: Bancroft-Whitney, 1886.

Thoreau, Henry David. *Walden and Civil Disobedience.* Edited by Owen Thomas. New York: Norton, 1966.

———. *A Week on the Concord and Merrimack Rivers.* New York: Thomas Y. Crowell, Apollo, 1961.

Tichi, Cecelia. *New World, New Earth: Environmental Reform in American Literature from the Puritans through Whitman.* New Haven: Yale University Press, 1979.

Tillson, Christiana Holmes. *A Woman's Story of Pioneer Illinois.* Edited by Milo Milton Quaife. Chicago: Lakeside Press, 1919.

Tompkins, Jane. *Sensational Designs: The Cultural Work of American Fiction, 1790–1860.* New York: Oxford University Press, 1985.

Trachtenberg, Alan. *The Incorporation of America.* New York: Hill & Wang, 1982.

Tully, James. *A Discourse on Property: John Locke and His Adversaries.* New York: Cambridge University Press, 1980.

Twain, Mark. *Adventures of Huckleberry Finn.* New York: Norton, 1977.

———. "Down the Rhone." In *Europe and Elsewhere.* Vol. 29 of *The Writings of Mark Twain.* Definitive ed. edited by Albert Bigelow Paine, 129–69. New York: Gabriel Well, 1923.

———. "How to Tell a Story." In *How to Tell a Story, and Other Essays,* 7–15. Vol. 22 of *The Writings of Mark Twain.* Hillcrest ed. New York: Harper, 1906.

———. *Life on the Mississippi.* New York: New American Library, Signet 1961.

———. *The Love Letters of Mark Twain.* Edited by Dixon Wecter. New York: Harper, 1949.

———. *Mark Twain's Correspondence with Henry Huttleston Rogers.* Edited by Lewis Leary. Berkeley: University of California Press, 1969.

———. *Mark Twain-Howells Letters: The Correspondence of Samuel L. Clemens and William Dean Howells, 1872–1910.* 2 vols. Edited by Henry Nash Smith and William M. Gibson, with the assistance of Frederick Anderson. Cambridge: Harvard University Press, Belknap Press, 1960.

———. *Mark Twain in Eruption.* Edited by Bernard De Voto. New York: Harper, 1940.

———. *Mark Twain's Letters.* Vol 1. Edited by Albert Bigelow Paine. New York: Harper, 1917.

———. *Mark Twain's Letters to Will Bowen: "My Oldest and Dearest Friend."* Edited and introduced by Theodore Hornberger. Austin: University of Texas Press, 1941.

————. *Mark Twain's Notebooks and Journals.* Vol. 2, 1877–1883. Edited by Frederick Anderson et al. Berkeley: University of California Press, 1975.

————. "What Paul Bourget Thinks of Us." In *How to Tell a Story, and Other Essays,* 141–64. Vol. 22 of *The Writings of Mark Twain.* Hillcrest ed. New York: Harper, 1906.

Unger, Irwin. *The Greenback Era: A Social and Political History of American Finance, 1865–1879.* Princeton: Princeton University Press, 1964.

Unger, Roberto Mangabeira. *False Necessity: Anti-Necessitarian Social Theory in the Service of Radical Democracy.* Part I of *Politics, a Work in Constructive Social Theory.* New York: Cambridge University Press, 1987.

————. *Knowledge and Politics.* New York: Free Press, 1976.

U.S. Congress. *Appendix to the Congressional Record.* Various volumes.

————. *Congressional Globe.* Various volumes.

————. *Congressional Record.* Various volumes.

————. *Gales & Seaton's Register of Debates.*

————. *The Statutes at Large of the United States of America.* Vol. 21.

U.S. Congress. House. "Letter from the Secretary of War." *House Executive Documents.* No. 153. 28th Cong., 1st sess., 26 February 1844.

————. "Preliminary Report of the Mississippi River Commission." *House Executive Documents.* No. 58. 46th Cong., 2d sess.

————. "Report of the Board of Engineers on the Ohio and Mississippi Rivers." *House Executive Documents.* No. 35. 17th Cong., 2d sess., 22 January 1823.

————. "Report of the Mississippi River Commission." *House Executive Documents.* No. 95. 46th Cong., 3d sess.

————. *Report of the Public Lands Commission. House Executive Documents.* 46th Cong., 2d sess. Vol. 22. Serial 1923. No. 46. Washington, D.C.: Government Printing Office, 1880.

————. "Report on the Improvement of the Navigation of the Ohio and the Mississippi Rivers." *House Reports.* No. 212. 20th Cong., 1st sess., 27 March 1828.

————. "Report on the Steamboat Act." By William M. Gouge. *House Executive Documents.* No. 10. 34th Cong., 1st sess., 6 November 1855.

U.S. Congress. Senate. "Report of the Committee on Commerce." By Mr. Barnett. *Senate Documents.* No. 137. 27th Cong., 3d sess., 9 February 1843.

————. "Report on the Mississippi and Ohio Rivers." By Charles Ellet, Jr. *Senate Executive Documents.* No. 49. 32d Cong., 1st sess., 1852.

U.S. Department of the Interior. Bureau of Land Management. *The Homestead Law: A Brief Sketch in United States History.* Washington, D.C., 1962.

U.S. Industrial Commission. "The Standard Oil Trust Agreement." *Preliminary Report on Trusts and Industrial Combinations, 1899–1900,* 1222–28. Vol. 1. Washington, D.C.: Government Printing Office, 1900.

Vanderbilt, Kermit. *The Achievement of William Dean Howells: A Reinterpretation.* Princeton: Princeton University Press, 1968.

Veblen, Thorstein. *The Engineers and the Price System.* New York: B. W. Huebsch, 1921.

————. *Essays in Our Changing Order.* New York: B. W. Huebsch, 1934.

————. *The Place of Science in Modern Civilization.* New York: B. W. Huebsch, 1919.

————. *The Theory of Business Enterprise.* New York: Scribner's, 1904.

————. *The Theory of the Leisure Class.* New York: New American Library, Mentor, 1953.

Vrooman, Henry C. "Methods and Devices." In "Gambling and Speculation: A Symposium." *Arena* 11 (1895): 416–26.

Walcutt, Charles C. *American Literary Naturalism: A Divided Stream.* Minneapolis: University of Minnesota Press, 1956.

Waldron, Jeremy, ed. *Theories of Rights.* New York: Oxford University Press, 1984. "Introduction," 1–20.

Walker, Franklin. *Frank Norris: A Biography.* New York: Doubleday, Doran, 1932.

Walsh, Harry. "Tolstoi and the Economic Novels of William Dean Howells." *Comparative Literature Studies* 14 (June 1977): 143–65.

Walters, Ronald G. *The Antislavery Appeal: American Abolitionism after 1830.* Baltimore: Johns Hopkins University Press, 1976.

Warner, Michael, "Franklin and the Letters of the Republic." *Representations* 16 (Fall 1986): 110–30.

Washington, Booker T. *Up from Slavery.* New York: Penguin, 1986.

Watkins, T. H. "Why a Biographer Looks to Muir." *Sierra Club Bulletin* 61 (May 1976): 16–19.

Weiskel, Thomas. *The Romantic Sublime: Studies in the Structure and Psychology of Transcendence.* Baltimore: Johns Hopkins University Press, 1976.

West, Cornel. *The American Evasion of Philosophy: A Genealogy of Pragmatism.* Madison: University of Wisconsin Press, 1989.

Westbrook, Max. "The Critical Implications of Howells' Realism." *Texas Studies in English* 36 (1957): 71–79.

Whipple, John. *Free Trade in Money, or Note-Shaving, the Great Cause of Fraud, Poverty and Ruin.* 1836. Reprint. Boston: Dayton & Wentworth, 1855.

White, Horace. "The Hughes Investigation." *Journal of Political Economy* 17 (October 1909): 528–40.

Whitman, Walt. *Democratic Vistas.* In *Leaves of Grass and Selected Prose,* edited by John Kouwenhoven, 460–516. New York: Random House, Modern Library, 1950.

Williams, Raymond. *Marxism and Literature.* New York: Oxford University Press, 1977.

Wolf, Bryan Jay. "A Grammar of the Sublime, or Intertextuality Triumphant in Church, Turner, and Cole." *New Literary History* 16 (Winter 1985): 321–41.

—————. *Romantic Re-Vision: Culture and Consciousness in Nineteenth-Century American Painting and Literature.* Chicago: University of Chicago Press, 1982.

Wood, Henry. *The Political Economy of Humanism.* Boston: Lea & Shepard, 1901.

—————. *Victor Serenus: A Story of the Pauline Era.* Boston: Lea & Shepard, 1898.

Wright, Benjamin Fletcher, Jr. *American Interpretations of Natural Law: A Study in the History of Political Thought.* Cambridge: Harvard University Press, 1931.

Wright, Conrad. "The Sources of Mr. Howells' Socialism." *Science and Society* 2 (Fall 1938): 514–17.

Yack, Bernard. "Towards a Free Marketplace of Social Institutions: Roberto Unger's 'Super-Liberal' Theory of Emancipation." *Harvard Law Review* 101 (June 1988): 1961–77.

Young, Mary E. "Congress Looks West: Liberal Ideology and Public Land Policy in the Nineteenth Century." In *The Frontier in American Development: Essays in Honor of Paul Wallace Gates,* edited by David M. Ellis, 74–112. Ithaca: Cornell University Press, 1969.

Index